Milestones in the Evolving Theory of Evolution

David Wool
Department of Zoology
Tel Aviv University
Israel

Language Editor
Naomi Paz
Department of Zoology
Tel Aviv University
Israel

Illustrations by
Dr. Leonid Friedman

CRC Press is an imprint of the
Taylor & Francis Group, an **informa** business
A SCIENCE PUBLISHERS BOOK

Cover credit: Lost forever: "Lonely George", the sole survivor of the Pinta tortoises, Galapagos, died in 2012.
Photographed at the Darwin Station, Santa Cruz, Galapagos:
David Wool 1987. Inset in the upper right corner: the same animal, photographed in 2010 by
R. Lamb & Patricia MacKay, 2010

CRC Press
Taylor & Francis Group
6000 Broken Sound Parkway NW, Suite 300
Boca Raton, FL 33487-2742

Version Date: 20200527

International Standard Book Number-13: 978-0-367-90333-6 (Hardback)

Library of Congress Cataloging-in-Publication Data

Names: Wool, David (Professor of zoology), author.
Title: Milestones in the evolving theory of evolution / David Wool,
Department of Zoology, Tel Aviv University, Israel.
Description: Boca Raton, FL : CRC Press, [2020] | Includes bibliographical
references and index.
Identifiers: LCCN 2020019400 | ISBN 9780367903336 (hardcover)
Subjects: LCSH: Evolution (Biology)--History.
Classification: LCC QH361 .W66 2020 | DDC 576.8--dc23
LC record available at https://lccn.loc.gov/2020019400

Visit the Taylor & Francis Web site at
http://www.taylorandfrancis.com

and the CRC Press Web site at
http://www.routledge.com

Dedication

In fond memory of my beloved grandparents,

Michael and Golda Einbinder,

and my parents,

Avraham (Avrum) and Esther (Tunia)

Preface

The present book emerged from my series of lectures on the history of the Theory of Evolution, delivered freely since my retirement in 2002, once a week for one or two semesters every year, at the School of Zoology, Tel Aviv University. In the audience were scientists, university students, and members of the general public.

I am a biologist, not an historian. I have taught courses related to evolution for almost 30 years. My book, "The Driving Forces in Evolution", was published in Hebrew in 1985 – and an extended version appeared in English in 2006. In the present book I have attempted to describe, at least briefly, the ideas that have contributed to the theory that changed the course of thinking about the biological world and impacted so many issues in human society, as these ideas have been expressed during the last 200 years.

My motivation to switch from active biological research to history – besides the need to give up my laboratory upon retirement – derived from a gift given to me of an 1874 copy of Darwin's book "The Descent of Man", by my late friend and colleague, Prof. Boaz Moav, following the publication of the Hebrew version of my book in 1985. The following year, when on sabbatical leave at Michigan State University, in East Lansing, I saw a large display of Darwin's books at the entrance to the library – and realized how little I knew about the contributions of Darwin to science (in particular, I was ignorant of Darwin's botanical books). In a used-book bookshop across the street from the university I purchased the two volumes of Darwin's book on domestication, as well as his book on the expression of emotions in animals and man, which fascinated me and stimulated my desire to collect more of his works – forming the nucleus of my personal library, which now includes most of Darwin's books as well as those of his supporters and other contemporaries.

The selection of events, and of scientists (below I shall use the term "savant", often used by Darwin and his contemporaries to refer to educated and thinking men) whose ideas are part of the history of the Theory of Evolution – and which are described in this book – are undoubtedly affected by my own impressions, pre-conceived ideas, and limits of knowledge. While preparing the material for my lectures, as well as for the present book, I devoted myself to reading and referring to the original works of each of the savants that I considered important. Until most of the material had been written, I avoided secondary materials apart from the biographies of these savants, which I gathered from sources such as the Encyclopedia Britannica and from databanks publicly available on the internet.

When reflecting on the contribution of these savants to the evolution of the Theory of Evolution, I wanted to illustrate how different their ideas were from the contemporary, accepted interpretation of the world at the time they had proposed them. I did not attempt to judge how closely their ideas agreed with the Theory of Evolution as accepted today. I did not intend to offer a progressive picture of the development of the theory evolution; but, rather, to evaluate the ideas according to the erratic chronological order in which they had been published.

History, like evolution, is a continuous process, and one that can be surveyed only in retrospect. The Theory of Evolution, however, seems to have been developed by saltations (to use Darwin's expression), marked by new points of view expressed by a savant – not entirely in agreement with the currently accepted opinion – or even directly opposed to it.

Chronologically, it is possible to mark the dawn of the 19th century as a reference point, when the great majority of persons who gave the matter any thought believed that the world had been created as we see it now, and as it has remained ever since[1]. Before that time, the biological world had been regarded as a chain or ladder, leading from the highest point – Man, the height of perfection created "in His image", down to the lower, degenerate, other animals – or, conversely, from simple organisms up to the higher and more complex, as argued by J.B. Lamarck in 1809[2]. The next landmark is 1859, the year that Darwin's book "The Origin of Species" was published and posited that the world had evolved gradually, with some forms becoming extinct and being replaced by others by natural selection. The chronological path from 1800 to 1859 – and then onwards to the "modern synthesis" and molecular evolution towards the end of the 20th century, was not paralleled by a linear progression of the theory. Lamarck's suggestion in his "Philosophie Zoologique"[3] that animal species were not fixed but have changed ("transmutation of species") and evolved gradually from simple to more complex organisms, was an idea that was later incorporated into Darwin's theory – but, at the time, Lamarck's eccentric ideas were ignored or ridiculed even by his colleagues at the French Academy of Sciences. In the 50 years that followed, biological thinking revolved around geology and fossils. The fossils, interpreted as the remains of extinct animals, were accepted as indications of previous, extinct forms of life – which had disappeared due to catastrophic changes on the face of the earth[4]. At the same time, the British geologist Lyell insisted that – given enough time – the normal, observable forces of erosion and volcanic activity were sufficient in themselves to account for the changes observed in the fauna between successive geological strata, and it was unnecessary to assume catastrophes to explain these changes.

Darwin adopted the ideas of transmutation and common descent[5] and offered a new mechanism which could explain how the transmutation was effected – Natural Selection. It took years before most biologists converted to the new theory, and some prominent ones even reverted to Creationism[6]. A significant advance occurred in 1900, when a paper on crosses of plants, written by Gregor Mendel in 1865, was rediscovered and began the era of genetics[7]. The modern Theory of Evolution came to life when genetics became amalgamated with the Darwinian theory.

In the absence of a linear trend, I decided that the best way to arrange the book would be in a series of four circles. The centers of the circles move by saltations along the temporal axis from 1800 to the end of the 20th century. Chapter 1 deals with theories of the origin of life on earth: clearly, no biological evolution was possible before there was life on the earth. The four circles commence in Chapter 2. The center of each circle is dedicated to a dominant figure. Around the centers, the work and ideas of distinguished savants are arranged irrespective of their more or less "advanced" views of evolution. Diverging ideas, sometimes contradicting each other, float within the circle – occasionally a "satellite" idea leaps away from the perimeter.

As the center of the first circle I have selected William Paley, a British clergyman who published, in 1802, a book called "Natural Theology"[8]. In this book – which Darwin claimed he knew almost by heart – Paley presented his outlook of the world as created by a Supreme Power – an interpretation continuous with that which had prevailed before him. These ideas still remain today as the fundamentals of Creationism, which rejects evolution. This outlook can be regarded as a "null hypothesis" in the context of the present work.

The center of the second circle moves from England to France. After the French Revolution in 1789, the French Academy of Sciences and the Museum of Natural History in Paris became the center of all sciences in early 19th-century Europe. Three great scientists made their mark on

---

1 "Milton's hypothesis": Huxley 1876, Science and the Hebrew Tradition. Lectures on Evolution I.
2 See Chapter 6.
3 See Chapter 6.
4 As advocated by Cuvier, Chapter 4.
5 See Chapters 8-9.
6 Chapter 10 on Gosse and Agassiz.
7 Chapters 20-25.
8 See Chapter 2.

the biological sciences: Georges Cuvier[9], Geoffroy Saint-Hilaire[10], and in particular, Jean Baptiste Lamarck[11]. In the same circle I have included the British (actually Scottish) geologist Charles Lyell[12].

The third circle, centered on Charles Darwin, encompasses a wide perimeter – dealing with the birth and publication of the "The Origin of Species", the central issues of Darwin's theory of natural selection[13] and his later books "The Descent of Man" and "Sexual Selection"[14]. It includes the contributions of Darwin's allies and supporters, Alfred Russel Wallace[15], Thomas Henry Huxley[16], and Ernst Haeckel[17]. The struggle of the new theory to survive against the opinions of the Church and other critics (Samuel Wiberforce, Robert Jenkin, and St. George Mivart in particular) is treated at length[18].

Towards the end of the 19th century, as if in anticipation of the interaction of heredity with evolution, we find the work of August Weismann[19]. Connecting the 19th with the 20th century we find Darwin's nephew, Francis Galton[20] and his friend and colleague Karl Pearson[21], laying the foundation of the science of statistics.

The widest circle is the fourth. It properly begins in the year 1900, with the discovery of the paper written 35 years previously by Gregor Mendel[22], and is dedicated to the contribution of genetics to evolution. The perimeter includes the American geneticist Thomas Hunt Morgan[23] – and a retrospective discussion of the inheritance of acquired characters, with social ideology and political interference causing a scientific rift between East and West[24]. It includes the three great founders of the mathematical Theory of Evolution – Ronald A. Fisher, John B.S. Haldane, and Sewall Wright[25], preceded by an introduction to population genetics theory[26]. The perimeter includes the great formalizers of the Modern Synthesis of evolution – Theodosius Dobzhansky and Ernst Mayr[27] – who put it all together, despite objectors like the "heretic" Richard Goldschmidt[28]. The following chapters discuss the study of natural selection in field populations ("Ecological Genetics") and Motoo Kimura's neutrality theory of molecular evolution[29], which dominated the study of evolution towards the end of the 20th century.

In this book I have attempted to provide readers who are interested in evolution with some knowledge – however brief and incomplete – regarding the major contributors to the theory. The fact that the book was originally written as a series of independent lectures necessitated my inserting in later chapters, notes referring to preceding ones, for the sake of continuity. I realize that the texts I referred to, some of them published over 100 years ago, are not readily available, and have included many quotations from the original works – if only to avoid misinterpretation.

My decision to publish this book was reinforced by the comments of colleagues:

---

9 See Chapter 4.
10 See Chapter 5.
11 See Chapter 6.
12 See Chapter 7.
13 See Chapters 8 and 9.
14 See Chapters 15 and 16.
15 See Chapter 11.
16 See Chapter 12.
17 See Chapter 19.
18 See Chapter 14.
19 See Chapters 18-19.
20 See Chapter 21.
21 See Chapter 22.
22 See Chapter 23.
23 See Chapter 25.
24 See Chapter 26.
25 See Chapters 28, 29, 30, respectively.
26 See Chapter 27.
27 See Chapters 32 and 33, respectively
28 See Chapter 34.
29 See Chapter 35.

As many teachers of science have noted, scientific problems are normally better understood from studying their history than their logic[30].

The history of science forces us to think more than science alone can reveal[31].

Finally, I cannot but sympathize with David Quammen[32], who expressed his disappointment that a PhD in evolutionary theory may be obtained at universities without the required reading of even one sentence of Darwin's works. What a shame, and what a loss for the students!

30 Sapp 2001, p. 10.
31 J.K. Nester, cited by Wood and Orel, 2001, p. IX.
32 Quammen 2006.

Introduction

Science is a human activity, motivated by hope, cultural prejudices, and the pursuit of glory, but stumbling in the erratic path towards a better understanding of Nature[33].

The American geneticist and evolutionist, Theodosius Dobzhansky, once wrote: "Nothing in biology makes sense except in the light of evolution"[34]. Darwin's introduction of the theory in 1859 made it a cornerstone in the history of human culture and a turning point in the approach to the study of nature, with important consequences for other sciences as well as for human society.

Like every other achievement in science, the present Theory of Evolution is not the product of a single active mind, Charles Darwin's, but has emerged from a background of ideas contributed by many savants; it also has not remained in the same form as that introduced by Darwin in 1859. The Theory of Evolution has evolved – and many great biologists have contributed to its evolution, as I seek to show in this book.

The Starting Point: Evolution versus Creation

Clearly, there could be no evolution before the appearance of life on earth. A living organism, as the name implies, is an ordered, organized system – be it as simple as a single cell or as complex as a man. The creation of order from disorder is opposed to the general tendency of things (entropy) and requires the investment of energy.

The biblical book of Genesis describes the creation of order from chaos. The process of creating the world, from disorder to the living world – man included, was accomplished in only six days. To accomplish this in such a short time certainly would require the immense energy of a Supreme Power. The Theory of Evolution suggests, in contrast, that the biological world may not have required an extraordinary source of energy if it had taken place gradually over a very very long time. The point of departure for the study of evolution is that the animals and plants, with which we are familiar, evolved gradually over a very long sequence of years, from some unknown non-living matter.

The Theory of Evolution has encountered, since the beginning, the aggressive opposition of those who considered the biblical stories as historical truth. The Creationists believe that organisms were created by the Almighty, in their present form, and that that form has not changed since Creation. An extreme example of the extent to which that Creationist belief can render a person blind to the facts, may be the book "Omphalos" [Greek: navel], published in 1857 by Philip Henry Gosse[35]. Gosse was a British biologist (he had lived for a few years in Canada as a young man) who published research articles on insects and marine organisms. In his book, published just two years before Darwin's "Origin", Gosse describes – in conventional terms – the geological evidence of the gradual accumulation of particles to form the layers of rocks, one on top of the other, a process that seemingly required millions of years. He describes the presence of different fossils in different rock layers, as indications of changes of the fauna in different ages. He describes the "annual rings" in the trunks of trees, with one or two being added every year as the tree grows – and similar well-known biological phenomena. The reader is taken by surprise when, at the end of each section, without

[33] Gould 1982a.
[34] Dobzhansky 1964: American Zoologist 44: 449 (413-452).
[35] Gosse 1998.

warning, he is faced with the conclusion: the scientific description you just read is not evidence for processes that took millions of years – it is an illusion – since we [represented by the author] know the truth beforehand, that the world was created in six days as documented in Genesis. The geological layers were not laid down gradually but created all at once. The fossils are not extinct animals: they formed with the rocks when the earth was miraculously created. Creation was a unique, singular event. Curiously, this is also how Gosse explains the presence of a navel in Adam[36] (as pictured in old church paintings) – although he was not born of a woman, he was created with a navel!

Considerable parts of the world population still believe that the biblical descriptions of Creation and the Flood are historical truths. In the United States, where the teaching of religion in schools is prohibited by the constitution, Creationists have repeatedly tried to ban the teaching of evolution in schools and universities as a heresy[37] (remember the "monkey trial" of 1925)[38]. In the last two decades, Creationists have tried to bypass the constitutional ban by devising a new disguise for God – that of the "Intelligent Designer" – and claiming "equal time" for their "new" theory in the science classes. In 2005, an editorial in the renowned scientific journal, Nature, suggested that allowing the publication of Creationist articles should convince potential religious readers that their belief can be reconciled with science. The editorial initiated a flood of angry responses from famous scientists, and the suggestion was dropped. Ironically, one of the founding fathers of Nature, Thomas Henry Huxley[39], fought all his life against the doctrines of the Church and against any claim of scientific support for the biblical stories:

> My belief… is, and long has been, that the pentateuchal story of creation is simply a myth. I suppose it to be an hypothesis respecting the origin of the universe which some ancient thinker found himself able to reconcile with his knowledge, or what he thought was knowledge, of the nature of things, and therefore assumed to be true[40].

The claim for "equal time" is not justified. Creationists assign the operation of the biological world to the wise planning and careful watching eye of the Almighty. Scientists explain biological phenomena by means of observation and experiments in the biological world itself, unaffected by supernatural forces. *The divergence between the two approaches is not about evolution at all: it is about the belief in God.* Belief is a personal option of individuals, not a scientific matter. Creationism, whatever its disguises, is not a scientific theory, and is in no way equivalent to the Theory of Evolution.

The Theory of Evolution offers a dynamic interpretation of the world around us. The theory itself has evolved. When new facts were discovered, the theory underwent change to accommodate them. This has happened in the 20th century, when genetics was incorporated into Darwin's theory, and later on also when molecular techniques provided new tools for the study of animals and plants. Evolutionists argue about the interpretation of the facts: there is no agreement as to whether new species arise gradually or by "saltations", nor about the phylogenetic origin of different groups of organisms. Even the term "species" has been given different meanings and definitions. All these disagreements lie within the framework of science, based on biologically observable facts

[36] See S.J. Gould, Adam's Navel. The Flamingo's Smile, p. 99-113. Norton, New York.

[37] See Futuyma, D.J. Science on Trial 1983, for a review. Also Ashley Montague (ed.), Science and Creationism. 1984.

[38] Reminder: John Scopes, a science teacher in Dayton, Tennessee, in 1925, was charged with breaking the law that prohibited the teaching of evolution. The trial became a national event and a famous lawyer, Clarence Darrow, was called to the defense. The judge refused to consider the issue of evolution and indicted Scopes for breaking the law. He was fined $100 but was acquitted on appeal.

[39] See Chapter 12.

[40] Huxley 1886: Mr Gladstone and Genesis. Collected Essays IV.

and processes. Creationism is not part of the scientific debates. One of the greatest 19th-century scientists – the geologist Charles Lyell[41], wrote in 1863:

> For what we term independent creation, or the direct intervention of the Supreme Cause, must simply be considered as an avowal that we deem the question to be outside the domain of science[42].

The dispute nonetheless will no doubt arise again and again in the future. The British scientist and philosopher, Karl Pearson, wrote in despair:

> If the reader questions whether there is still war between science and dogma, I must reply that there always be, as long as knowledge is opposed to ignorance[43].

And the paleontologist, George G. Simpson, questioned whether it is worthwhile to argue with the Creationists:

> It is a human peculiarity... that no amount of proof suffices to convince those who do not want to accept the truth... and reiteration for those who do want to know the truth is quite unnecessary, because they already know it[44].

---

41 See Chapter 6.
42 Lyell 1873.
43 Pearson 1900, preface.
44 Simpson 1961.

Contents

1

The Origin of Life

Life is a sexually-transmitted disease which is always fatal (anonymous)

In "The Origin of Species", Darwin did not discuss the origin of life – he apparently took it for granted that life originated somewhere, in the remote past, from non-living matter, in primitive form. In a letter to his friend, Joseph Hooker, he explained that this question is not worth worrying about:

> It is mere rubbish, thinking at present of the origin of life. One might as well think about the origin of matter[45].

Perhaps Darwin considered the problem too complex to comprehend: organisms differ vastly from inorganic matter:

> The most humble of organisms is something much higher than the inorganic dust under our feet. And no one with an unbiased mind can study any living creature, however humble, without being struck with enthusiasm at its marvelous structure and properties[46].

At the turn of the 20th century, the philosopher and mathematician Karl Pearson[47] wrote that from physical considerations of the conditions at the time of the formation of the earth, the physicist

> …is compelled to postulate a period, distant it is true, many millions of years back, in which owing to the conditions of fluidity and temperature, no life such as we know it now …could exist on earth[48].

If animals were not created as they are now, life had to begin somehow from non-living matter. T.H. Huxley later stated the problem:

> …I have translated the term "protoplasm" …by the words "The physical basis of life" …suggesting that there is some kind of matter which is common to all living things, and that their endless diversities are bound together by a physical, as well as ideal, unity. In fact, when first apprehended, such a doctrine as this appears almost shocking to common sense[49].

---

45 Darwin to Hooker, 20.3.1863. In Darwin, F., Darwin's Life & Letters III: 18.
46 Darwin in "Descent of Man (1871); 1952: 528.
47 See Chapter 21.
48 Pearson 1900, The Grammar of Life, p. 347-348.
49 Huxley 1868, The Physical Basis of Life. Collected Essays I, p. 130.

Spontaneous Generation

I admit that spontaneous generation, in spite of all the vain attempts to demonstrate it, remains for me a logical necessity (August Weismann, 1881 and 1891, p. 34).

Since the time of the Ancient Greeks, people have believed that some animals are formed spontaneously – fish and frogs are formed from the mud at the bottom of ponds and streams, etc. Similar ideas persisted for centuries. At the beginning of the 18th century, Van Helmont offered the following recipe for the artificial production of mice:

> Close the mouth of a pot containing wheat grains with a dirty shirt [and place it in a warm place behind the stove]. After about 20 days, a ferment from the dirty shirt mixes with materials emitted by the wheat, and the grains turn into mice of both sexes, which can mate among themselves or with other mice. The mice from the wheat... are perfect... and do not have to be suckled by their mother like other mice[50].

Insect larvae found in plant galls were believed to be formed spontaneously by the plant. John Ray, dealing with the induction of galls by insects, in 1691 encouraged biologists to study the reproduction of "all sorts of insects" because the belief in their spontaneous generation stands against faith in Creation:

> If it be demonstrated that all creatures are generated univocally by parents of their own kind, and that there is no such thing as spontaneous generation in the world, one main prop and support of atheism is taken away, and their strongest hold demolished. They cannot then exemplify their foolish hypothesis of the generation of Man and other animals at first by the like of frogs and insects of the present day[51].

Objections and Proofs to the Contrary: Proof that spontaneous generation did not occur was provided by the Italian, Francesco Redi, in 1668. He put the carcasses of a snake, fish, eels, and a piece of raw meat in four jars, and closed them tightly with paper. Four other jars with the same contents were left open. Shortly afterwards, maggots appeared in the open jars, but not in the closed ones. This proved that the maggots had not been formed "spontaneously" from the rotting flesh in the jars, but had arrived from the outside. Redi repeated the experiment, this time closing the jars with thin cloth rather than paper. The results were the same, but he could see where the maggots had come from: flies hovered around the jars and laid eggs, from which the maggots hatched[52].

With further improvement of the microscope, attention turned to minute creatures ("animalcules") which could be seen in every drop of water: these were surely formed spontaneously from inorganic matter? Another Italian, Spallanzani, then showed that after boiling the broth and sealing the vessels, no "animalcules" appeared spontaneously.

Nonetheless, and almost a century later, J.B. Lamarck dedicated an entire chapter in his "Zoological Philosophy" (1809) to the spontaneous generation of life from non-living matter:

> Nature, by heat, light, electricity and humidity, creates directly and spontaneously [living bodies] at the fringes of each of the animal kingdoms, where the simplest organisms are found[53].

[50] Glass et al. 1968, Forerunners of Darwin, p. 40.
[51] Cited by M. Redfern, 2011, p. 484.
[52] Glass et al. 1968, Forerunners of Darwin, p. 40.
[53] Lamarck (1809) 1917, Zoological Philosophy, p. 244.

> This Nature is daily engaged in the formation of the elementary rudiments of animal and vegetable life, which corresponds to what the ancients termed spontaneous generation. She is always beginning anew, day by day, the work of creating monads, or rough drafts, which are the only living things she gives birth to directly[54].

Darwin's grandfather, Erasmus Darwin, accepted Lamarck's proposal of spontaneous generation – and ridiculed those who opposed it:

> From the misconceptions of the ignorant or the superstitious, it has been thought somewhat profane to speak in favour of spontaneous vital production, as if it contradicted holy writ, which says that God created all things that exist[55].
>
> To suppose that the eggs of the former microscopic animals float on the atmosphere and pass through the sealed glass phials, is so contrary to apparent Nature as to be totally incredible! And as the latter are viviparous, it is equally absurd to suppose that their parents float universally in the atmosphere to lay eggs in paste or vinegar[56]!

Louis Pasteur, in the middle of the 19th century, proved experimentally beyond doubt that "animalcules" – known today as bacteria and yeast (fungi) – do not form spontaneously. He boiled the broth in vessels with curved but open necks, which allowed the access of air into the broth but captured solid particles of dust floating in the air [to counter claims that the exclusion of air killed the animalcules]. No animalcules appeared in the boiled broth after it had cooled. Certain of the generality of his conclusion, Pasteur wrote:

Pasteur's Claim

> Never will the doctrine of spontaneous generation recover from the blow of this simple experiment[57].

Pasteur won the admiration and support of the growers of silk worms – a major economic industry in the 19th century in France – when he demonstrated that epidemic mortality of the moth larvae was not spontaneous but resulted from infection by microbes, and could be avoided by simple sanitary procedures. He later saved the sheep growers from great losses by demonstrating that anthrax, a lethal disease, is also caused by infection, and its spread can be prevented by properly managing the carcasses. Pasteur claimed that this was true of all cases of apparently spontaneous generation. True to his conclusion, he suggested that sterile procedures be introduced in military hospitals: if infection could be prevented, the lives of many hundreds of wounded soldiers who were dying from gangrene could be saved. Surprisingly, the French medical profession rejected this idea. The French physicians insisted that gangrene and other complications arose spontaneously in the body of the soldiers.

---

54 Summary of Lamarck's idea by Charles Lyell, Principles of Geology, 1853 p. 574.

55 Erasmus Darwin, The Temple of Nature, additional notes p. 1.

56 Erasmus Darwin, The Temple of Nature, additional notes, p. 3.

57 Vallery-Radot 1930, p. 109.

In 1874, Pasteur received a letter of praise from a Scottish doctor, Joseph Lister. Lister reported that he was applying sanitary procedures in his hospital and many lives were being saved[58]. After surveying the evidence, T.H. Huxley (1870) concluded that life only issues from previous life:

> Spontaneous generation could only have occurred – if at all – in the remote past, under conditions which no longer exist on this planet[59].

Not everybody concurred with this conclusion. Huxley's student, Henry C. Bastian, caused a stir when he published that Pasteur was wrong: animalcules were found in his vessels after boiling to 150°C. He claimed that life can form spontaneously under current conditions. In 1876, the German biologist and evolutionist Ernst Haeckel wrote,

> The fundamental idea… is that of a gradual development of all… organisms out of quite simple and imperfect original beings, which came into existence... by spontaneous generation… out of inorganic matter[60].

Lamarck believed that Nature is ever producing simple organisms – monads – spontaneously, all the time. Most investigators, however, believed like Darwin, that life had evolved only once in the history of the globe, and if new life was still forming spontaneously somewhere, we are unlikely to find any evidence of it:

> If we could conceive, in some warm little pond, with all sorts of ammonia and phosphoric salts, light, electricity &c present, that a protein compound was chemically formed ready to undergo still more complex changes, at present such matter would be instantly devoured or absorbed – which would not have been the case before living creatures were formed[61].

The question was raised: if a transition from non-living to living matter could have occurred in the remote past, why cannot we repeat the process experimentally and form life from inorganic materials? Two savants addressed this question:

> The reply probably lies in the statement that we seek to reverse a process that is irreversible. In five or ten minutes we convert living to lifeless substance, but there is no reason for asserting that the reverse can be gone through even in the lifetime of a man[62].

> When we pour sulphuric acid on a piece of chalk, we only change the form, but the inorganic matter remains. But when we pour sulphuric acid on a worm, or when we burn an oak tree, those organisms are not changed into some other animal and tree, but they disappear entirely as organized beings and are resolved into inorganic elements[63].

Experiments in the 20th Century

The successful synthesis of organic compounds from simple, inorganic precursors was first reported in the 1940s[64]. An external source of energy was applied to a mixture of gases – methane (CH_4),

---

58 Vallery-Radot 1930, p. 238. Sterilization procedures ["Pasteurization"] have since enabled the preservation of food-stuffs for human consumption, and the sterilization of surgical equipment in hospitals prevails in modern medicine.

59 Chapter 12. Huxley 1870, Biogenesis and Abiogenesis.

60 See Chapter 19. Haeckel 1876, History of Creation, 1876, p. 75.

61 Darwin, F., Charles Darwin Life & Letters III: 18.

62 Karl Pearson, Chapter 22. The Grammar of Life 1900, p. 349.

63 August Weismann, Chapter 20. 1889, The Duration of Life, p. 35.

64 Miller 1953, 1954.

ammonia (NH_3) and water vapor – a mixture that could have existed in the atmosphere of the primitive earth (in different experiments, electric sparks, ultraviolet light, and other sources of energy were used). Among the organic compounds obtained in these experiments were amino acids and even short peptides! The accumulation of organic molecules could have created an "organic primeval soup", in which processes of improving the affinities, replication ability, and organization of some molecules could have taken place.

These experiments proved that, given the right conditions, organic materials could be formed from non-living materials. Organic molecules, however, complex as they may be, are still not a living organism.

> How can we conceive that dead organic matter could come together in such a manner to form living protoplasm[65]?

A number of hypotheses were suggested for the evolutionary steps from the "organic soup" to the formation of a living cell. In the "soup", molecules interacted at random and could combine by their chemical affinity. Some selective mechanism was required to eliminate molecule combinations that may be chemically possible, but biologically unfit[66]. A replicating mechanism was essential, in order to increase the frequency of the "fit" molecules in the "soup": self-replicating molecules must have formed at some point in the process. In today's living world, self-replicating molecules are the "nucleic acids", DNA and RNA. These molecules however, are synthesized and replicated in cells by the activity of other molecules – specific enzymes – which are proteins, themselves the products of enzymatic processes on an RNA or DNA template. This creates a "chicken and egg" dilemma: which was the primeval molecule?

Most of the modern theoretical research on the origin of life revolves around the properties of the first self-replicating system. An auto-catalytic system was sought, one that self-replicates without the need for enzymes. DNA could not be replicated experimentally without enzymes. Some RNA molecules fit this requirement[67], but only short sequences are produced. An enzyme – RNA polymerase – was required for the precise replication of longer sequences.

If such RNAs were formed spontaneously, each with a "domain" that could recognize a specific amino acid, amino acids could group together to form short peptides, which in turn could unite to make proteins. Their ability to self-replicate would give these RNA molecules an advantage in the primeval soup (the "RNA World" hypothesis)[68]. An alternative theoretical model had been suggested[69].

Further, a mechanism is still required for the assembly of different molecules together to create the first living cell. Some common cellular organelles – mitochondria and chloroplasts – originally evolved as independent, unicellular organisms, which were acquired as symbionts by later-forming multicellular cells. Evolutionist Lynn Margulis suggested that the nucleus of the eukaryotic cell acquired many genes from the incorporated primitive bacteria that by a process of "symbiogenesis". This process she described as "the inheritance of acquired genomes":

---

65 Weismann, The Duration of Life, 1891, p. 35.
66 Eigen 1982; in Evolution Now, ed. J. Maynard-Smith.
67 Eigen 1982; in Evolution Now, p. 18.
68 Gilbert, W. 1986: The RNA world. Nature 319: 618.
69 Lahav 1991, 1993.

> Random changes in DNA alone do not lead to speciation. Symbiogenesis – the appearance of new behaviors, tissues, organs, organ systems, physiologies or species as a result of symbiont interaction – is the major source of evolutionary novelty in Eukaryotes – animals, plants, and fungi[70].

Further Developments in the 21st Century: New discoveries in deep-sea research in the 21st century breathed new life into the study of the origin of life. A rich bacterial community was discovered associated with geothermal vents of ocean-floor thermal springs[71]. These events occur worldwide and emit vapor at temperatures up to 400°C, which is cooled down by the surrounding seawater. Some of the vents are large and have received names such as "Lost City" and "Black Smoker". Since these vents have probably existed since water first accumulated in the primitive planet more than 4 billion years ago, they may have been sites where life first started. The associated fauna contains anaerobic bacteria which can replicate at 120°C. Some generate methane (CH_4) and live as Archaea in symbiosis with other, sulfur-reducing bacteria. They obtain their energy from chemical reactions, not photosynthesis, and may represent the early stages of life.

A different conclusion was reached by comparing the chemical composition of living cells with that of seawater. Based on the principle that the traits of organisms are more conservative than the environment in which they live, the discovery that all cells contain more K ions than Na ions (K^+/ Na^+ ratio greater than 1), and other differences in contrast to seawater, make it unlikely that protocells originated in the ocean – in particular because protocells probably did not have thick membranes and active pumps to regulate their contents like modern live cells[72]. The authors suggested instead that life had originated in small vapor-dominated geothermal ponds on earth, in which the ion composition is compatible with that of cells – reminiscent of Darwin's "warm little pond"[73].

Extra-terrestrial Origin of Life?

Although the prevalent opinion is that life evolved on our planet, some scientists believe that life evolved elsewhere and later arrived on Earth, e.g., via meteorites[74]. Inter-planetary probes are still sent to explore the possibility of life existing elsewhere. This however offers no solution to the problem of how life evolved, but instead makes the problem unanswerable as long as we cannot access the unknown site where it occurred (if at all). T.H. Huxley wrote in 1870:

> Whether they [i.e., life] have originated in the globe itself, or whether they have been imported by, or in, meteorites from without, the problem of the origin of those successive faunas and floras of the earth... remains exactly where it was. For I think it will be admitted that the germs brought to us by meteorites, if any, were not ova of elephants, nor of crocodiles; not cocoa nuts nor acorns; not even eggs of shellfish and corals, but only those of the lowest animal and vegetable life. Therefore... the higher forms of animals and plants... either have been created, or they have arisen by evolution[75].

[70] Margulis, in Shermer, 2005-7.

[71] W. Martin, J. Baross, D. Kelly and M.J. Russell, 2008. Nature reviews microbiology 6: 805-814.

[72] A.Y. Mulkidjanian, A.Y. Bychkov, D.V. Dibrova, M.Y. Galperin and E.V. Koonin 2012, PNAS Feb 2012 E821-E838.

[73] P. 21.

[74] Bernal 1967; Hoyle and Wickramasinghe, 1978; Crick, 1981.

[75] Huxley 1870, Biogenesis and Abiogenesis. In Critiques and Addresses, Macmillan, London, pp. 218-250.

The First Circle

The savants in the first circle represent the "null hypothesis" for the Theory of Evolution: they believed that the world did not evolve, but was created as we see it now – in all its complexity. Even when they described the world as an *ordered* system – the "ladder of life " or the "chain of beings" – they did not visualize it as a dynamic system but as a map of the plan by which the Creator constructed His world.

The six savants in this circle (Chapters 2 to 7) lived and worked in the 18th and early 19th century. In the center of the circle I chose to place **William Paley,** a clergyman whose book "Natural Theology" (1802) describes enthusiastically the marvelous foresight of the Creator and His unlimited ability, as demonstrated by his observable creations. Next is **Carolus Linnaeus,** who constructed the classification of the biological world in his "Systema Naturae" (1735, but mainly 1758) and initiated the methodology for nomenclature and systematics in biology, but he did not consider the hierarchical groups he described to be connected by descent. **Georges–Louis Leclerc, Count de Buffon,** was perhaps the most famous and prolific biologist in the 18th century, and affected biological thinking well into the 19th century. **Charles Bonnet,** in his biological researches, described parthenogenetic reproduction in aphids, studied the development of the (chick) embryo to the adult form, and described the "chain of beings" – but did not consider the possibility of change from one species to another, and postulated that all "intermediate forms" were created as they are now.

Thomas Malthus, in his influential "Essay on Population" (1798), considered almost exclusively the regulation of population size of a single species only – the human species – but had a strong influence on Darwin and Wallace in introducing the concept of the struggle for existence. **Erasmus Darwin** – Charles Darwin's grandfather – did mention in his poem "Zoonomia" (1794?) the possibility of a common origin for the entire biological world, but he was more a poet than a biologist, and his ideas were not considered by his grandson as a contribution to the theory of the Origin of Species.

2

Early Theories: Creation, Pre-formation and Order in the Biological World

William Paley

More than two hundred years ago, a British theologian, William Paley (1743-1805), published a book called "Natural Theology" (1802). The book apparently became very popular – even Darwin, an atheist later on in life, was enthusiastic about it in his youth and admitted to having known it almost by heart[76].

Paley held a Master's and a PhD degree in theology from Cambridge University and occupied senior religious posts in Carlisle and other communities in the UK. His published works[77] include "Evidences of Christianity", but also a large book on "Moral and Political Philosophy", ranging from interaction among individuals to moral behavior of princes (rulers), to interactions among nations – a true treatise of political sciences from the standpoint of morality. Paley's books were mandatory reading for Cambridge students in 1787-1820[78].

Among the immoral actions of princes, which Paley condemns, are the destruction of forests and fields cultivated by one's adversary, unnecessary killing of animals including livestock, netting of salmon in unlawful nets (which do not allow young fishes to escape), and the distillation of spirits from bread corn[79]. The reasoning is theological:

> It appears to be God Almighty's intention that the products of the earth should be applied to the sustenance of Man. Consequently, all waste… of these productions is contrary to Divine intention, and therefore wrong.

"Natural Theology": "Natural Theology", published three years before his death, appears to be Paley's last book. The full title of the book describes its purpose: "Natural Theology, or the existence and attributes of the Deity, collected from the appearances of nature".

Paley sought to convince his readers of the existence of the Deity, by demonstrating the beautiful adaptation of biological structures to the functions they fulfil: every natural adaptation is a result of very careful and detailed planning, illustrating the talents of the Creator.

The example which opens the discussion is that of a watch. Nobody denies that a watch is designed and meticulously built by an engineer for the purpose of keeping time. What is true for the watch, claims Paley, must also hold true for the observable adaptations in Nature.

---

[76] Darwin, Autobiography, p. 47. Darwin to Lubbock, Darwin, F.: Charles Darwin Life and Letters II, p. 219.

[77] Paley 1844.

[78] Sulloway 2009, J. Biosci. 34: 173-183. It is noted that Darwin, as a student at Cambridge, occupied the same rooms as Paley had done fifty years before.

[79] Ibid. pp. 32, 33.

> Every observation which was made… concerning the watch, may be repeated with strict propriety concerning the eye; concerning animals; concerning plants; concerning, indeed, all the organized parts of the works of nature[80].

The eye was designed for the same purpose as the telescope: for taking in light and showing us the beauty of the world around us. It is a contrivance planned in great detail to overcome various optical difficulties and enable good vision. There cannot be a contrivance without a contriver, argues Paley. This adaptation must necessarily be the result of planning by the Creator, a "super-planner", whose ability knows no boundaries.

Paley was well-read and familiar with the structures of different body parts in many different organisms. The bodies of man and other animals, in which all parts are so well adjusted to each other to fulfill their part in the activities of the animal, provide many examples of adaptation of structure and function. The eye, the ear, and any other organ in each case show modifications in structure to fit the differences in ways of life, on land or in water. All of these are demonstrations of planning and design, and wherever there is a design there must be a designer. This point is reiterated, in many scores of pages, with endless examples from biology.

Some of the processes in natural organisms were not understood in Paley's time – as he admits; especially the "chymical" processes of digestion, blood circulation, the excretion of different glands, the operation of the nervous system – but:

> True fortitude of understanding consists in not suffering what we know to be disturbed by what we do not know. If we perceive a useful end, and means adapted to this end, we perceive enough for our conclusion… All these are fair questions, and no answer can be given, but what calls in intelligence and intention[81].

Sometimes the observed planning in nature surpasses everything imaginable by man. In a section entitled "The succession of plants and animals" (in modern terms, their reproduction) Paley argues for the foresight of the Divine designer,

> The parent is not the contriver... Neither [the plant nor the bird] have this sort of relation to what proceeds from them [seed or eggs] which a joiner has to the chair he makes. Now a cause, which bears this relation to the effect, is what we want for the suitability of means to an end[82].

The production of milk in the female mammal and its perfect timing with the birth of the baby is a wonderful demonstration of the foresight and planning of the Deity. Paley writes with admiration:

> It is not very easy to conceive a more evidently prospective contrivance than that which, in all viviparous animals, is found in the milk of the female parent. At the moment the young animal enters the world, there is its maintenance ready for it… We have, first, the nutritious quality of the fluid, unlike… every other excretion of the body, and in which nature hitherto remained uninitiated: neither cookery nor chymistry having been able to make milk out of grass[83].

Why Paley Chose to Publish his 1802 Book: People of Paley's generation accepted the Scriptures as true, and the story of Creation in Genesis as an unquestioned historical record. Theologians

---

[80] Ibid. p. 14.
[81] Ibid. pp. 18, 23.
[82] Ibid. p. 14.
[83] Paley 1844 (1802), p. 64.

of the 17th century had even calculated the exact time of Creation – 4,004 years before Christ[84]. Few members of the public in 18th-century England doubted that God created and ruled the world. Therefore, why did it seem necessary to Paley to make such an effort to convince the public of this doctrine?

In the wake of the French revolution of 1789, the atheism of some of the greatest British scientists of the 19th century, like T.H. Huxley, Francis Galton, and Darwin himself, signifies a weakening of the interpretation of the story of Genesis as a description of an historical fact. Only a few years after the publication of Paley's book, Lamarck's important book "Zoological Philosophy" came out in France (1809)[85]. Lamarck suggested a plausible alternative to the belief in Creation: spontaneous generation of tiny "animalcules" from inorganic matter, followed by a gradual development, from the simplest to the most complex. The Anglican Church and the governmental circles were worried about possible effects of the French Revolution, and were suspicious of anything French. Paley's "Natural Theology" may have been an attempt to prevent the French atheistic teachings from gaining a hold in Britain.

Creationism: Paley's example of the watch is still brought up today in arguments for "creation science" – often by people who have never heard of Paley and seem to know much less about biology than did Paley, more than 200 years ago. The creationist "argument from design" has not lost its appeal[86]. The American philosopher, E. Sober, wrote about the "design" argument as follows:

> It was obvious to Paley and to other purveyors of the organismic design argument, that if an intelligent designer built organisms, that designer would have to be far more intelligent than any human being could ever be... When the Spanish conquistadores arrived in the New World, several indigenous peoples thought these intruders were gods, so powerful was the technology that the intruders possessed. Alas, the locals were mistaken. They did not realize that these beings with their guns and horses were human beings.
>
> The organismic design argument for the existence of God embodies the same mistake. Human beings in the future will be the conquistadores, and Paley will be our Montezuma[87].

Carolus Linnaeus (Carl von Linne') and the Systema Naturae

The great reformer of plant and animal nomenclature, the naturalist who exerted the most incontestable influence on his century[88].

The Swedish scientist, Carolus Linnaeus (1707-1778), wrote his books in Latin. Interesting details about him and his work are scattered in a book on the classification of fishes – to which Linnaeus made important contributions. The book was written by Georges Cuvier[89] in the 19th century, but only translated into English and published in 1995[90].

Linnaeus studied medicine and specialized in botany, which was an important part of the medical curriculum. He studied at the universities of Lund and Uppsala, where he worked for many years, finally receiving his Doctor's degree at the University of Leiden in the Netherlands, where he first published his book – "Systema Naturae" – in 1735. He collected and studied plants and made several field trips, as far north as Lapland. In 1741 he was appointed Professor of Botany at Uppsala, where he continued his botanical studies and classified plants sent to him by his students

---

[84] Sapp 2003, p. 11.
[85] See Chapter 4.
[86] See Introduction, p. 13.
[87] Sober 2002.
[88] Georges Cuvier on Linnaeus, 1825. Cuvier 1995.
[89] See Chapter 4.
[90] Cuvier 1995.

from remote parts of the world. In later years he was appointed private physician to the Swedish Royal Family.

Linnaeus was a keen biologist. He grew up in a small village, where biological phenomena were abundant. Swallows nested in the village and he wondered where they disappeared to in winter (he assumed that they dived to the muddy bottom of lakes to overwinter). As a doctor, he studied Ague (=malaria) and tried to produce drugs from Indian plants. He introduced temperature measurements into his students' experiments and prepared conversion tables from the Fahrenheit scale to the decimal scale prepared by his friend and colleague, Andreas Celsius, who died young (1701-1744). He was involved in a successful project to acclimatize the Spanish Merino sheep to Sweden (the Merino sheep were smuggled out of Spain by his friend) and the establishment in 1746 of the study of "Applied Economics" in Swedish universities, with sheep and silk-worm cultivation as mandatory topics[91]. His attempt to acclimatize tea, cocoa and bananas in Sweden, and to domesticate the yak and the elk, were unsuccessful. His private collection of plants, insects, mollusk shells, fossils, and minerals was turned into a formal museum in 1768.

His major work was the "Systema Naturae", which he frequently edited and enlarged (the 1758 edition is perhaps the most comprehensive).

Linnaeus established *order* in the bewildering variety of plants and animals that the biological world presents. His plant classification was based on the number and arrangement of parts in the flowers[92]. In animal classification he referred to the visible characteristics shared by each group, such as the skeleton, limbs, and skin cover, but also internal characteristics such as the blood vessels and the color of the blood, and the location of the nervous system. Individuals similar to each other he grouped into *species*. Similar species were grouped into a *genus*, similar genera into a *family*, similar families into an *order*, similar orders in a *class*, and similar classes in the two great biological *Kingdoms* – vegetable and animal. The hierarchical system has been adopted by biologists worldwide and is the basis today of biological classification.

Linnaeus also established order in the system of giving names to organisms. Each species was given a specific Latin name, preceded by the name of the genus to which it belonged (i.e., the domestic dog was named *Canis familiaris*, since it belonged to the genus Canis, together with other species such as the wolf, *C. lupus*, and the jackal, *C. aureus*. This "binomial" system persists in biology to this day.

Each new specimen was compared with those already described and, if dissimilar to them, was given the designation "type" and a new specific name.

As is implied by the title he gave to his book, Linnaeus regarded his system as a reflection of the plan that the Creator had in mind when He designed the world. Therefore species cannot be subtracted from the world (by extinction), because this would disturb the harmony of nature. The borders between species are maintained by the absence of interspecies mating. If new forms are produced in nature by hybridization, the new form remains as a fixed entity – [hence, perhaps, the giraffe – *Camello pardalis* – could have been of a hybrid origin between the camel (*Camellus camellus*) and the leopard (*Panthera pardus*).]

Cuvier (1825) approved of Linnaeus' decision to use easily-remembered *common* names for every species (such as dog, cat, etc.) in addition to the cumbersome Latin names, and commends the accuracy of his description of the characters he used, the clear definitions of the terms. The utility of the Linnaean system is proven by the fact that scientists all over the world have adopted it[93]. Since the forms of organisms were considered as fixed at the time of their Creation and could never change, this system became the scientific justification for the building of the huge natural history museums in European capitals in the 19th century.

---

[91] Wood & Orel 2001, footnote p. 28.

[92] More on Linnaeus' contribution to classification, see W. Whewell, History of inductive science, 1866, IV History of Botany, p. 388 ff.

[93] Cuvier 1995, pp. 102-103.

Cuvier noted a few errors in Linnaeus's classification of fishes: the cartilaginous fishes (sharks) had been grouped together with reptiles, under the assumption that they breathed with lungs. On the other hand, Linnaeus corrected the erroneous classification – prevalent since Aristotle – of the marine mammals (dolphins, whales, and sea lions) together with the fishes, and assigned them correctly to the Mammalia.

Linnaeus noted the structural similarity between Man and the other vertebrates and found no reason to exclude Man from his animal classification. He placed Man in a subgroup *Primates* (meaning first, or best) within the class Mammalia, together with the chimpanzee, the gorilla, and the orangutan (reflecting the common belief that Man was created in the image of God; and, as is the culmination of Creation, Man was placed at the top of the biological hierarchy).

Linnaeus's classification was based on morphological and anatomical similarity and bore no notion of a common descent! Nevertheless it aroused strong religious opposition. The Lutheran Archbishop of Uppsala accused him of "impiety". Linnaeus wrote apologetically,

> It is not pleasing that I place Man among the primates, but ...let us not quibble over words. It will be the same to me whatever name we use. But I request from you and from the whole world to bring to my knowledge any generic difference between Man and Simian [i.e. monkey], and this from the principles of natural history. I certainly know of none. If only someone might tell me just one! If I called Man a simian or vice versa, I would bring together all the theologians against me. But perhaps I ought to [do so], in accordance with the law of the discipline [of natural history][94].

Georges-Louis Leclerc, Count de Buffon

Buffon (1707-1778) was perhaps the best-known and most influential biologist in Europe in the 18th century. He had trained as a lawyer but studied mathematics in Dijon. Forced to leave the college after killing an officer in a duel, he traveled widely in France and Italy in the company of a young Englishman. Returning to France, he studied medicine and botany. In 1733 he submitted a paper to the French Academy of Sciences, introducing differential and integral calculus into probability theory. The paper was greatly applauded by the mathematicians.

Buffon later translated two of Newton's works, as well as an important botanical essay, from Latin into French, and consequently was accepted as a member of the Academy of Sciences, aged only 27. In 1738 he was appointed "overseer" of the Royal Botanical Gardens. He was impressed by J.B. Lamarck's[95] book on the Flora of France, and hired him as a "king's botanist" after Lamarck's release from the army (he also hired Lamarck as a tutor and companion for his son for a two-year tour of botanical gardens in Europe).

Buffon greatly expanded the public awareness of biological sciences, writing in French rather than Latin. He published a series of 36 books on minerals and especially on the fauna of the world – birds, reptiles, fishes, and mammals, with many illustrations. Eight additional volumes were published after his death[96].

In the first of his books, published in 1749, Buffon presented a "Theory of the Earth". He suggested that the earth was formed when a large meteorite hit the sun. The collision produced a cloud of blazing-hot particles, one of them being the earth. The red-hot planet later cooled down

---

[94] Linnaeus to J.G. Gmelin, February 25, 1847; translated from Latin by A. Ionescu.

[95] See Chapter 6.

[96] **Note:** One of his biological statements caused what amounts to an international incident. He claimed that the fauna of the New World was inferior to that of Eurasia, because in North America there were no large animals like the elephant or rhinoceros. This upset the American president, Thomas Jefferson, who dispatched a company of soldiers to New Hampshire to hunt a large moose bull, which he sent to Buffon to prove him wrong. The individual bull did not have large enough antlers, and the antlers of a still larger bull were attached to it!

enough to be suitable for life, and the process of gradual cooling has been going on ever since, from the poles toward the equator.

His theory further explained the distribution of plants and animals on earth: the faunas are the product of the environment into which they are born: "The earth makes the plants; the earth and the plants make the animals"[97]. Life began in the north of the planet, because the south was too hot for life. The finding of remains of "tropical" animals – the mammoth, elephants and rhinoceroses – in northern Europe supported his idea of a continuous gradual cooling[98].

Buffon encouraged research into the history of the biological world, and his ideas stimulated savants well into the 19th century.

> Just as in civil history we refer to deeds, seek for coins, or decipher antique inscriptions in order to determine the epochs of human revolutions and establish the dates for moral events – so, too, in natural history, we must search the archives of the world, recover old monuments from the bowels of the earth, collect their fragmentary remains, and gather into one body of evidence all the signs of physical changes that may enable us to look back onto the different ages of nature. This is our only means of fixing some points in the immensity of space, and of setting a certain number of waymarks along the eternal path of time[99].

Classification: The "Chain of Beings": Buffon accepted the concept of the ideal world as reflected in the writings of the ancient Greek philosophers, with a continuum of forms, changing insensibly from the most to the least perfect. Wherever there seems to be a gap, there must be intermediate forms that have not yet been found[100]. Nature contains only individuals, not species or genera. Classification is therefore a human artifact:

> The notion of species is an artificial one. The error consists in a failure to understand nature's processes, which always take place by fine gradations ...it is possible to descend by almost insensible degrees from the most perfect creature to the most formless matter.
>
> In general, the more one increases the number of one's divisions... the nearer one comes to the truth. Since in reality, individuals alone exist in Nature[101].

Buffon disagreed strongly with Linnaeus about the order of nature. He rejected Linnaeus's idea that the systematic order reflects the Divine plan in creating the world, perceiving the systematic order was a purely human, artificial arrangement of things for convenience. Buffon considered the relationships among animal species according to the similarities in the form of the limbs and their function – the basis of the study of 19th-century comparative anatomy[102]: carnivores had canines to kill their prey and cutting molars; herbivores had flat teeth for grinding, etc.

---

97 Cited in Mayr 1982, p. 441.

98 Buffon studied the process of cooling in a factory he built on his estate at Montbard. He built an industrial park with an iron forge at its center. He produced and tested different designs of guns for the French army and navy. The workers in the forge were housed on the premises and had access to a garden, a bakery, and a chapel.

99 Cited by Huxley, "The Progress of Palaeontology" 1881, p. 35.

100 Lovejoy 1960. In his later writing, Buffon did allow discontinuities – in the form of species – resulting from the reproduction system: the contributions of both parents are molded into a form which remains constant, "a whole which was counted as one among the works of nature". (Ibid. p. 230).

101 Cited in Lovejoy, 1960, p. 230. (e.g. measuring length in millimeters is closer to the true length than measuring in centimeters).

102 Chapters 4 and 5.

A.R. Wallace (1900) cites a long section from Buffon's first volume, to demonstrate the latter's criteria for classifying organisms – and, in contrast to Linnaeus, considering their descent:

Buffon, Chain of Beings

> The horse, for example – what can at first sight seem more unlike mankind? Yet when we compare man and horse, point by point and detail by detail, our wonder is excited rather by the resemblances than by the differences between them [in the parts of the skeleton]... If we regard the matter thus, not only the ass and the horse, but even man himself, the apes, etc might be regarded forming members of the same family... If we once admit that there are families of plants and animals, so that the ass may be of the family of the horse, and that the one may only differ from the other by degeneration from a common ancestor, we might be driven to admit that the ape is of the family of man, that he is but a degenerate man, and that he and man had a common ancestor.
>
> If it were once shown that we had right ground for establishing these families, and ...among plants and animals there have been even a single species which have been produced in the course of direct descent from another species, then there is no further limit to the power of Nature, and we should not be wrong in supposing that with sufficient time, she could have evolved all other organized forms from one primordial type[103].

His theory of the earth almost caused him to be condemned by the Church as heretical. He tried to avoid confrontation with the Church by using terms which could be understood in more than one way. Directly following the text cited above, he added:

> But this is by no means a proper representation of Nature. We are assured by the authority of Revelation that all organisms have participated equally in the grace of direct Creation and the first pair of every species issued fully from the hands of the Creator.

The church changed its decision in 1781.

Buffon was familiar with Voltaire and other intellectuals who were later involved in the 1789 French Revolution. He was instrumental in the planning of the future Museum of Natural History, and his biological ideas were discussed well into the 19th century[104].

---

103 Wallace 1900: Evolution, p. 8. In the 1901 collection "The Progress of the Century". Also cited by Mayr 1982, p. 332.

104 See Eiseley, 1961, pp. 39-45.

Charles Bonnet: Pre-formation and the Chain of Beings

Charles Bonnet (1720-1793), a native of Switzerland and a lawyer by profession, was deeply interested in biology. He discovered and described the tracheal respiration in insects (1742) and the parthenogenetic reproduction in aphids[105].

The discovery that some aphids develop from unfertilized eggs, which become embryos within the mother's body, supported the belief in *pre-formation*: when the biological world was created as described in Genesis, all individuals of all species to be born from then to eternity were packed and ready – all that happens with time is the unfolding and growth of the miniature forms.

The concept of pre-formation was given further support when Von Leewenhoek – who became famous as the inventor of the microscope – or one of his students, reported that he had observed a complete miniature man ("homunculus") in the head of the human sperm[106]. Other people of the same period objected, contending that the "homunculus" should be housed in the egg cell, not in the sperm. The physiologist Haller calculated that on the 6th day of Creation, 6,000 years ago, the Creator made 200 billion (!) humans, and packed them in his wisdom in Eve's ovaries for future use. Even the philosopher Leibnitz accepted this idea[107].

The discovery that an unfertilized egg can develop into a viable individual suggested to Bonnet that the chick embryo was, perhaps, folded up in the egg before fertilization: the sperm then stimulates the heart of the embryo and induces it to grow[108].

But, Bonnet wondered, if the chick was present in the egg before fertilization, perhaps the foal too was present in the mare's ovaries "if it is proved that mammals develop from eggs"[109]: the difference is that the chick begins to develop after the egg has left the mother hen's belly, while the foal develops inside his mother. But if the foal was pre-formed, how is it that a mare impregnated by an ass gives birth to a mule – a very different animal in many aspects – and not to a horse? The embryo must be affected by external conditions during its development.

The Chain of Beings: Bonnet expanded the concept of the "Chain of Beings" to include the entire universe. Everything – from the stars to animals, plants, and man, was created by the same designer, forming a continuous "Chain of Beings", including all the intermediate forms between different organisms[110].

> Between the lowest and the highest degree of corporeal and spiritual perfection, there is an almost infinite number of intermediate degrees. The result of these degrees composes the universal chain. This unites all beings, connects all worlds, comprehends all spheres. One sole being is out of this chain, and that is He who made it.

Bonnet's writings contain detailed descriptions of the living world. He claimed that there is no precise distinction between the vegetable and animal classes: we cannot clearly discern where the vegetable terminates and the animal commences[111]. Intermediate forms close the apparent gaps in the continuum: asbestos rocks are intermediates between minerals and plants due to their fibrous texture. Colonial polyps and sea-anemones (Coelenterata) are intermediates between plants and

---

105 The parthenogenetic development of aphids was described in greater detail by T.H. Huxley 100 years later (Huxley 1858).

106 S.J. Gould suggested that this "observation" was nothing but a way of bringing the idea of preformation "ad absurdum". Gould, Ontogeny and Phylogeny, 1977, p. 19-28; "The Flamingo's smile", p. 145.

107 Haeckel, The Riddle of the Universe, 1876, p. 55.

108 The French term *Evolve,* meaning growth and differentiation of an individual, was adopted 100 years later to describe the changes of species by natural selection.

109 Fifty years later, K.E. von Baer discovered the mammalian egg.

110 Charles Bonnet, Abridgment of the contemplation of Nature, p. 7.

111 Ibid. p. 10.

invertebrates[112]. Eels are intermediates between fishes and reptiles, flying fish – between fish and birds, ostriches and bats are intermediate stages between birds and mammals. The savages (primitive people) in Africa and South America are intermediates between monkeys and humans.

> The ape is this rough draught of man, this rude sketch, an imperfect presentation, which nevertheless bears a resemblance to him and is the last creature which serves to display the admirable progressive [ness] of the works of God.

The "beings" in the chain are arranged according to the degree of their "perfection". There are two general classes of perfection – the corporeal perfection which is peculiar to bodies, and the spiritual perfection which is peculiar to souls. At the summit of the scale of our globe is placed Man, the masterpiece of earthly creations[113].

> If corporeal perfection corresponds with spiritual perfection, as there is reason to believe it does, Man – as he is superior to other animals by understanding, so he is likewise by organization. Whence we may infer that those animals, whose structure most nearly resembles Man, ought to be considered as most elevated in the scale. The faculty of generalizing ideas, or abstracting from a subject what it has in common with others and expressing it by arbitrary signs, constitutes the highest degree of spiritual perfection. And therein consists the difference between the human soul and the soul of brutes.

Belief in the Chain of Being continued as a model of the structure of the world for a long time.

[112] The "polype" was a sensation in 18th - century biology. Abraham Tremblay, a tutor on the estate of a rich Dutchman, discovered a small green thing in a pond, resembling a coral polyp (later named Hydra). He noticed that it moved toward the light like an animal, but was green and reproduced itself from cuttings like a plant! Tremblay reported his observations to the French Academy and to Bonnet (a family relation) but was met with disbelief. In despair, Tremblay mailed specimens to many scientists in Europe to convince them (R. Stott, Darwin's Ghosts, 2012).

[113] Ibid. pp. 15 and 23.

Bonnet, Complete Ladder

3

More Theories

Historical Background

Shortly before the publication of Paley's "Natural Theology" a very influential book came out – Thomas Malthus's "Essay on Population" (1798). Both publications had a common background: a few years earlier, the French Revolution had broken out (1789). This event had a profound effect on British intellectuals. T.H. Huxley wrote 100 years later: "An electric shock affected the nation".

Thomas Malthus (1766-1834), a member of the Royal Society, was educated at Jesus College, Cambridge, and was appointed professor of history and political economics at a college founded by the East India shipping company for training their employees. His "Essay on the Principle of Population" opens with a description of the French egalitarian ideas.

> The French Revolution... like a blazing comet, seems destined either to inspire with fresh hope and vigour, or to scorch up and destroy the shrinking inhabitants of the earth[114].

Thoughts of the approach to the future "ideal society", with no conflict or struggle between people, were being circulated among British intellectuals. In Malthus's own words:

> The great question now at issue is, whether man shall henceforth start forward with accelerated velocity towards illimitable, hitherto unconceived development, or be condemned to a perpetual oscillation between happiness and misery... The advocate for the present order of things is apt to treat the sect of speculative philosophers either as a set of artful and designing knaves, who ...draw captivating pictures of society only... to enable them to destroy the present establishment, or as wild and mad-headed enthusiasts whose silly speculations are not worthy of the attention of any reasonable man.
>
> The advocate of the perfectability of man, and of society, retorts on the defender of the establishment with a more than equal contempt. He brands him as the slave of the most miserable and narrow prejudices of civil society, only because he profits by them.

Malthus does not believe in the perfectability of man, and responds with the following metaphor:

> A man may tell me that he thinks Man will ultimately become an ostrich. I cannot properly contradict him. But before he can expect to bring any reasonable person over to his opinion, he ought to shew that the necks of mankind have been gradually elongating, and the hair is beginning to change into stubs of feathers[115].

Several British clubs and societies were very supportive of the new ideas. Among them was the "Lunar Society" in Birmingham, gathered around Darwin's grandfather Erasmus (see below)

---

[114] Malthus 1798, p. 1.
[115] Ibid. pp. 2-3.

and a liberal fraction of the English Church (the "dissenters"). A prominent member of the Lunar Society was the clergyman and chemist, Joseph Priestley[116], who stated clearly, years before, the circumstances when a revolution may be justified: in an essay on "The first principles of government" he wrote in 1771,

> ...if, instead of considering that they are made for the people, they should consider the people as made for them; if the oppressions and violation of rights should be great, flagrant and universally resented... If, in consequence of these circumstances it should become manifest that the risk which should be run in attempting a revolution would be trifling, and the evils which might be apprehended from it were far less than those which were actually suffered and which were daily increasing – in the name of God, I ask, what principles are those which aught to restrain an injured and insulted people from asserting their natural rights, and from changing or even punishing their Governors[117]?

The French Revolution worried the government as well as the Church. On the second anniversary (Bastille Day, 14 July 1791), members of the Lunar Society gathered in Birmingham to celebrate the event. An agitated mob ransacked their homes and laboratories and set them on fire, crying "Away with the philosophers!" and "The King and Church". Priestley left the town and eventually immigrated to America[118].

Malthus's "Essay on Population"[119]: The full title of Thomas Malthus's book fully reflects its purpose: "An essay on the principle of population, as it affects the future improvement of society". Malthus's book is about the human population and the conduct of society from an economic point of view.

Economically, Malthus makes two basic assumptions: a, that food is necessary for the existence of man; and b, that the attraction between the sexes is necessary, and will remain so. He states that

> The power of population is infinitely greater than the power of the earth to produce subsistence for Man.
>
> Population, if unchecked, increases in a geometric ratio. Subsistence increases only in an arithmetic ratio. This implies a strong and constantly operating check on population, from the difficulty of subsistence[120].

The evidence that Malthus provides to support his theory is statistical. Published data showed that the population in the 13 colonies in North America (the founding nucleus of the future United States) had doubled in 25 years, while the population of Europe had hardly changed in size. Malthus argues that, in America, the population was initially small and it had vast areas available for agriculture

---

[116] Priestly is known in biology for his demonstration of the role of oxygen in animal respiration. In a well-known experiment, he confined a mouse in a glass flask. When the mouse died, Priestley noticed that the volume of air in the flask was reduced by 20%. He called the missing component of the air "de-phlogistonated air" [combustion was thought to add "phlogiston" to the air]. At about the same time, Lavoisier in France discovered the properties of this gas and called it oxygen. In further experiments, Priestley introduced a potted green plant into the flask with a mouse. When the flask was positioned in sunlight, the mouse did not die. [As we now know, oxygen is released by photosynthesis in the green plant].
Incidentally, Priestley is the inventor of the common drink, soda water, produced by passing CO_2 through water. This was a by-product of his studies of fermentation by yeast.

[117] Cited by Huxley, 1874: "Joseph Priestley", pp. 126-127.

[118] Huxley 1874.

[119] The term "population" is used in the sense of "the act of populating" (Webster's dictionary).

[120] Malthus 1798, p. 5. Malthus illustrates numerically the difference in growth at the two ratios; after 10 generations the difference will be as 512 to 10.

– whereas in Europe the same agricultural land had been exploited for hundreds of years, while hardly any new land had been added in the same period.

The "check on population" is universal in Nature, and Man cannot escape its consequences:

> Among plants and animals, its effects are waste of seed, sickness, and premature death. Among mankind, misery and vice (ibid.)

This has clear, unavoidable social consequences:

> The actual distress of some of the lower classes, by which they are disabled from giving the proper food and attention to their children, act as a positive check to the natural increase of the population. The distresses they suffer from want of sufficient food, from hard labour and unwholesome habitations, must act as a constant check to incipient population. [To this] may be added vicious customs with respect to women, great cities, unwholesome manufactures, luxury, pestilence and war[121].

Malthus argues against the "poor laws", which required every community in Britain to allocate considerable sums of money for helping the poor in various ways, from providing employment, housing, and education to direct financial support. According to Malthus, increasing wages will not reduce the number of hungry people and their misery; but, on the contrary will increase them. The increased income will increase the standard of living of those employed, and enable them to support larger families. As a consequence, populations and the competition for the available jobs will increase, and the wages will consequently decrease. The demand for food will increase food prices, and with them the number and misery of the hungry people.

According to Malthus, the only positive factor for population increase is an increase in agricultural production. Expanding industry (which produces consumer goods) provides new jobs and increases the standard of living, but enables an increase in population with the same ill consequences as that of direct support of the poor!

Malthus and the Struggle for Existence: As stated above, Malthus's book deals with human economics, and the great majority of the text has nothing to do with the subject of the present book. However, in one paragraph, Malthus considers the problem of shortage of food as a mechanism for limiting population growth–in the entire biological world, not only that of humans:

> Through the animal and vegetable kingdoms, nature has scattered the seeds of life abroad with the most profuse and liberal hand. She has been comparatively sparing in the room and the nourishment necessary to rear them.
>
> The germ of existence contained in this spot of earth, with ample food and ample room to expand in, would fill millions of worlds in the course of a few thousand years.
>
> **Necessity**, that imperious all-pervading law of nature, restrains them within the prescribed bounds. The race of plants and the race of animals shrink under this restrictive law. And the race of man cannot, by any effort of reason, escape from it.

[121] Malthus 1798, p. 31.

The concept of the struggle for existence attracted the attention of both Darwin and Wallace, who realized its importance for their theory. The concept became a cornerstone in the Theory of Evolution[122].

Erasmus Darwin

Charles Darwin's grandfather, Erasmus (1731-1802), was a successful and well-known physician. He lived most of his life in a small town, Lichfield, in central England. His patients ranged for miles around the city and he traveled long distances in his carriage to visit them – some patients even came to live in his house while being treated. King George III wished to appoint Erasmus to the post of royal physician, but he declined, refusing to live in London[123].

Erasmus's first wife, Mary, gave birth to three children. The oldest, Charles, who became a highly promising medical student in Edinburgh, died of blood poisoning from an accidental cut during a post-mortem operation. The second son, Robert Waring, married Suzannah, daughter of Erasmus's close friend Josiah Wedgwood – founder of the famous British ceramics industry – and fathered Charles Darwin.

When Mary died, Erasmus married a friend's widow, Elisabeth Pole, who brought three of her own children into the household. Erasmus also supported two daughters he had sired by another woman out of wedlock. His second wife bore him six children; their daughter, Violetta, married the industrialist Samuel Galton, and gave birth to Darwin's cousin, Francis Galton[124]. Elisabeth lived long after Erasmus's death, and for 30 years took good care of all the children.

The Darwins and the Wedgwoods were further interrelated: Charles Darwin's wife Emma was the daughter of Josiah Wedgwood II, his mother's brother.

Erasmus Darwin belonged to the circle of British intellectuals called "The Lunar Society", because they met once a month on the day of a full moon, to discuss science and philosophy. Among the members of the Society were Josiah Wedgwood, the industrialist James Watt (who developed the steam engine), Joseph Priestley, and others. Benjamin Franklin, scientist and inventor of the lightning rod, then serving as the American ambassador to France, was in touch with the Society. All of them admired the ideas of the French Revolution – and most of them also supported the struggle for independence of the American colonies, in defiance of King George III and his government.

In addition to treating patients, Erasmus was concerned with public health. He recommended opening the windows in homes to allow in fresh air – contrary to the prevalent customs – and advocated activity in the open air, in particular for young women who customarily spent their lives indoors. He recommended that good drinking water should be supplied to city dwellers and condemned the pouring of sewage into the rivers [sewage should be separated and used for agriculture][125].

Erasmus was well versed in classical literature and art, and gained a reputation as a poet. He published three long poems – "Zoonomia", "The Love of the Plants", and "The Temple of Nature" – in verse, accompanied by extensive explanatory notes (see below). He was also well known as an inventor. Among his inventions was a steering device for carriages, correcting their tendency to overturn on sharp curves. He designed a wind-powered mill for grinding together raw materials

[122] In a recent article (2016) both Darwin and Wallace are blamed for treating Malthus's theory as scientific although he presented no proof. In fact, the author claims that as concerned the human population, Malthus was wrong. The human population size does not decline when food is scarce – it is affected by the difference between the cost of food production and its price to the customers, which is controlled by land owners and affected by politics. The notorious famine in Ireland was not due to over-population, but to the political regulation of land and imports by the English rulers of Catholic Ireland. The population was left with but a single source of food – potatoes – and when the potato fields were destroyed by fungus, famine resulted (Remoff 2016).

[123] King-Hele 1977.

[124] Chapter 22.

[125] In Charles Darwin's biography of his grandfather, Darwin 2003.

for Wedgwood's ceramics industry. He understood the phenomenon of artesian wells, and designed perhaps the first ever system for bringing water into his house by gravity. He was involved in planning and financing parts of the canal transport system in England.

Erasmus Darwin as a Botanist: Knowledge of the properties of plants was a required part of the training of medical doctors, who had to manufacture for themselves many of the drugs they prescribed for their patients (one drug that Erasmus often prescribed was opium!). Erasmus translated into English two of Linnaeus' works on plant classification, replacing the Latin terms with English ones [listing as the author of the translations "The botanical society of Lichfield."] He explained why the classification of plants was so important:

> The great use of the distribution of plants into natural classes is not only for the purpose of more readily distinguishing them from each other, and discovering their names, but also for that of more readily detecting the virtues or uses of them in diet, medicine, or in arts, as for the purposes of dyeing, tanning, architecture or ship building[126].

Erasmus published a large volume – more than 600 pages – on the structure and vital processes of plants. The book, titled "Phytologia", was published in Ireland in 1800 and became the standard botanical text in British universities in the early 19th century[127]. More than half of "Phytologia" deals with the practical issues of agriculture and horticulture (including a description of agricultural machinery that Erasmus himself had designed).

The first half of the book describes the basic properties of plants, which were little known at the time. Erasmus interpreted the anatomy of plants by analogy with the better-known anatomy of animals: all the vital systems of animals must be present, in one form or another, in plants as well. Erasmus noted some similarities between plants and colonial animals such as corals:

> Vegetables are, in reality, an inferior order of animals. This evinces that every bud on a tree is an individual vegetable being, and that the tree therefore is a family or a swarm of individual plants, like the polypus with its young growing out of its sides, like the branching cells of the coral insect[128].

But the plants differ from animals in that they are fixed to the ground and cannot move, and they obtain all their nourishment from the earth and the air.

> It follows that in examining their anatomy, we are not to look for muscles of locomotion, as legs and arms, nor for organs to receive and prepare their aliment, as a mouth, throat, stomach and bowels, by which animals are enabled to live many hours without new supplies of food from without[129].

Like animals, plants must breathe. The *respiratory* system – like the lungs or gills in animals – requires a large surface area, and is therefore located on the upper surface of the leaves. Erasmus objected to the idea, perhaps expressed by Priestley, that the leaves acquire some phlogistic materials from sunlight: if their role was to absorb light, they should have been black and not green, since Benjamin Franklin had shown that black color is the most effective in absorbing light.

In looking for the *circulatory* system in plants, the analogy with animals led Erasmus to suggest that the "blood" of plants is the liquid in the leaf veins. The "blood" of plants is oxygenated on the upper surface of the leaves, its color changes, and it returns to the trunk and roots on the leaf

---

126 Erasmus Darwin 1800, p. 514.
127 Kutschera 2009.
128 Phytologia, 1800, p. 2.
129 Ibid. p. 5.

underside (Erasmus interpreted the milky substance excreted by some plants as the oxygenated blood). He was impressed by the fact that the circulatory system works so well – without a heart:

> In some of the experiments of Dr Hales, who fixed glass tubes to vine stumps in the spring, the sap rose about thirty feet; and in some trees must rise still higher.

Plants, like animals, *excrete* unwanted materials. Some of these excretions are compounds which are very useful for animals and are utilized by man – such as starch, turpentine, cauchuk and that important resource, sugar:

> If sugar could be made from its elements, without the assistance of vegetables, such abundant food might be supplied as might tenfold increase the number of mankind!

Interestingly, Erasmus regarded the nectar of flowers ["honey"] as an important resource for the plant, which needs to be protected from predation by insects by means of deep and concealed nectaries[130].

Erasmus and his contemporaries looked in vain for a *nervous* system in plants, and could not explain phenomena such as the "sleep" of plants and sensitivity to touch (in Mimosa). In considering plant *reproduction*, Erasmus saw the seed as an equivalent of the animal egg, and imagined a system of vessels similar to the umbilical vessels, for bringing the nutrients from the cotyledons to the growing plant. He recognized the asexual reproduction in plants (as by cuttings), which produces offspring identical to their parent plant – unlike reproduction from seeds, which produces variable offspring.

Common Origin and the Evolution of Life: In his poem "Zoonomia", published in 1794 with many "footnotes" in prose, Erasmus catalogued human diseases and the appropriate remedies [derived from various plants] for treating them, arranged against the background of variation in nature. In doing so, he traced the evolution of life from microscopic specks in a primeval ocean, through plant life, fishes, amphibians, reptiles, land animals, and birds, and described the struggle for existence – fifty years before his grandson was to do so[131].

There is no doubt that Erasmus contemplated a common origin of the entire living world, and thought about the possibility of changes with time. In a "footnote" to the poem Erasmus writes,

> Would it be too bold to imagine, that in the great length of time since the earth began to exist – perhaps millions of years before the commencement of the history of mankind – would it be too bold to imagine, that all warm-blooded animals have arisen from one living filament, which the Great First Cause endowed with animality, with the power of acquiring new parts… thus possessing the faculty of continuing to improve by its own inherent activity[132].

In the notes appended to his last poem, "The Temple of Nature" (1803), Erasmus describes a scenario for the evolution of life on earth. He begins with "spontaneous generation" of the simplest creatures: mold, microscopic vegetable organisms, Monas and Proteus (amoeba) are spontaneously produced wherever suitable conditions exist. Then,

> After islands or continents were raised above the primeval ocean, great numbers of the most simple animals would attempt to seek food at the edges or shores of the new land, and might thence gradually become amphibious – as it is seen in the frog. At the same time,

130 Sixty years later, his grandson Charles assigned to nectar the important role of attracting insects for pollination.
131 King-Hele 1977.
132 Zoonomia 1794; in King-Hele, 1977, p. 244.

> new microscopic animalcules would immediately commence wherever there was warmth and moisture and some organic matter, that might induce putridity... and by innumerable successive reproduction for some ...millions of ages may at last have produced many of the vegetable and animal inhabitants which now people the earth[133].

Erasmus suggests the possibility that organisms could change with time – but notes that the idea is not his own. He cites another researcher, who is not mentioned by name.

> A naturalist, who had studied this problem, thought it not impossible, for example, that the first insects were the anthers and stigmas of flowers... and that other insects in process of time had been formed from these – some acquiring wings, others fins, still others claws, from their ceaseless efforts to procure food or secure themselves from injury. He [that investigator] contends that none of these changes are more incomprehensible than the transformation of caterpillars into butterflies[134].

Who was the mysterious author of those outstanding ideas? There is reason to believe that it was J.B. Lamarck in France[135]. Erasmus spoke French, and the members of the Lunar Society probably often discussed ideas emanating from France.

His poem "The Loves of Plants" was enthusiastically received by some of the readership, but stirred much criticism. In this poem, pollination and fertilization in plants are described as a love affair between the male and female parts of the flower. This analogy irritated some readers, who accused Erasmus Darwin of causing moral damage to young women, because pollination in plants is not restricted to one partner and is in some cases carried out freely by the wind – unlike the marriage system advertised by the Church – and might give the young women immoral ideas. A junior minister went so far as to publish a satirical poem under the title "The loves of the triangles", imitating Erasmus's style[136].

Erasmus apparently refrained from revealing the source of his ideas openly, and preferred to add them as "notes", hoping that most readers would stick to the poetry and miss the "notes". He had much to lose were he to be labeled liberal. He delayed the publication of the notes in "Zoonomia" as much as he could, as these were sufficient to turn the establishment against him. Erasmus certainly preferred not to advertise his French connections.

Erasmus Darwin and his Famous Grandson: Charles Darwin did not know his grandfather – Erasmus had died seven years before his grandson was born. When Charles was 70 years old, a German publisher asked his permission to publish a monograph on his grandfather. Charles agreed, and promised to take care of its translation into English and append an introduction. That introduction was to become a 100-page volume[137].

After listing the family connections and some details of Erasmus's life, Charles praises his grandfather as both an inventor and a physician. He mentions his grandfather's discovery of an heritable component in epilepsy and consumption (tuberculosis), and his concern with public health: as noted earlier, Erasmus had advocated the need to supply good drinking water to the population and to avoid pouring sewage into the rivers (sewage water should be diverted for use in agriculture). Erasmus had advocated activity in the open air, especially for young ladies who were traditionally confined to the home, and the opening of windows to let fresh air in (traditionally windows were shut to ward off diseases...)

---

133 The Temple of Nature, notes, p. 29.
134 The Temple of Nature, notes, p. 66.
135 See Chapter 6.
136 Connolly 2016.
137 Darwin 2003.

Charles evaluates Erasmus's biological contributions in his poems, albeit with little enthusiasm. He reports that "Zoonomia" had been translated into French, Italian, and German, and that had the Pope included it in the list of books forbidden by the Church! (Charles praises his grandfather's physiological ideas, which indicates that he had read "Zoonomia" carefully.) "The loves of the plants" had been received with enthusiasm at the time, but "in our own times nobody reads this kind of poetry". Charles commends his grandfather's declared objection to slavery, years before an anti-slavery law was passed in the British parliament[138].

Charles does not mention his grandfather's thoughts about evolution, to which historians of evolution were attracted[139], but he does cite the dramatic comment by his grandfather about the interaction between animals – in contrast to the then postulated ideal harmony in nature:

> The stronger locomotive animals devour the weaker without mercy. Such is the condition of organic nature, whose first law might be expressed in the words 'eat or be eaten', which would seem to be one great slaughter house, one universal scene of rapacity and injustice[140].

In a letter to his friend Charles Lyell (Chapter 7), Charles refers to Erasmus – but not as a precursor of his own Theory of Evolution. True, Erasmus had thought about a common origin of all animals, but this was not a new idea: Charles considered natural selection, not the common descent, to be the essence of his theory.

> Plato, Buffon, my grandfather before Lamarck, and others, all propounded the obvious view that if species were not created separately, they must have descended from other species[141].

Charles disregards the detailed description by his grandfather of competition between males for the possession of females – a subject upon which Charles would later expand in an entire book[142]. An important desire in part of the animal world [e.g. in male mammals], states Erasmus, is to acquire complete possession of the females. And the males have acquired weapons for the purpose. The male boar has long and sharp canine teeth, which have no other function except for fighting other males, since the boar is not a carnivore. It also has a heavy protective shield of skin at the front of its body. The male strikes its opponent obliquely, and the shield protects him from the blows of his opponent. Following is a poetic description of the selection scene in Erasmus's words:

> Contending boars with tusks enameled strike
> And guard with shoulder-shield the blow oblique
> While female bands attend in mute surprise
> And view the victor with admiring eyes[143].

James Hutton and "The Theory of the Earth"

James Hutton (1726-1797) is rarely mentioned in the evolutionary literature, but was perhaps the first true geologist[144].

---

138 Darwin 2003, pp. 88-89.
139 King-Hele 1977.
140 Darwin 2003, p. 41.
141 Darwin to Lyell. Darwin, F., Charles Darwin's Life and Letters, II: 198.
142 Sexual Selection, 1871. See Chapter 16.
143 The Temple of Nature, 1803, p. 68.
144 The title "the father of geology" is often given to Charles Lyell, who was born the same year that Hutton died.

Born in Scotland, Hutton studied medicine and received his degree at Leiden, the Netherlands, but then turned to agriculture and successfully managed an estate he had inherited from his father. His observations during the years of farming led him to wonder about the origin of the soil. He noticed the great destructive force of weather elements on the earth, and turned his thoughts to their long-lasting effects. In a paper presented to the Royal Society of Edinburgh in 1788, he wrote:

> A soil is nothing but the materials collected from the destruction of solid land. Therefore the surface of this land, inhabited by man, and covered with plants and animals, is made by nature to decay… and this soil is necessarily washed away, by the continual circulation of the water, running from the summits of the mountains toward the general receptacle of that fluid[145] [the "unfathomable" depth of the ocean].

But the earth had been made for Man, and man, the plants and the animals need the soil for subsistence. The world (seen as a machine) must have a mechanism to repair the damage done by the elements, lest it will be destroyed forever. The forces that cause the damage must be countered by the forces of rebuilding new soil. This requires two different operations: forming new rocks and lifting them up above the sea level.

> Consequently, besides an operation by which the earth at the bottom of the sea should be converted into an elevated land… there is required, in the operation of the globe, a consolidating power by which the loose material that had subsided from water should be formed into masses of the most perfect solidity…[146]. It is thus, upon chemical principles… that all solid strata of the globe have been condensed by means of heat and hardened from a state of fusion.

Hutton experimented with different chemical methods and concluded that the consolidating power must be heat, like the melting of iron or gold.

He found proof of his theory by observing the geological strata, near Edinburgh and later elsewhere. The phenomena he observed supported his theoretical expectations. He reasoned as follows,

> The strata formed at the bottom of the ocean are necessarily horizontal… and continuous in their horizontal direction or extent… But, if these strata are cemented by heat of fusion, and erected with an expansive power acting from below, we may expect to find every species of fracture, dislocation and contorture… and every degree of departure from an horizontal toward a vertical position[147].

The upward moving force is that which operates in volcanoes: after volcanic eruptions such as that of Mount Etna in Sicily, the layers of rock at the bottom of the sea are elevated. This is how the land is renovated.

> The world which we inhabit is composed of the materials, not of the earth which was the immediate predecessor of the present, but of the earth which… had preceded the land that was above the level of the sea, while our present land was yet beneath the waters of the ocean. Here are three successive periods of existence, and each of these is, in our measurement of time, a thing of infinite duration.

Hutton concludes that the cycle of destruction and renovation is a perpetual system:

---

[145] Hutton, The Theory of the Earth, p. 6.
[146] Ibid. p. 13.
[147] Ibid. p. 45.

Nature has contrived to productions of vegetable bodies, and the sustenance of animal life, to depend upon the gradual but sure destruction of a continent: that is to say, these two operations must go hand in hand. But with such wisdom has nature ordered things in the economy of this world, that the destruction of one continent is not brought about without the renovation of the earth in the production of another...[148].

The result, therefore, of our present inquiry is that we find no vestige of a beginning – no prospect of an end[149].

148 Ibid. p. 67.
149 Ibid. p. 74.

The Second Circle

The center of the second circle is located in France. Following the French Revolution in 1789, the greatest scientific minds in Europe – in mathematics, physics, chemistry, zoology, and botany, gathered at the French Academy of Sciences in Paris[150].

> Twenty years ago, if a stranger penetrated the sanctuary of the Academy of Sciences, he would have found himself struck to the depth of his soul by a respect amounting to awe, at finding assembled around him so many rare geniuses, the pride of France and the shining lights of their century.
>
> Geometricians bowed before Laplace, Ampere, Legendre, Poisson; physicists came to honor Hauy, Berthollet, Chaptal, Vauquilin; and naturalists proudly added to those illustrious and popular names those of Jussieu and Desfontaines, those of Cuvier and of Geoffroy Saint-Hilaire, long consecrated also by the veneration of Europe[151].

The Natural History Museum and the Royal Botanical Gardens became the property of the new Republic. By special decree, the new Natural History Museum was established with 12 professors, appointed by name, to be replaced only following retirement or death – and with each in charge of a specified scientific field. Among the founding professors were J.B. Lamarck, as Professor of "insects, worms, and microscopic animals", and Etienne Geoffroy Saint-Hilaire, as Professor of vertebrate zoology. It was Geoffroy who invited the young Georges Cuvier to join the Museum, first as a temporary appointment, and then, when one of the professorial chairs became vacant, as a member of the permanent staff.

The end of the 18th and early 19th century was a period of great biological discoveries. Biological data were being gathered from distant parts of the globe by scientific expeditions. James Cook traveled to Australia and the accompanying biologists described strange plants and animals unknown in Europe. Von Humboldt traveled in South America and the South Seas and described the distribution of plants and animals. Napoleon's armies briefly occupied North Africa and Palestine and scientists accompanying his army brought back mummies from Egyptian tombs. At the Natural History Museum in Paris, zoologists Georges Cuvier, Etienne Geoffroy Saint-Hilaire, and Jean Baptiste de Lamarck were at work. Young Charles Darwin started his voyage around the world in 1832, carrying with him the first edition of the monumental book, "The Principles of Geology" by Charles Lyell. Lyell, a Scot, had spent several months in his youth at the Museum in Paris, associated with Lamarck and Cuvier, and admired their work. He may be thought of as a biological bridge between the scientists of the two countries. The contributions of these scientists formed a solid basis for the Theory of Evolution that would later develop.

Until the mid-1830s, the discussions and scientific publications of the Academy of Sciences and the Museum were published only in French. Many of these publications remained only in French in the 150 years that followed, even though English had replaced French as a global scientific language (Lamarck's "Zoological Philosophy" was translated into English as late as 1912, more than 100 years after its first publication). Some of these important publications have been more recently translated into English[152].

[150] Le Guyader, 2004, p. 19.

[151] From the eulogy by the chemist Dumas at the funeral of Geoffroy, 1844.

[152] Rudwick 1997; Le Guyader 2004.

4

Georges Cuvier and the Theory of Catastrophes

Georges Cuvier (1769-1832) was a central figure in 19th-century biological science. Born in France but educated in Stuttgart, Germany, he was invited by Etienne Geoffroy Saint-Hilaire, then head of zoology in the newly-established Museum of Natural History in Paris, to join the staff of the Museum. He specialized in comparative vertebrate anatomy, was appointed Professor of Anatomy in 1802, and served as a permanent secretary to the Academy of Sciences.

Cuvier survived the dramatic changes in the government of France – from monarchy to revolutionary republic, and to the Napoleonic administration – and fulfilled administrative roles in the government as Minister of Education and Deputy Minister in the Home Office. He married a widow [whose husband had been executed during the Terror] with four children – one daughter became his secretary – but all their mutual offspring died in childhood.

Cuvier studied and classified the large numbers of fossils that were being discovered in the quarries around Paris (or brought to Paris from various collections in European capitals occupied by the armies of Napoleon). Previously collected as mere curiosities, Cuvier recognized them as the remains of extinct animals[153]. He eventually became a world authority on fossils.

The "primary discussion" of his book on fossils was almost immediately translated into English (by Jameson in Edinburgh in 1813), under the title "Essays on the Theory of the Earth" and became popular – three editions were printed[154]. In addition to four volumes on fossils, Cuvier prepared a series of 21 volumes on the taxonomy of fishes, many of which were published after his death.

Among the first discoveries published by Cuvier was the description of a large fossil from South America, known as "the animal from Paraguay" (it was collected in Argentina, then under Spanish rule, and the skeleton was assembled in Madrid). Cuvier identified it as an (extinct[155]) relative of the sloth, and named it Megatherium (= giant animal). This find was important, because its size made it "scarcely probable that, if this animal still exists, such a remarkable species could hitherto have escaped the researches of naturalists"; therefore, it must have entirely disappeared from the face of the earth.

---

[153] T.H. Huxley stated that the first person to identify fossils as animal remains was the Italian, Steno, in 1669. Steno looked at strange, triangular objects that were common in the rocks and were called "glossopterae" (shining wings). He observed that they looked very much like the teeth of a shark, which he dissected for comparison. He wrote that bodies which are closely similar must have been produced in the same way – "like effects imply like causes" (Voltaire's Principle of Zadig).

[Huxley, The progress of palaeontology; p. 30 in "Science and the Hebrew Tradition", 1876].

Fortley (2009) noted that the German mathematician Leibnitz recognized in 1749 that fossils were extinct animals (R. Fortley, Nature 455(4)).

[154] The translation somewhat distorted Cuvier's views on the subject: the translation emphasized the presence of fossils as evidence in support of the biblical story of the Flood – while Cuvier questioned the credibility of all the old stories, and founded his theory of catastrophes [see below] purely on geological and fossil evidence, devoid of any religious motives; Rudwick 1997.

[155] Believers in the Linnaean Systema Naturae could not accept the possibility that species became extinct. Since the world was planned and created by the Deity, the removal of any species would disturb the harmony of nature. The missing species were believed to exist in some unexplored corner of the earth.

Comparative Anatomy

Cuvier maintained that every organism is an integrated system, in which the parts interact with each other. An organ cannot change independently, without causing changes in other organs. The recognition of this principle enabled the reconstruction of the form and structure of an entire organism from the study of any part of it, however small:

> Every organized being forms a whole, a unique and closed system, in which all the parts correspond mutually and contribute to the same definitive action by a reciprocal reaction. None of the parts can change without the others changing too[156].
>
> This is because the number, direction and shape of bones that compose any part of an animal's body are always in a necessary relation to all other parts, in such a way that – up to a point – one can infer the whole from any part of them, and vice versa[157, 158].

For example, from the structure of the teeth one can infer the type of food the animal ate and the structure of the alimentary canal. From the structure of the leg bones one can infer the mode of locomotion. It is possible to identify the points of attachment of the muscles to the bones, and draw them in place:

> The flesh being once reconstructed, it would be straightforward to draw them covered by skin and we would thus have an image not only of the skeleton, that still exists, but of the entire animal as it existed in the past[159].

Cuvier published drawings of such reconstructions of a number of extinct mammals.

The comparative anatomy of existing and fossil animals provided evidence that the fossils were anatomically different and thus must be extinct species. Cuvier's confidence in his knowledge is illustrated by the following incident, which was published by Cuvier himself. A split slab of rock containing a rare, unknown fossil collected near Paris was brought to the Museum. Cuvier examined the exposed jaw and teeth and identified the fossil immediately as an opossum, a marsupial – and marsupials were unknown from Europe. To prove that he was right, Cuvier assembled a crowd of spectators and, in their presence, he slowly worked his way into the stone using a knife – risking the destruction of the precious specimen – to reveal the two characteristic "marsupial bones" that were embedded deeper in the rock.

History of the Globe: Geological and Climate Changes

Theories about the history of the globe varied among different writers. Cuvier devoted many pages in his "Discourse on the Revolutionary Upheavals on the Surface of the Globe" (1825)[160] to a review of the older theories. He concludes that the estimated ages of different civilizations have been greatly inflated, and that no evidence exists of civilizations older than four to six thousand years. One item is common to all the old traditions: the story of a flood. This, Cuvier is convinced, reflected a real event.

---

156 In Le Guyader 2004, p. 14.

157 Cuvier Text 5, Rudwick 1997, p. 36).

158 The American zoologist Henry F. Osborn denied that this conclusion is justified. He wrote that teeth are correlated with skull and neck structure, but not with limb and body structure, and therefore diverge independently in evolution. "The correlation is not morphological, as Cuvier supposed, but physiological, function always preceding structure. It becomes closest when teeth and feet combine in the same function as in prehensile canines and claws of the Felidae, and most diverse where the functions are more diverse, as in the teeth and paddles of the Pinnipedia". (Osborn 1902).

159 Cuvier text 5, Rudwick 1997, p. 40. Ibid. Cuvier text 7.

160 On the internet:www.mala.be.ca/~Johnstoi/Cuvier-e.htm

> Thus, I am of the opinion ...that if there is something confirmed by geology, it is that the surface of our world has been the victim of a great and sudden upheaval, whose date cannot go back much beyond five or six thousand years – and that this revolutionary upheaval pushed down the countries where human beings, and the species of animals best known to us today, previously used to live and made them disappear – that it, by contrast, made dry land of the bottom of the most recent sea and from it created the countries now inhabited – that since this revolution the small number of individuals which it spared have spread out and propagated throughout the territories recently made dry land, and consequently that it is only since this time that our societies have... formed institutions, raised monuments, collected facts about nature, and put together scientific systems.
>
> But these countries inhabited today... had already been inhabited previously, if not by human beings, at least by terrestrial animals. Consequently at least one previous revolution has put them under water[161].

All the geologists of the period recognized that the earth had been occupied, in different periods, by different species of animals. Fossil remains of mammals that were known to live in tropical countries – elephant, hippopotamus, rhinoceros, hyena, lion and leopard – were discovered in northern Europe. Geologists believed that various catastrophes had caused extreme changes in the surface of the earth. Some believed (the "neptunists", named after Neptune, the god of the sea) that the earth was originally covered by water, which had disappeared into subterranean spaces. Others believed (the "vulcanists", after Vulcan, the god of fire) that volcanic eruptions and earthquakes had changed the surface of the world.

As noted above, Buffon maintained that as the globe was slowly cooling, warmth-loving species – whose fossil remains are found where they had once lived – were forced towards the equator when the north and south began to become too cold for them.

Cuvier did not accept this interpretation. In a paper he read to his colleagues at the Museum early on in his career[162], he compared the dentition and skull of the African and Asian elephants with those of the extinct northern-Europe mammoth, all of which were formerly thought to belong to one and the same species. Cuvier strongly argued that the extinct mammoth was not a tropical animal, and could not be used as evidence that a tropical climate had once dominated the North. Rather, the mammoths could have been adapted to a cold climate:

> The mammoth is different from the elephants more than the dog is different from the jackal or the hyaena", and the two species of elephants differ from each other "more than the ass is different from the horse, or the goat from the sheep[163].

Fossils

Cuvier organized the collection of fossils for the Museum, and even hired people to look for them in the quarries to ensure their arrival in good shape. He declared that he intended to study the anatomy of all the fossilized animals, because

> The antiquities of nature, if they may be so termed, will provide the

Cuvier, Reconstruction from Tooth

[161] Conclusions. Ibid. p. 57.
[162] Cuvier text 3, Rudwick 1997.
[163] Rudwick 1997, p. 24.

> physical history of the globe with monuments as useful and as reliable as ordinary antiquities provide for the political and moral history of nations[164].

According to Cuvier, the geological layers had accumulated gradually, one on top of the other. The order of the layers is chronological and important. In the oldest, granitic layers there are no fossils, because these rocks were formed before the appearance of life on the earth. In later layers, each stratum has a distinct assemblage of fossils. The first (earliest) fossils are invertebrate – mainly mollusks – then fishes are encountered – the remains of mammals appear only in the upper (most recent) strata. Cuvier considered the vertebrate fossils more important than the mollusks, because vertebrates were definitely terrestrial animals, and the fact that the fossil species had not been found living anywhere in the present was proof that they were extinct. Like the Megatherium, it was unlikely that such animals could exist unseen.

Striking differences in the fossil fauna occurred abruptly between consecutive layers of rock, with no intermediate stages. This, thought Cuvier, must reflect a series of sudden changes – catastrophes – because the forces in operation in his time, like local volcanic eruptions, earthquakes, and floods – which he reviewed at length – were too slow and not strong enough to account for the sudden extinction of entire faunas.

> For a long time we thought we could explain earlier revolutionary upheavals by present causes, just as we readily explain past events in political history. But we are going to see that unfortunately things are not the same in the history of physics. The thread of the processes is broken. Nature's march has changed, and none of the agents which she uses today would have been sufficient to produce these ancient works[165].

> All these facts… seem to me to prove the existence of a world previous to ours, destroyed by some kind of catastrophe[166].

The Catastrophes

Cuvier believed that the history of the earth was short – just a few thousand years. He scorned the ideas of some of his colleagues – particularly J.B. Lamarck – that the age of the world was very much greater:

> I know that some naturalists rely a great deal on the thousands of centuries which they add up with the stroke of a pen. But in those matters we can hardly judge what a lengthy time would produce, except by multiplying mentally what a lesser time produces[167].

Cuvier examined the descriptions of animals in the writings of Greek and Roman authors, and found them to be very similar to extant animals. In particular he examined the collection that his colleague, Geoffroy Saint-Hilaire, had brought from 3000-year-old Egyptian tombs[168].

> We certainly do not observe more differences between these creatures and those which we see today, than between human mummies and today's human skeletons. There is nothing which can in the least support the public opinion that the new genera I have discovered or established among the fossils… could have been the ancestors of some animals today[169].

---

[164] Cuvier text 5, Rudwick 1997, p. 35.
[165] Ibid. p. 7.
[166] Rudwick 1997, p. 24.
[167] Cuvier 1825, p. 30.
[168] See Chapter 5.
[169] Cuvier 1825, p. 30.

Cuvier was impressed by the dramatic differences in the fossil fauna of the different geological strata. These differences proved to him that there had been sudden geological changes. Each catastrophe had destroyed the entire existing fauna, and the earth had to be repopulated in some way. The biblical Noah's Flood was considered the most recent of these catastrophes, as well as the most dramatic.

> Thus life on earth has often been disturbed by terrible events... living organisms without number have been the victims of these catastrophes. Some were destroyed by deluges, other were left dry when the sea bed was suddenly raised. Their races are even finished forever[170].

These changes are unlikely to have occurred slowly – time was too short – and there is no evidence for slow changes in the structure of animals – as suggested by Cuvier's colleague and adversary, J.B. Lamarck[171] (for Cuvier each species had remained the same for as long as it existed):

> ...if the species have changed by degrees, we ought to have found traces of these gradual modifications – that we ought to have discovered certain intermediate structures between the Palaeotherium and today's species – and that up to the present time this has not happened at all. Why have the depths of the earth not preserved monuments of such a curious genealogy, unless it is because the earlier species were as unchanging as our own[172]?

In his public lectures in 1805, Cuvier suggested that there were six important geological eras, parallel to the Six Days of Creation. The changes in the fauna between eras were sudden and not gradual: strata containing fossils of marine and terrestrial organisms alternate without any discernable transitional changes. The thickness of the strata was seen as proof that the same conditions had persisted for a long, stable period after each catastrophe.

> One is thus soon disposed to believe, not only that the sea invaded all our plains, but also that it stayed there long and peacefully, so as to form deposits so thick and extensive... containing such well-preserved remains[173].

The catastrophes were mainly floods of seawater, following volcanic events with the emergence of mountain chains – as shown by layers containing marine mollusks at 14,000 feet elevation in the Alps[174].

Cuvier's book, "Recherches sur les Fossiles du Quadrupeds" (Researches on the vertebrate fossils), was published in several parts by the Museum. The most important part, entitled "Preliminary discussion" (Discours préliminaire) was printed repeatedly in the 19th century, years after Cuvier's death. In this book Cuvier describes his ideas about the catastrophes that had shaped the world:

> Thus the great catastrophes of our planet have not only caused the different parts of our continents to emerge by degrees from beneath the waves: it has also happened several times that areas made into dry land have been covered again by water... and the particular ground that the sea left during its last retreat had already been dried once before, and had sustained quadrupeds, birds, plants and all kinds of terrestrial productions[175].

---

170 Cuvier text 19, Rudwick 1997, p. 190.
171 See Chapter 6.
172 Cuvier 1825, p. 28.
173 Rudwick, 1997, p. 187.
174 Rudwick 1997, p. 86.
175 Rudwick 1997, Cuvier text 19, p. 189.

> I do not claim that a new creation was needed to produce the existing species. I only say that they did not exist in the same places, and they must have come there from elsewhere[176].

Large mammals, discovered frozen in Siberia – like the mammoth that was discovered intact with skin and hair – were proof that the catastrophe had been a sudden event. These animals had not been swept into Siberia with flood waters from tropical areas (as some claimed): they had frozen to death where they had lived and had not been able to escape.

Cuvier wondered about the absence of *fossil human bones* in the geological strata. Human bones should have been preserved like the bones of vertebrates:

> It is certain that no one as yet found human bones among the fossils… everything therefore leads to the belief that, in the same time period as the upheavals which buried the fossil bones, the human species did not exist at all in the countries where the bones are discovered [they could have survived the catastrophe in some "regions of small extent, from where they repopulated the earth after these terrible events"][177].

Classification

Cuvier suggested a classification system for the animal kingdom, based on phenotypic appearance – mainly the pattern of body plan. His system of classification recognized no linear, hierarchical order as in the Linnaean system, nor any lines of descent, as suggested by his colleague and adversary, J.B. Lamarck. Cuvier arranged the organisms in four mutually-exclusive "branches": Vertebrata, Mollusca, Articulata (arthropods), and Radiata (including coelenterates and worms).

> The true classification sees each organism in the midst of others. It shows all the variations or lineages by which organisms are linked more or less closely in this immense network which makes up organized nature.
>
> Therefore let it not be thought that, because we shall be placing one genus or family before another, we actually consider it superior, or more nearly perfect in the system of organisms.
>
> Only he could make that claim, who pursues the widely fanciful project of ranking organisms in a straight line – and this is an approach we renounced long ago[178].

Cuvier's system of classification was not accepted by his contemporaries, but was later revived and praised by Louis Agassiz[179].

Concluding Notes: From the standpoint of the Theory of Evolution, Cuvier is regarded as a conservative because he believed that all species had been created as we now see them, and are immutable; and because he opposed Lamarck's idea of similarity by descent and transmutation. But he had his merits.

> Fixism, that is the error of which he [Cuvier] is often accused. But history often preserves only a caricature of the thought of men of science, insisting as if with pleasure on the points they have consequently been shown to be mistaken. Granted, Cuvier was a fixist. But we can only admire, on the one hand, the (almost) perfectly objective manner in which he carried out his palaeontological work, and, on the other hand, the tenacity he displayed in finding the only likely hypothesis that could at the time be opposed to transformism[180].

---

176 Rudwick 1997, p. 229.
177 Cuvier 1825, p. 32.
178 Cuvier 1828, p. 281.
179 See Chapter 10.
180 Le Guyader 2004, p. 16.

5

Etienne Geoffroy Saint-Hilaire: The Theory of Comparative Anatomy

Etienne Geoffroy Saint-Hilaire (1772-1844) was a brilliant child. He studied law in Paris and obtained his first degree when he was only 16 years old. After the Revolution he chose to study medicine and then specialized in crystallography. At the age of 21 he was invited – as one of 12 professors – to join the staff of the new Museum of Natural History, where he was appointed Professor of Zoology. Cuvier accepted Geoffroy's invitation to join the institute, and the two became close friends. Both were interested in comparative anatomy and published several articles together. The friendship broke down, however, when a scientific controversy over philosophical principles turned into a bitter personal conflict (see below).

Geoffroy was appointed Head of Zoology at the Museum in 1810 (he had politely offered the job to Lamarck, who declined). Geoffroy was accepted into the Academy of Sciences twelve years later than his younger colleague. In 1840, having lost his eyesight, he resigned, and died four years later, aged 70. He was given an impressive funeral.

Unlike Cuvier, Geoffroy refused to take on any administrative positions outside the Museum. Instead, he joined the scientific expedition that accompanied Napoleon's armies into Egypt, bringing back precious finds from ancient Egyptian graves, including mummified animals[181]. Cuvier opened the wrappings of the mummies and discovered that they were identical to extant species: their form was the same although they were definitely thousands of years older.

> It is impossible to control our flight of imagination on seeing still preserved with its smallest bones and hair, perfectly recognizable, an animal which two or three thousand years ago had in Thebes or Memphis its priests and altars… This part of Mr Geoffroy's collection shows that these animals are exactly similar to those of today[182].

The Philosophy of Comparative Anatomy

Like Cuvier, Geoffroy was an enthusiastic comparative anatomist. But while Cuvier used the comparative methods as a practical instrument for species recognition and classification, Geoffroy developed comparative anatomy as a scientific philosophy. Cuvier discussed the similarities and differences of species only within his four "embranchements" (circles), while Geoffroy believed that there was a unifying pattern of structure that connected between animals from different circles, even if the pattern is not always obvious at first glance. He considered this "unity of pattern" as a

[181] Le Guyader 2004. The British navy blockaded the shores of Egypt and prevented Napoleon's fleet from sailing home – but nevertheless let the ship carrying Geoffroy's collections sail to France.

[182] Lamarck 1984 – Zoological Philosophy, pp. 41-42. Cuvier interpreted this similarity as proof of the fixity of species. Lamarck questioned this interpretation, suggesting that the time was too short to expect a change: the climate in Egypt may not have changed much in 3,000 years, therefore animals did not have to adapt to new environments and their structures did not change (Lamarck 1984, p. 43).

philosophical principle, "the principle of analogues", which should be the basis for understanding the living world.

> Naturalists have returned to the doctrine of analogies. They begin to glimpse this fact of great importance for theory: that an organ, varying in its conformation, often passes from one function to another. They can follow the front foot just as well in its different uses as in its numerous metamorphoses, and see it successively applied to flight, to swimming, to leaping, to running, etc. To being here an instrument for digging, there hooks for climbing, elsewhere offensive or defensive weapons; or even becoming, as in our species, the principal organ of touch, and consequently one of the most efficacious vehicles of our intellectual faculties[183].

In the introduction to his book "Anatomical Philosophy", Geoffroy explains that his theory is based on four principles[184], three of them important: *The theory of analogy* states that in order to establish similarity between organs in different animals, one should look for correspondence between all component parts of the organs in question – because the parts determine the characteristics of the whole organ (he warns that analogy is only an advisory tool and should not be used without discrimination[185].

The principle of balancement between organs, maintained during development, guided Geoffroy in identifying reduced organs of which only traces could be detected, by applying *the principle of connections.*

> Nature constantly uses the same materials, and is ingenious only in varying their forms... we see her always tending to make the same elements reappear, in the same number, in the same circumstances, and with the same connections. If one organ takes on extraordinary growth, the neighboring parts... are nonetheless conserved, although in minimal degree that often leaves them without utility. They become, as it were, so many rudiments that testify... to the permanence of the general plan[186].

Geoffroy suggested that his philosophical approach was very useful for teaching anatomy to students. Rather than examining in detail every species, this principle requires the student of biology to study in detail only representatives of the vertebrates – man, a quadruped (ruminant), a bird (pigeon), and a fish – in order to know the basic structure of the whole class[187].

The Unified Body Plan

In his first paper on the subject (1820) Geoffroy applied the principle of a unifying pattern to the vertebrates. The skeleton, the nervous system, and the respiratory system of all vertebrates seemed to be built on the same model. This paper was praised by Cuvier. Later, however (1822), Geoffroy made a bold attempt to apply the same principle to the insects: their chitinous external skin (integument) supports the internal organs – like the internal bony skeleton of the vertebrates, and the limbs (legs) in both groups are built on a similar pattern. In the insects the internal organs are enveloped in the chitinous skin – analogously, in the tortoises and turtles the internal organs are enclosed by their bony skin. The wings of the insect may be equivalent to the swim-bladder in fish:

---

183 Anatomical philosophy 1818, preliminary discourse, in Le Guyader, 2004, p. 29.

184 Geoffroy 1922, in Le Guyader 2004, p. 45.

185 Geoffroy 1820, in Le Guyader 2004, p. 60.

186 In Le Guyader 2004, p. 22. The application of this principle enabled the tracing of the auditory bones in the mammalian ear – to the ancestral position as parts of the reptilian jaws.

187 In Le Guyader 2004, p. 34.

> I have a beetle which I preserved in ecdysis, and it is clearly seen that the wing is nothing but a bladder.

The skeleton of insects is located outside their body, while in the vertebrates the skeleton is inside. It thus seemed reasonable to Geoffroy to form a rule:

> Every animal lives around or inside its skeleton[188].
>
> Insects live within their vertebral column, like molluscs in the interior of their shells.

Geoffroy extended his analogy between insects and vertebrates in a detailed study on the vertebral column. The skeleton of a young fish served as a vertebrate model. As a model of the Articulata, Geoffroy used the tail of a lobster: when viewed from its side it looked similar to a fish's tail. Each vertebra begins its development as a bony ring, or tube, and is gradually filled with concentric rings of bony material [later, projections appear to form arches, protecting the nerve cord and the blood vessels]. Each segment in the insect – and other Articulata – is then analogous to the body of a vertebra, which encloses the internal parts in the segment and then adheres to the skin.

Since in vertebrates the nervous system is dorsal and in the insects it is ventral, Geoffroy saw the insects as "inverted vertebrates", and illustrated them in this form in his paper[189].

The Conflict with Cuvier

The attempt to combine two of his "circles" – Vertebrata and Articulata – upset Cuvier, who responded furiously:

> Your memoir on the skeleton of insects lacks logic from beginning to end, said M. Cuvier, expressing himself on my first work, with an extreme vivacity, before the professors of the Museum. You compare things that are in no way susceptible of comparison. There is nothing in common, absolutely nothing, between the insects and the vertebrates – at most one point: animality[190].

Geoffroy was offended by this response. He tried unsuccessfully to assure his friend – and their colleagues at the Museum – that their friendship should not be affected by the differences in opinion. But Geoffroy refused to yield to the authority of his former friend.

> It is clear that I am not reporting these observations so that they may profit persons who are of mature age: a person who has received the lessons of a long experience is immune from all seduction. I am addressing myself to youth, naturally avid for novelties. My honesty in the sciences, my love for the truth, and the misgivings that I just refused to dissimulate commit me to warning this interesting youth against my own results... the reason is an

---

[188] Geoffroy 1822; in Le Guyader 2004, pp. 57-58.

[189] The presentation of insects as inverted vertebrates sounded ridiculous. Recently, however, molecular evidence was published to suggest that there may be an element of truth in the idea. Sander and Schmidt-Ott (2004) examined the genes which control dorso-ventral differentiation in embryos. The homologues of some genes which were expressed on the ventral side of the insect embryo (Drosophila) were expressed on the dorsal side of the embryo in the vertebrate (frog). Injection of mRNA from Drosophila into frog eggs reversed this differentiation. The authors suggested that Geoffroy had had the right idea, but he wrongly thought that the inversion took place in the evolution of insects, while in fact it occurred in the evolution of the vertebrates! (Sander and Schmidt-Ott 2004, p. 86).] See also Le Guyader 2004, Chapter 7, for an evaluation of the conflict.

[190] Le Guyader 2004, p. 50. Geoffroy was at a disadvantage in the verbal exchange. He is described as graying, carelessly dressed, and slow in expressing himself. Cuvier looked younger (although actually three years older than Geoffroy), was ginger-headed, impeccably dressed, and fluent in speech.

> absolute condemnation of these same views, pronounced (with a little violence, doubtless too much) by the greatest of the naturalists of our age.
>
> No, I have not wanted to wound an old friend. I have remained the same with regard to him – always equally devoted as at the time everything was on good terms with us and we shared everything. What M. Cuvier is condemning at this moment is the totality of my views, my whole philosophy. But what does this divergence of opinions really prove? Only that M. Cuvier and I think differently about theories[191].

The conflict intensified when two fossil crocodiles were discovered in Normandy. Cuvier, with some hesitation, classified them as Gavials (Asiatic crocodiles). But Geoffroy – who after the Egypt campaign was considered an expert on crocodiles – found important distinguishing features in the skulls that enabled him to place the new finds – classified as a genus Teleosaurus – as intermediates between the Gavials and crocodiles. The characteristics of the fossil crocodiles led him to think about the possibility of descent from older ancestors:

> To show that these animals are in natural order, we shall say that they must follow the genus Crocodilia at some distance… the [low?] degree of probability that the Teleosauruses... animals of antediluvial times, are the source of the crocodiles prevalent today in the hot climates of the two continents.
>
> I can propose that it is by no means repugnant to reason, that is, to physiological principles, that the crocodiles of the present epoch might descend by a succession without interruption from antediluvial species, recovered today in the fossil state in our territory[192].

These carefully expressed statements were contradictory to Cuvier's Theory of Catastrophes, and must have widened the gap between the two former friends. The geologist Charles Lyell extended Geoffroy's position to include the entire animal world:

> It is said, the stability of a species may be taken as absolute, if we do not extend our views beyond the narrow period of human history. But, let a sufficient number of generations elapse… and the characters of the descendants of common parents may deviate indefinitely from the original type. Accordingly, M. Geoffroy Saint-Hilaire has declared his opinion, that there has been an uninterrupted succession in the Animal Kingdom, and that the ancient animals whose remains have been preserved in the strata, however different, may nevertheless have been the ancestors of those now living[193].

In 1830, the conflict was brought to a head when two young scientists presented a paper to the Academy on the structure of mollusks, in particular Cephalopods (squid). They attempted to illustrate that the structure of these animals agreed with the unified "general plan". The paper was accepted enthusiastically by Geoffroy, but Cuvier was furious because it meant breaking through another of his independent "circles" – the Mollusca.

> It is not I who will suppose that even the most vulgar naturalists could have employed these words, Unity of composition, Unity of plan, in their ordinary sense of *identity*. No one of them would maintain for a moment that a polyp and a man had in this sense one single composition, one single plan. For the naturalists …*unity* does not mean *identity*, but is given a twisted sense to mean resemblance, analogy. Unity of composition [of man and

---

[191] On the organization of insects, 1820, In Le Guyader 2004, p. 62.

[192] In Le Guyader 2004, p. 93-94. The idea of evolution by common descent had been suggested by their colleague J.B. Lamarck in his book of 1809, but Lamarck had few supporters at the Museum (see Chapter 6).

[193] Lyell 1859, p. 567. See Chapter 7.

> whale] means only "a great resemblance in composition", or [man and snake] "a certain resemblance in composition"[194].

This started a much-publicized debate, which lasted several weeks. A series of scientific papers was exchanged between the contestants (and read before the Academy members) within a short period in 1830. The conflict was also publicized in England and Germany – even Goethe took sides[195].

Other Scientific Contributions

Embryo Development and Teratology: Between 1811 and 1830, Geoffroy – with his colleague Etienne Serres – studied the development of embryos. They were interested in the disturbances during development that result in a "monster" when the embryo reaches the adult stage. Geoffroy thought that their study might clarify the mechanism behind the formation of new species[196]. He considered the claims that an "advanced" organism (on the "chain of being") passes during its development through stages resembling "lower" organisms, and pondered upon the possibility of finding some suitable conditions during the incubation of a hen's egg that would arrest the development of the chick and leave it at a "reptile" stage[197].

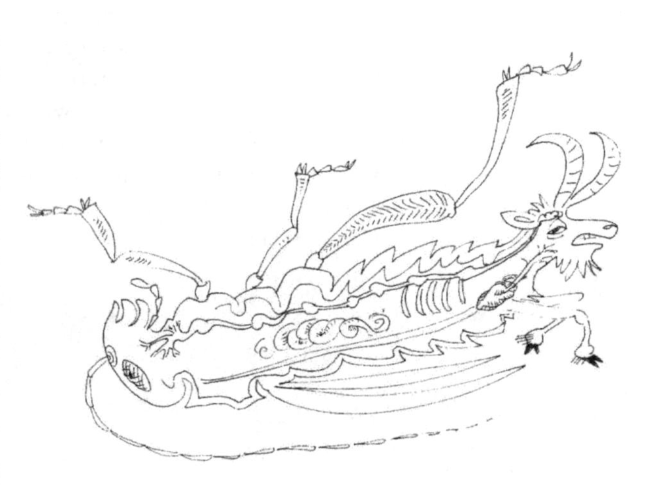

Geoffroy st. Hilaire, Insect as Inverted Mammal

Ape Skulls: Two late papers by Geoffroy deal with the skulls of a young orangutan that had died in the Paris zoo[198]. Geoffroy studied the skulls of orangutans in the collection of the Museum, and misidentified a young specimen as belonging to an entirely different genus. After examining the new young specimen, he admitted that he had been wrong: he could not have guessed that the museum skulls represented two developmental stages of the same species, because the differences between them were greater than those between *ursus* (bear) and *canis* (dog). The skull of the young orangutan resembled that of a human child, exhibiting tenderness and endearing features (which agreed with the observations on the animal when alive) – while the features of the skull of the adult orangutan were savage and frightening.

Geoffroy thought that in the ape, the brain stops growing early and does not fill the entire cranial cavity; whereas in man, the brain grows and fills the entire space because the human skull grows more slowly than does the brain.

The Platypus: A controversial problem among zoologists in the early 19th century was: Does the platypus reproduce live offspring – or does it lay eggs?

---

194 Le Guyader 2004, p. 140.

195 For translations of the papers see in Le Guyader 2004; E. Haeckel, History of Creation, pp. 86-88.

196 Gould 1977.

197 These ideas upset Cuvier – who strongly believed in the fixity of species. Cuvier tried to use his authority at the Home Office to forbid his colleague from carrying out experiments in this direction. Apparently he was not sure that these experiments would fail! Le Guyader 2004, p. 95.

198 Gould 1977, pp. 354-357.

When the first skins and skeletons of the platypus (*Ornithorhynchus*) arrived in England from Australia, zoologists were utterly confused. Some did not believe that this was a genuine animal: there had been cases of eager dealers "manufacturing" exotic animals by grafting together parts of different animals and selling them to gullible travelers. The beak-like mouth of the new arrival seemed to have been grafted onto a rodent body[199].

Observations then revealed that the female platypus produces milk, on which the pups feed. The opinion that this animal is thus viviparous prevailed. The anatomists then discovered that the female reproductive organs were unlike those of a normal mammal: there was no uterus, and the twin oviducts merged and opened into a single cloaca, together with the urinary and digestive excretory ducts – like the system in reptiles and birds. Based on his philosophical theory of anatomy, Geoffroy argued strongly that the platypus laid eggs: "The reproductive system of the platypus cannot produce anything but an egg." When another exotic animal with a similar reproductive system was brought from Australia – the Echidna – Geoffroy arranged them in a separate vertebrate class, parallel to fishes, amphibians, reptiles, birds and mammals, and coined the term *monotremata* for the group.

It was not until the end of the 19th century that Geoffroy was proven right, following direct observations of Platypus reproduction.

[199] Robert Chambers (Chapter 10) in 1844, placed the platypus as an intermediate between birds and mammals!

6

Jean Baptiste de Lamarck: The Theory of Descent

His Life

Jean Baptiste Pierre Antoine de Monet, Chevalier de Lamarck (1744-1829), aged 16, volunteered for the Royal French army. His initiative and courage in battle earned him the rank of an officer. He was stationed in the south of France and served seven years. During this period he assembled and published (1778) a 3-volume guidebook to the plants of his country (Flora Francaise). Released from the army following a training accident, his botanical book attracted the attention of Buffon – then Director of the Royal Botanical Gardens – who employed Lamarck as a "king's botanist" as well as a private tutor, to accompany his son for a two-year tour of botanical gardens in Europe. Following the French Period of Terror in 1793, when the Royal Gardens became part of the Natural History Museum, Lamarck was appointed one of its 12 Professors, responsible for "insects, worms, and microscopic animals".

Lamarck worked and lectured for many years in the Museum, contemporary with Cuvier and Geoffroy Saint-Hilaire. His ideas about the origin and development of the biological world, as expressed in his 1809 book "Zoological Philosophy", were truly revolutionary, and Cuvier in particular became Lamarck's bitter opponent (see below).

Lamarck's fortune changed for the worse as he got older. His lost his eyesight and most of his possessions. His eldest daughter took care of him to his last day. He died aged 85, impoverished and blind. His daughter erected a monument to her father in the Botanical Gardens in front of the Museum in Paris, with the epitaph: "Posterity will admire you, and avenge you my father".

Contribution to Biology

In the Museum, Lamarck organized the display of the collections in a scientific order, under the supervision of expert curators. He reclassified Linnaeus's class Vermes (worms) which included all the invertebrates – very different organisms – into ten different classes. His book on "Animals without Backbones" was published in several volumes (1815-1822). The main divisions in his classification of the invertebrates still stand today[200].

Lamarck's book "Philosophie Zoologique" (Zoological Philosophy, 1809) eventually became a landmark in the history of biology and the Theory of Evolution. His greatest biological legacy is his statement that the entire living world had descended from a common source: from very simple organisms, which are generated spontaneously by nature from non-living material. The more complex organisms developed from simpler forms which preceded them:

---

[200] Lamarck (1809) 1914, pp. 131-133.

> I do not mean that existing animals form a very simple series, regularly graded throughout. But I do mean that they form a branching series, irregularly graded and free from discontinuities[201].

In a generation that believed in Creation and the "Scale of Beings" – that "lower" creatures were degraded forms of the ideal, God or Man – these ideas were undoubtedly revolutionary. Indeed, Lamarck was recognized as such by contemporaries like Charles Lyell[202], despite the latter's criticism of Lamarck's ideas. Ernst Haeckel, the greatest supporter of Darwinism in Germany[203] wrote in 1876:

> To him [Lamarck] will always belong the immortal glory of having for the first time worked out the theory of descent as an independent scientific theory of the first order, and the philosophical foundation of the whole science of biology.
>
> The portion of the theory of development [= evolution] which maintains the common descent of all species of animals and plants from the simplest common original forms might, therefore, in honour of its eminent founder and with full justice, be called *Lamarckism*, if the merit of having carried out such a principle is to be linked to the name of a single distinguished naturalist[204].

Lamarck should have been recognized as one of the giants of the life sciences in the 19th century (for one thing, he was the first to use the term "biology", in 1802). This was not to be. His "Zoological Philosophy" was rejected and greeted with scorn, both in his own Institute and outside France. When Lamarck died in 1829, his colleague (and bitter opponent) Georges Cuvier wrote a eulogy, summarizing Lamarck's achievements. The eulogy was translated into English, and may have been the source of information about Lamarck's ideas for the majority of British intellectuals, who did not read French[205]. It gave the impression that Lamarck had been a fantasy writer, who had published theories with no foundation in the real world.

> Lamarck belongs to the class of naturalists who have laboriously constructed vast edifices on imaginary foundations, resembling the enchanted palaces of old romances, which vanished into the air on the destruction of the talisman to which they owed their birth[206].

Zoological Philosophy: Lamarck's Evolutionary Ideas

"Philosophie Zoologique", published in 1809, represents a true revolutionary change in biological thought, compared with the then prevalent beliefs. With few exceptions, the scientific and philosophical men of the period believed, led by Cuvier, that all species were created as they now appear, and can never change ("fixity of species"). Lamarck, on the contrary, wrote that Nature cannot directly produce animals and plants of all classes, and that the species can – and do – change form with time. Nature must proceed very slowly, beginning with the simplest forms. From these ("monads"), nature goes on to make more and more complex organisms (animals and plants) according to their required conditions of life. The book[207] opens with the following statement:

---

[201] Ibid. p. 37.
[202] See Chapter 7.
[203] See Chapter 20.
[204] Haeckel, The History of Creation I: 149-150.
[205] The full text of the Eulogy is printed as an annex to the 1984 reprinting of the English translation of Zoological Philosophy.
[206] Ibid. p. 434.
[207] Only the first of three parts of the book is relevant to the Theory of Evolution. The other two parts – fully two-thirds of the 400-odd pages – are usually ignored, as part of Lamarck's "scientific rubbish" (Graur, Gouy, & Wool, 2009), but nevertheless reveal Lamarck's approach to biology.

> How could I avoid the conclusion that nature had successfully produced the different bodies endowed with life, from the simplest upwards? For in ascending the animal scale, starting with the most imperfect animals, organisms gradually increase in complexity in an extremely remarkable manner[208].

> It is quite clear that both animal and vegetable organizations have, as a result of the power of life, worked out their own advancing complexity. Beginning with that which was the simplest and going on to that which represents the highest complexity, the greatest number of organs and the most numerous faculties. It is also clear that every special organ and the faculty based on it, once obtained, must continue to exist in all living bodies after those which possess it in the natural order[209].

Lamarck reversed the accepted direction of the traditional "Ladder of life", which started with man at the top and worked its way downwards in increasing degrees of degradation: Nature started at the bottom, and worked its way upwards.

> I do not hesitate to say, however, that our natural classifications of animals up to the present have been in inverse order from that followed by nature, when bringing her living productions successively into existence.

> The existing custom of placing at the head of the animal kingdom the most perfect animals, and terminating the kingdom with the most imperfect and simplest in organization, is due on the one hand to that natural prejudice toward giving the preference to the objects that strike us most[210].

The Mechanism of Species Change ("Transmutation")

The face of the earth, wrote Lamarck, is not in a stable state. The face of the earth is changing by erosion – high ground is washed away and silt is carried by the rivers to the lakes. This changes the conditions of life for living organisms, and they must likewise change.

> It is not a futile purpose to decide definitely what we mean by the so-called species among living bodies, and to enquire if it is true that species are of absolute constancy, as old as Nature, and have all existed from the beginning just as we see them today; or if, as a result of changes in their environment, albeit extremely slow, they have not in the course of time changed their characters and shape.

> [Any collection of like individuals which were produced by others like themselves is called a species.] To this definition is added the allegation that the individuals composing a species never vary in their specific characters... It is this allegation which I propose to attack, since clear proofs drawn from observation show that it is ill-founded[211].

Lamarck went one step further by suggesting a mechanism for the transmutation of species. All animals and plants are descended from monads, which are formed spontaneously from non-living material. More complex organisms descend from simpler ones according to the requirements of their conditions of life:

---

[208] Lamarck 1912, Zoological Philosophy, Preface, p. 1.
[209] Lamarck (1809) 1984, Zoological Philosophy, p. 295.
[210] Lamarck (1809) 1984, p. 128.
[211] Ibid. p. 35.

> Every fairly considerable and permanent alteration in the environment of any race of animals works a real alteration in the needs of that race. Every change in the needs of animals necessitates new activities on their part for the satisfaction of those needs, and hence new habits. Every new need... requires the animal either to make more frequent use of some of their parts which it previously used less... or else to make use of entirely new parts. To which the needs were imperceptibly given birth by efforts of inner feelings[212].

The transmutation mechanism obeys two laws:

> **First Law.** In every animal that has not passed the limit of its development, a more frequent and continuous use of any organ gradually strengthens, develops and enlarges that organ... while the permanent disuse of any organ imperceptibly weakens and deteriorates it until it finally disappears.
>
> **Second law.** All the acquisitions or losses wrought by nature on individuals, through the influence of the environment in which their race has long been placed, and hence through the predominant use or permanent disuse of any organ, – all these are preserved by reproduction to the new individuals that arise, provided that the acquired modifications are common to both sexes, or at least to the individuals which produce the young[213].

The "first law" provides an explanation for the presence of rudimentary organs such as the appendix and the rudiments of the tail in the primates and in Man, and the elongation of the neck in the giraffe. The "second law" is the explanation – or rather a statement – that these changes, acquired in the life of an individual, are transmitted and accumulated over time: the effort causes a change, and the change is passed on to the progeny.

The geologist Charles Lyell summarized Lamarck's views for the benefit of his countrymen who did not read French:

> It is not the organs, or in other words, the nature and form of the parts of an animal, which have given rise to its habits, its manner of living and its particular faculties. But on the contrary, its habits, its manner of living and those of its progenitors, have in the course of time determined the form of its body, the number and condition of its organs – in short, the faculties which it enjoys[214].

Lamarck's revolutionary ideas were rejected and greeted with scorn by his contemporaries. One reason for this was Lamarck's suggested driving mechanism for transmutation: in particular, that the "wishes" and "aspirations" of organisms to fulfill their needs, as required by the change in the external environment, affected heritable, adaptive changes in their anatomy and physiology.

The Examples: Lamarck listed several examples to illustrate possible scenarios explaining the formation of familiar characters of animals according to the mechanisms described by his two laws. The examples were received with ridicule.

> We find in the same way that the bird of the waterside which does not like swimming and yet is in need of going to the water's edge to secure its prey, is continually liable to sink in the mud. Now this bird tries to act in such a way that its body should not be immersed in the liquid, and hence makes its best efforts to stretch and lengthen its legs.

[212] Ibid. p. 112.
[213] Ibid. p. 113.
[214] Lyell, Principles of Geology, 1853. See Chapter 7.

> The long-established habit... of continually stretching and lengthening its legs, results in the individuals of this race becoming raised ...on long, bare legs denuded of feathers...
>
> If some swimming birds like the swan and goose, have short legs and yet a very long neck, the reason is that those birds while moving about on the water, acquire the habit of plunging their heads as deeply as they can into it in order to get the aquatic larvae...[215].
>
> It is interesting to observe the result of habit in the peculiar shape and size of the giraffe (*Camelo pardalis*). This animal, the largest of the mammals, is known to live in the interior of Africa, in places where the soil is nearly always dry and barren, so that it is obliged to browse on the leaves of trees and to make constant efforts to reach them. From this habit long maintained in all of its race, it has resulted that the animal's fore legs have become longer than its hind legs, and that its neck is lengthened to such a degree that the giraffe, without standing up on its hind legs, attains a height of six meters (nearly 20 feet)[216].

Lamarck had imagined a scenario for the evolution of erect posture and bi-pedalism in man. "If a race of monkeys" (Quadrumana, "possessing four hands" – Linnaeus's term), "due to a change in their habitat, were forced to abandon living in trees and had to use their legs for walking only – their feet would lose the ability to grasp branches, and they would become apes" (Bimana, "having two hands"). And if they wished to command a wide and distant view of their surroundings – they would have endeavored to stand upright and continually adapt this posture. The faculty of standing and walking would be transmitted from generation to generation. That race would ultimately have gained supremacy over other races of animals. This change in posture would involve changes in food, hence also in the skull and teeth, and eventually would have led to a communication system (language)[217].

Classification

Lamarck declared that the genera, families, orders, and classes are not natural groups; and that the classification of organisms is an artificial system worked out by human taxonomists for convenience. Moreover, even species are not permanent units and their characters can change. The only things Nature produces are individuals – which are also temporary units with limited duration. People imagined at that time that every species was as old as nature and created as we see it now. But as more and more creatures were being collected and brought together in museums, it became more and more difficult to distinguish among species, and the decisions of biologists become increasingly arbitrary.

Lamarck considered that it was possible to group animals according to their "affinity"[218], which he defined as the similarity between two individuals when all their characters are compared – in particular the more "essential" parts of their bodies: the nervous system (particularly the brain if present), the respiratory system, and the circulation of the blood. In plants, it is the reproductive organs that are used for classification. He describes his experience in classifying the invertebrates in his collections:

---

[215] Lamarck (1809) 1984, Zoological Philosophy, pp. 119-120.

[216] Ibid. p. 122. Charles Darwin, 50 years later, offered a different mechanism to account for these phenomena – natural selection (see Chapters 8-9).

[217] For many years biologists accepted the notion that an increase in brain size and capacity preceded the bipedal propagation in the development of the human line. Lamarck was correct in suggesting that erect posture and bipedalism was the first step – as confirmed when early human fossils were discovered in 170 years later.

[218] Geoffroy and Cuvier defined affinity more precisely (Chapters 4 and 5); but chronologically, Lamarck's book "Zoological Philosophy" (1809) was published years before the publications of his younger colleagues and opponents. They worked under the same roof for many years, and the ideas were certainly shared and/or discussed in the Museum.

> How great is the difficulty now of studying and satisfactorily deciding on species among the multitude of every kind of polyps, radiarians, worms and especially insects… what a swarm of mollusc shells are furnished by every country and every area, eluding our means of distinction and draining our resources[219].

Linnaeus had divided the animals into six classes: Mammalia, Aves, Amphibia (including reptiles), Pisces, Insecta, and Vermes. Lamarck retained the first four as Vertebrata, and grouped together animals that did not have a bony skeleton as Invertebrata. The latter he divided into ten separate classes – including the Insecta. As early as 1800 he removed the Crustacians (crabs) and the Arachnids (spiders and scorpions) from Vermes (worms). Then, based on reports by Cuvier on the circulation of blood in some "Vermes", Lamarck separated the Annelida (e.g., earthworms) and gave the Mollusca, Coelenterata (e.g. medusas and corals), Cirripedia (barnacles[220]), and Infusoria (microscopic animals), class status[221].

Lines of Descent: The Earliest Phylogeny

Lamarck provided a long and detailed table[222] of all major animal groups and their distinguishing characteristics, and added a summary graph illustrating the natural order of their formation, according to his principle of development from the most simple to the more complex. This seems to have been the first attempt to connect all animals into a network of their lines of descent (phylogeny).

At the base of the animal world he placed the protozoa, which live in water, and the worms – some of which are aquatic. From the worms, branches lead to the insects (some of them, like the mosquitoes and mayflies, have aquatic larval stages). The arachnids are derived from the insects. The crustaceans are derived from certain arachnids which gradually abandoned life on land and returned to live in water.

Another branch of the worms gave rise to the annelids and mollusks. All these transitions Lamarck considered as gradual changes.

The next transition required a major, drastic jump in body organization. From the Mollusca to fishes (Vertebrata) there is a structural hiatus with no intermediate stages known. From this point on the changes are again gradual: amphibians were derived from fishes, reptiles from amphibians. Reptiles were the point of origin for diverse groups of vertebrates: turtles (reptiles) were the progenitors of birds [both groups have a toothless mouth and a horny cover of the jaws]. The crocodiles were the progenitors of the marine mammals, from which developed three lines of descent – one led to the whales (Cetacea), one to the ungulates, and the third to all other mammals.

Lamarck Measuring Size

Some of the lines of descent that Lamarck proposed are incorrect[223], but his phylogeny was nevertheless a bold first-ever attempt to connect the animal world by lines of descent (Lamarck did

[219] Ibid. p. 38.

[220] The barnacles – like *Balanus* – were an enigma. They have a calcareous shell like mollusks, but do not resemble mollusks in other traits. Many years later, Charles Darwin discovered their true affinities as crustaceans – when he studied their larval stages [published in four volumes in 1874].

[221] Lamarck (1809) 1984, p. 66.

[222] Ibid. pp. 128-129.

[223] In particular his idea that marine mammals were the primitive stage in mammalian evolution was incorrect: the evidence compiled since shows that mammals developed on land, and it was only later that some of them returned to water.

not use the term "evolution'). It is noteworthy that he believed that such changes in "development" occurred gradually:

> If the procedure of nature is attentively examined, it will be seen that in creating, or giving existence, to her productions, she has never acted suddenly or by a single leap, but has always worked by degrees towards a gradual and imperceptible development[224].

Appeasing the Church: Lamarck assigned the formation and control of the biological world to nature itself, independent of Divine intervention. But despite the French Revolution and its atheist perspective, the French continued to believe in divinity and the position of the Church in France was too strong to ignore. Lamarck tried to smooth over the problem:

> Doubtless nothing exists but by the will of the sublime author of all things. But can we set rules for him in the execution of His will, or fix the routine for Him to observe? Could not His infinite power create an *order of things* which gave existence successively to all that we see, as well as all that exists but that we do not see[225]?

Cuvier's Eulogy for Lamarck

> In sketching the life of one of our most celebrated naturalists, we have conceived it to be our duty, while bestowing the commendation they deserve on the great and useful work which science owes to him, likewise to give prominence to such of his productions in which too great indulgence in a lively imagination has led to results of a more questionable kind[226].

After giving a review of Lamarck's personal life, Cuvier emphasized that the deceased had published books on physics, chemistry, meteorology, and psychology in the 30 years before entering the Academy as a botanist. In all these fields he had not hesitated to express opinions contrary to established doctrines. None of his ideas had been considered seriously by his contemporaries. Cuvier mentions in particular Lamarck's two volumes on chemistry, which were contrary to the teachings of Lavoisier (by then accepted by all European chemists) – without presenting a single experiment to support his statements. For 11 years, Lamarck had published weather forecasts for the Paris area, which had never eventuated, "as if Nature enjoyed demonstrating his follies". In particular, Cuvier had this to say about Lamarck's idea of transformation of species:

> It will be easily perceived, that nothing is wanting but time and circumstances to enable a monad or a polype gradually and indifferently to transform into a frog, a stork, or an elephant. But it will also be perceived that M. de Lamarck could not fail to come to the conclusion that species do not exist in Nature. And, if mankind thinks otherwise, they have been led to do so only from the length of time which had been necessary to bring about these innumerable varieties of form… This result ought to have been very painful to a naturalist, nearly the whole of his long life has been devoted to the determination of what had hitherto been believed to be species.
>
> Unlimited time, which plays such an important part in the religion of the Magi, is no less necessary to Lamarck's physics, and it was to it that he had recourse to silence his own doubts and to answer all the objections of his readers[227].

[224] Lamarck (1809) 1984, p. 323.
[225] Lamarck, Physiological Zoology (1809) 1984, p. 36.
[226] Ibid. p. 434.
[227] Lamarck (1809) 1984, pp. 446, 448.

Only towards the end of his long eulogy did Cuvier return to a positive description of Lamarck's biological achievements.

> He finally left all those fruitless subjects and concentrated on the one subject he excelled and should not have strayed away from – the classification of the invertebrates, a subject where he earned uncontested praise and appreciation, and the right for gratitude of future generations[228].

Lamarck's Self-reflection (1809): "In spite of the errors into which I may have been led, the work may possibly contain ideas and arguments that will have a certain value for the advancement of knowledge, until such time as the great subjects, with which I have ventured to deal, are treated anew by men capable of shedding further light on them"[229].

Later Appreciation (1876)

The great German supporter of Darwin, Ernst Haeckel[230] explicitly noted that acquired characters are heritable. He suggested that Lamarck should be credited for his the theory of common descent from primitive forms[231].

Haeckel's suggestion was not accepted. In the 20th century, "Lamarckist" became a derogatory name for those biologists who did not follow the accepted trends – especially as regards the inheritance of "acquired" characters – not the least because of the adoption of Lamarck's views by the Soviet biologists, led by Lysenko, one hundred years after Lamarck's death[232]. In 1984, the American historian David Hull offered this explanation for why Lamarck's contribution had been ignored: He was French!

> Few British or American scientists got their views of Lamarck by reading translations of Lamarck, let alone Lamarck's original works in French. What Lamarck really said has not been all that important. Scientists had Lamarck say what they needed him to say. At times, Lamarckism became a pejorative term used to characterize the views of one's opponents[233, 234].

Lamarck and "Philosophie Zoologique": A Critical 21st-century Note: The description so far of Lamarck and his book is incomplete, and consequently biased: it relates only to Part I of the book [Parts II and III seem to have been traditionally ignored by scientists].

Part II[235] deals with the properties of life. It is announced as "an enquiry into the physical causes of life; the conditions required for its existence; the exciting force of its movements; the faculties which it confers in bodies possessing it, and the results of its present in those bodies". These include a comparison between living and non-living matter, the "orgasm" (meaning turgor?) and irritability of living matter, and the structure of cells and tissues.

---

228 Ibid.

229 Ibid. p. 405.

230 See Chapter 19.

231 Haeckel, The History of Creation, 1876, pp. 149-150.

232 See Chapter 26.

233 Hull 1984.

234 NOTE: Research on the teaching of evolution in the 20th century showed that, *before they were taught about Darwin and Natural Selection,* children tended to believe that acquired changes are heritable (e.g. Kargbo et al. 1980); and high-school as well as first-year medical students naturally tended to explain adaptive phenomena using the Lamarckian mechanism of use and disuse (e.g. Brumby 1979, 1984). Thus Lamarck should not be blamed for suggesting the only explanation (save Creation) for these phenomena more than 200 years ago, before the nature of heredity was discovered!

235 Lamarck (1809) 1984, p. 181ff.

Part III[236] is dedicated to the nervous system. It is announced as "an enquiry into the physical causes of feeling, into the force which produces actions, and lastly into the organ of the acts of intelligence observed in various animals". The subjects included are the formation of the nervous tissue, the nervous "fluid" and its activity, and the physical basis of "inner feeling", thoughts, understanding, intelligence, and the formation of ideas[237].

Very little evidence existed on these matters in Lamarck's time – the book was published in 1809! Lamarck filled in the large gaps in knowledge by using his imagination. He imagined some sort of fluid, with electrical properties, flowing through the nerves[238]. He invented a solution for every unknown, and was apparently carried away by the idea that his speculations had solved all the difficulties. Nevertheless, even in his speculative chapters Lamarck was ahead of his time. He insisted that all the phenomena assigned to the "Mind" or "Soul" – as thought and feeling – had a physical basis[239].

With the benefit of 200 years of research that followed, we now realize that Lamarck's suggestions were incorrect. When the book is taken as a whole, it appears that Cuvier was not too unfair in his eulogy when he described Lamarck's writings as a fantasy. But in doing so he wrongly also dismissed Part I of the book as well, in which Lamarck's fantasy turned out to be a true introduction to the evolution of the living world.

> In fact, the amount of scientific rubbish that Lamarck put on paper certainly exceeds the quantity of good science in his scientific oeuvre... But by writing about evolution directly rather than en passant (as did dozens of philosophers from Empedocles to Count Buffon), and by tackling the subject of evolution in scientific rather than poetical terms (as did Erasmus Darwin), Lamarck is without doubt the father of evolutionary theory[240].

Lamarck, Cuvier's Eeulogy

[236] Ibid. p. 283 ff.
[237] Ibid. pp. 314, 344.
[238] Ibid. p. 317.
[239] Ibid. pp. 284, 286.
[240] Graur, Gouy & Wool, 2009.

7

Charles Lyell – "The Father of Geology"

Biographical Notes

Charles Lyell (1797-1875) was born in Scotland to a wealthy family. Trained as a lawyer, he gave up practicing law because of his poor eyesight. Lyell's interest in geology and in fossils led him to spend several months in the Museum of Natural Sciences in Paris. In his letters, he expresses his admiration for Cuvier and Lamarck:

> The labours of Cuvier in comparative osteology, and of Lamarck in recent and fossil shells, has raised these departments of science to a rank of which they never previously been deemed susceptible[241].

Until the age of 70, Lyell spent much of his time traveling. He crisscrossed Europe, from southern Italy to northern Scandinavia and from Spain to eastern Germany – on horseback, in carriages (later, trains), by boat – and on foot. He climbed the Alps and ascended the volcanoes Vesuvius and Etna to gather geological evidence. After his marriage, his wife joined him on his excursions. Twice in his career – in 1842 and again in 1852 – he traveled to America and spent some months studying rock formations in geologically-interesting areas, from Canada and New York to Alabama and the Mississippi Valley. Wherever he went he arranged to meet with the best local authorities on fossils. His monumental book "The Principles of Geology" soon earned him a leading position among geologists.

Lyell was twice elected President of the Royal Geological Society, and for a few years taught geology at Cambridge, but refused a permanent official position. He was socially close to the royal circles – Queen Victoria and Prince Albert – and attended many royal dinners and other functions, which are mentioned very frequently in his letters[242]. Particularly noteworthy are his letters about his travels in America, in which he advocated support for the North, not the South – contrary to the official foreign policy of his government at the time.

Lyell was buried in Westminster Abbey, the resting place of England's most honored individuals.

Geology, Fossils, and Time: "The Principles of Geology"

> *Part, at least, of the ancient memorials of Nature were written in living language*[243].

The first volume of the "Principles" was published in 1830, followed by the second in 1832. A third volume followed later. At least nine editions were issued over the next 20 years – with the three parts of the ninth, 1853, expanded edition covering nearly 900 pages.

---

[241] Lyell 1853, p. 60.

[242] Two volumes of his "Life and Letters", edited by his sister-in-law, contain many letters but little information about the man himself. See also Wool 2001, in Oikos 94: 385-391.

[243] Lyell 1853, p. 60.

The "Principles" is not simply a textbook of geology, with systematic descriptions of the geological formations[244]. It deals with global issues such as climate, volcanic eruptions, erosion of land, and transport of soil to river deltas, and erosion of sea coasts by waves – as well as the distribution of plants and animals across the globe. The book was unique for its time – in that it explains all the physical and biological changes observable in the geological formations without resorting to catastrophes and miracles.

In the 19th century geology became an applied science. Huge quantities of metal ore and coal were required by expanding industry. There was a need for trained people to identify the rock formations where such commodities might be found. The identification of rock formations by means of chemical analysis was slow, difficult and often inaccurate. It was much easier to identify them according to their fossil content. In 1841, the Museum of Practical Geology was opened in London, serving as a technical school of mining and a center for public scientific instruction[245].

All the geologists agreed, as did Cuvier, that fossils were extinct animals, and that the fossils in the upper (most recent) rocks were the most similar to extant animals. They observed that Europe had been colonized, in successive periods, by different species of animals – some marine, some terrestrial. The finding in northern Europe of fossil animals that were now limited to tropical climates – elephants, rhinoceroses, hippopotamuses, hyenas, lions, and tigers – suggested that a drastic climate change had taken place. The differences in fossil fauna between epochs were interpreted by Cuvier as evidence of repeated deluges destroying all life on earth. Because of Cuvier's great authority, his "catastrophe theory" continued to prevail in Europe for many years. Lyell, however, disagreed with this interpretation.

> A close comparison of the recent and fossil species, and the inference drawn in regard to their habits, accustomed the geologist to contemplate the earth as having been, at successive periods, the dwelling place of animals and plants of different races – some terrestrial, and others aquatic – some fitted to live in seas, others in the waters of lakes and rivers. By the consideration of these topics, the mind was slowly and insensibly withdrawn from imaginary pictures of catastrophes and chaotic confusion, such as haunted the imagination of early cosmologists[246].

Changes of the Surface of the Earth

Lyell opens his book with an extensive review of the beliefs about the creation of the world in different ancient cultures. His conclusion is that the belief in disasters such as the Flood, shared by old traditions, was based on real, local events that had destroyed large tracts of land, and in which thousands of people and untold numbers of livestock and other animals had perished[247]. Reports of local floods, of eruptions of volcanoes and earthquakes, became magnified as they were transmitted from generation to generation, and finally reached the proportions of the imagined destruction of "the whole world".

Lyell dedicated a considerable part of his book to evidence of the importance of erosion and volcanic activity in shaping the surface of the earth[248]. He considered that the earlier need to explain the evidence as resulting from catastrophes was a consequence of squeezing the history of the world

244 To this subject Lyell later dedicated his "Elements of Geology" (1874), which even appeared as a special student edition with many good illustrations of typical fossils.

245 Bibby 1960, pp. 108-9.

246 Lyell 1853, p. 60.

247 Lyell 1853, p. 160. The Tsunami disaster in south-east Asia (Thailand) in 2005, and the earthquake in Haiti in 2010, which caused hundreds of thousands of casualties, are sad examples of such local events.

248 The same forces were discussed by Cuvier (1825, p. 11-14) but declared too slow to account for past geological events.

into a period of 5,000 to 6,000 years since Creation – as the priests had calculated and taught for generations. If this restriction is removed, the present rates of erosion and deposition of sediment are sufficient to explain past events. The observed distribution of fossil forms may be explained – *given enough time* – by the slow, cumulative effects of the same natural forces that can be observed at work today. Lyell wrote to a friend,

> These principles, as you know, are neither more nor less than that no causes whatever have from the earliest time to which we can look back, to the present, ever acted, but those now acting. And that they never acted with different degrees of energy from that which they now exert[249].
>
> Until we habituate ourselves to contemplate the possibility of an indefinite lapse of ages having been comprised within each of the modern periods of the earth's history, we shall still be in danger of forming most erroneous and partial views of geology[250].

Volcanic eruptions and earthquakes cause mass destruction, as Lyell himself had seen in Pompeii and Herculaneum – but also bring about the accumulation of new fertile soil: despite the repeated disasters, people recolonized the foot-slopes of the volcanoes Vesuvius and Etna. Lyell lists a number of old cities and forts (in Italy for example) that had been built at river mouths in historical times, but whose remains were now located miles inland, due to the slow buildup of the deltas into the sea. The cover plate of the 9th (1853) edition of Lyell's "Principles" shows a dramatic proof of repeated submergence and uplifting of the ground near Naples within historical times: the standing marble pillars of a Roman temple (Temple of Seraphis) are perforated, at about a meter or two above ground, with tunnels made by marine mollusks – proof that this part of them had been submerged for a long-enough period in the past, and was then uplifted with the next volcanic activity. Lyell cites Darwin's observations in Chile and the Andes as evidence of the gradual elevation of the land. While visiting Norway and Sweden, he calculated the slow rate of elevation of Scandinavia with the aid of marks chiseled in the rocks at low tide 20 years previously. In his visits to America he studied the Niagara Falls, and calculated that the immense power of the water had eroded the falls at the rate of 30 cm a year: the original falls had been twice as high as he now saw them, and have taken 35,000 years to retreat to their present position.

Lyell described an important force in shaping major features of the northern continents: glaciers. A glacier is a mass of ice, pushed downwards by the weight of accumulating snow and ice on the high ground. Lyell found evidence that glaciers had shaped alpine valleys and Scandinavian fjords in the geological past. Glaciers also break up huge rocks and transport them, buried in the ice, for long distances, releasing them where the ice finally melts. This was his explanation for the presence of blocks of granite in Swiss valleys, which had previously received supernatural explanations. Glaciers fit well into Lyell's concept of time: they move extremely slowly and generations or epochs must pass before the ice melts and their effects on the landscape become visible.

Lyell considered the presence of different fossils not only as identification markers for different rock strata, but as indicators of the type of environment in which they had lived. After considering – and rejecting – various other theories, Lyell explains that climatic changes in the geological past were a result of changes in the proportion of land and sea on the surface of the globe, caused by the erosion and accumulation of matter over time. Climate is determined by a balance between the areas covered by sea and land, aided by sea currents and winds (he obtained data on the latter from the records of captains of the British Royal Navy and merchant navy ships, who routinely accumulated them). Land warms up faster than water, in particular near the equator, and loses heat faster. In the northern hemisphere, more than half of the surface is land, while in the southern

[249] Lyell to Murchison. Life and Letters I: 234. This principle is referred to in the literature as the doctrine of Uniformitarianism.
[250] Ibid.

hemisphere only 1/16th of the surface is land. This explains the differences in climate between the hemispheres at comparable latitudes. Some parts of continents are continually being elevated, while others sink. Erosion transfers huge quantities of sediment into the sea, and their cumulative actions determine local climate, and with it the composition of the flora and fauna. A gradual change in proportion of land and sea could, eventually, have led to a global climate change – as documented in the fossil record.

Lyell and Lamarck: Creation versus Transmutation

Lyell knew Lamarck, and read the "Philosophie Zoologique" carefully, if critically. Some aspects of it he found appealing. He wrote to a friend in 1827,

> I devoured Lamarck en voyage… his theories delighted me more than any novel I ever read, and in much the same way – for they address themselves to the imagination, at least for geologists who know the mighty inferences which would be deducible were they established by observation…
>
> I confess I read him rather as I hear an advocate on the wrong side, to know what can be made of the case in good hands…
>
> I am glad he has been courageous enough and logical enough to admit that his argument, if pushed as far as it must go, if worth anything, would prove that men may have come from the orangutan.
>
> That the earth is quite as old as he suggests has long been my creed[251].

Unlike Cuvier, Lyell considered Lamarck's ideas seriously. He dedicated a chapter in his "Principles" to an objective summary account of Lamarck's theory, thus exposing it to English readers. He then proceeded to criticize the theory, point by point, in the next two chapters.

Lyell was a strong believer in God's role in Creation; at the end of the "Principles" he states his belief:

> But in whatever direction we pursue our researches, in time or in space, we discover everywhere the clear proof of a Creative Intelligence, and of His foresight, wisdom, and power[252].

He could not therefore accept Lamarck's theory of "spontaneous generation"[253]:

> We must be on our guard, not to tread in the footsteps of the naturalists of the middle ages, who believed in the doctrine of spontaneous generation to be applicable to all those parts of the plant and animal kingdom which they least understood… When at length they found that the insects and cryptogamous plants were also propagated from eggs or seed, [they] still persisted in retaining their old prejudices respecting the infusory animalcules and other minute beings, the generation of which had not then been demonstrated by the microscope to be governed by the same laws[254].

As a Creationist, Lyell also rejected the theory of progressive development from simple to more complex organisms – the suggestion of common descent:

251 Lyell to G. Mantell; Life and Letters, I: 168.
252 Lyell 1853, concluding remarks, p. 799.
253 See Chapter 1.
254 Ibid. p. 581.

> It had not been evidently as part of the plan of nature, to hand down to us a complete or systematic record of the former history of the earth. The popular theory of the successive development of the animal and vegetable world, from the simplest to the most perfect forms, rests on a very insecure foundation[255].

He criticized the laws suggested by Lamarck[256] for the transmutation of species, and found no proof for the appearance of new organs in response to the needs imposed on them by the environment.

> No positive fact is cited to exemplify the substitution of some *entirely new* sense, faculty or organ, in the room of some other suppressed as useless… all the instances adduced go only to prove that the dimensions and strength …may be lessened… by disuse, or… be matured and augmented by exertion. When Lamarck talks about the effect of "internal sentiment", "the influence of subtle fluids", and "acts of organization" as causing where animals and plants may acquire new organs, he substitutes names for things[257].

Responding to stories that the natives of Borneo train monkeys to climb the trees and bring down the coconuts they need, Lyell comments,

> It is for the Lamarckians to explain how it happens that those savages of Borneo have not themselves acquired, by dint of longing for many generations for the power of climbing trees, the elongated arms of the Orang. We should have naturally anticipated that their wants would have excited them to efforts, and their continuous efforts would have given rise to new organs[258].

Many years later Lyell had second thoughts about Lamarck. He wrote to Darwin:

> When I came to the conclusion that, after all, Lamarck was going to be shown to be right… I re-read his book and, remembering when it was written, I felt that I have done him injustice. Lamarck's belief in the slow changes in the organic and inorganic world, in the year 1800, was surely above the standard of his time, and he was right about progress in the main – though you have vastly advanced that doctrine[259].

Lyell and Darwin

The first edition of Lyell's "Principles of Geology" was one of the few books that the young Charles Darwin took with him on board the "Beagle" in 1832[260]. Thirteen years later, on the front page of the second edition of his "Journal of Researches" during the voyage of the "Beagle", we find the following dedication:

> To Charles Lyell, esq., F.R.S., this second edition is dedicated, with grateful pleasure, as an acknowledgement that the chief part of whatever scientific merit this journal, and the other works of the author, may possess, has been derived from studying the well-known and admirable "Principles of Geology[261].

[255] Ibid. p. 146.
[256] See Chapter 6.
[257] Lamarck 1853, p. 571.
[258] Ibid. p. 599.
[259] Lyell's Life and Letters, p. 365.
[260] See Chapter 8.
[261] The full title of the book was "Researches into the natural history and geology of the countries visited during the voyage round the world of H.M.S. "Beagle" under the command of Captain Fitz Roy, R.N." Darwin 1845.

What did Darwin find in Lyell's book that had prompted him to relate the scientific value of all his work to ideas derived from the "Principles"? Darwin's greatest contribution to science, the idea of "descent with modification" by means of variation and natural selection, was certainly *not* derived from Lyell – who at the time opposed the idea that species could change at all. But *the concept of unlimited time* was a major contribution to Darwin's ideas on the origin of species: it was not possible to consider natural selection as the driving force in evolution unless unlimited time was allowed for its operation[262].

Darwin provided Lyell with much useful information on the geology of South America[263]. Darwin's observations on the elevation of vast areas of land and of mountain chains on the western shores of Chile and the Andes provided supporting evidence for Lyell's geological theories. On one issue they disagreed: while Lyell believed that coral atolls were formed on the rims of craters of submerged volcanoes, Darwin explained their presence by the building up of coral belts around slowly sinking continents or islands – thus including in one theory atolls, fringing reefs, and barrier reefs[264]. Lyell wrote to a friend,

> I am very full of Darwin's new theory of coral islands, and have urged Whewell to make him read it at our next meeting. I must give up my volcanic crater theory forever, although it costs me a pang at first... Coral islands are the last efforts of drowning continents to lift their heads above water[265].

Lyell was one of two friends – the other was the botanist Joseph Hooker – whom Darwin approached in June 1858 for advice, after receiving by mail an article by Alfred Russel Wallace, then in Ternate in the Malay islands, expressing the same ideas about natural selection that Darwin had been deliberating on for the previous twenty years. Lyell and Hooker presented Wallace's and Darwin's theory to the Linnaean Society in July, 1858 (in the absence of both authors) and encouraged Darwin to publish his book, which finally came out in November 1859[266].

Lyell's letters after 1845 reveal that he had endeavored to convince others that there was much to be appreciated in Darwin's books, but despite letters from and discussions with Darwin and his supporters, like Huxley and Hooker, Lyell was reluctant to accept Darwin's Theory of Evolution. He did *not* accept the idea that species could change. In particular, he could not agree that Man is part of the same process:

> I feel that progressive development or evolution cannot be entirely explained by natural selection. I rather hail Wallace's suggestion[267] that there may be a Supreme Will and Power which may... guide the forces and laws of nature[268].

---

262 It may not be trivial to note that Darwin had thought out the explanation for the formation of coral islands while on board the "Beagle", when one of the few books available to him was Lyell's "Principles": the formation of the reef combines the very slow rate of submergence of the land with the very slow growth rate of the corals. In his last book, on the activity of earthworms (Darwin 1985), Darwin presented interesting examples of Roman artifacts, some of considerable dimensions, buried under the earth and brought up from beneath by the slow activity of earthworms over hundreds of years.

263 Darwin 1890 (1842).

264 Ibid.

265 Lyell to J. Herschel, May 1837. In the 9th (1853) edition of the "Principles", Lyell included both theories, his own and Darwin's. He added a note that the latter is perhaps more correct, but he did not remove his own incorrect explanation from the new edition of the book!

266 See Chapter 8.

267 See Chapter 11.

268 Lyell to Darwin, 5 May 1869.

> The sudden passage, from an irrational to a rational animal, is a phenomenon of a distinct kind from the more simple to the more perfect of animal organization. To pretend that such a step, or rather leap, can be part of a regular series of changes in the animal world is to strain analogy beyond all reasonable bounds[269].

Fixity of Species versus Transmutation

Before attempting to explain the observed differences in fossil remains in the different geological formations, Lyell had to answer a very basic problem: Is a species a true natural unit?

> …Whether species have a real existence in Nature, or whether they are capable – *as some naturalists pretend* – of being indefinitely modified in the course of a long series of generations[270]?

By "some naturalists" Lyell refers to supporters of Lamarck's idea of the transmutation of species[271]. Lyell was convinced that species were real entities and immutable. Each species stems from a single source – a single pair, or a single individual when this was sufficient – and cannot be mixed with individuals of other species. Some morphological changes can be caused by the environment, and some of these may be heritable, but they are not sufficient to change one species into another. Each species will endure for an appointed long time, and then become extinct, but never change[272]. Even when the environment changes, the species need not change their characteristics:

> We must assume that when the Creator made an animal or a plant, he took into account all the conditions in which its offspring may find themselves, and made sure that their structure and organization will enable them to survive and reproduce in these conditions[273].

Lyell presents three kinds of evidence to support his perceived permanence of species[274]. First, from domestication: the efforts to modify animals for human purposes did not result in the creation of new species. Despite the difference in structure and other characteristics of different races of dogs, they are still dogs. The cat was transferred by man to all parts of the globe and established itself everywhere, but still remains the same animal deified by the ancient Egyptians. The Creator in his wisdom gave some species the flexibility to adapt to different circumstances. The ability of the horse, the dog, the sheep, the cat and many kinds of domestic fowl to live in almost every climate, was given them explicitly so that man could benefit from their service wherever he lives[275]. This flexibility removes the need to change.

The second kind of evidence is from attempts to hybridize species. The sterility of interspecies hybrids proves that species boundaries cannot be crossed.

The third kind of evidence is that of the mummified animals from the ancient tombs of the Pharaohs – cats, monkeys, dogs, crocodiles, mongooses and even an ibis – collected by scientists who accompanied Napoleon's armies into Egypt[276]. The preserved animals were identical with extant species, although they were thousands of years old. This is conclusive evidence that species did not change with time.

---

269 Lyell 1853. p. 148.
270 Lyell 1853, p. 582.
271 Although the core of Darwin's theory is that species did change with time, Charles Darwin was still an unknown student at Cambridge when Lyell wrote his book.
272 Lyell 1853, p. 582.
273 Ibid. p. 665.
274 Lyell 1853, pp. 595-597.
275 Ibid. p. 597.
276 Chapter 4.

But if species are constant and did not change their form, how is it that we find different faunas in successive geological strata? Why are no fossils of mammals, for example, found in older geological formations? Lyell first suggests that we should be more careful in interpreting the paleontological finds:

> We must not, therefore, too hastily infer from the absence of fossil bones of Mammalia in the older rocks, that the highest class of vertebrate animals did not exist in remoter ages... the casualties must always be rare by which land quadrupeds are swept into the sea[277].

On the other hand, finding some animal remains at a site may be accidental and does not prove that the animals had really existed where the bones were found:

> Supposing that our mariners were to report that, on sounding in the Indian Ocean near some coral reefs, and at some distance from the land, they drew up on hooks attached to their line, portions of a leopard, an elephant, or tapir. Should we not be sceptical as to the accuracy of their statements? And if we had no doubt of their veracity, might not we suspect them to be unskillful naturalists? Or, if the fact were unquestioned, should we not be disposed to believe that some vessel had been wrecked on the spot[278]?

Faunal Changes in Successive Geological Strata

Despite these perceptions, the changes of species composition with time at different localities could not be denied. Lyell offers an explanation: the fauna changed as a consequence of the ever-changing environment. Yet transmutation – even if it were possible – is too slow to explain the faunal differences between geological periods.

> The disposition of the seas, continents, and islands, and the climates, have varied. The species likewise have changed[279].
>
> It is idle to dispute about the abstract possibility of the conversion of one species into another, when there are known causes so much more active in their nature, which must always intervene and prevent the actual accomplishment of such conversion[280].
>
> [For example], if a tract of salt water becomes fresh by passing through every intermediate degree of brackishness, still the marine molluscs will never be permitted to be gradually metamorphosed into fluviatile species, because long before any such transformation may take place by slow and insensible degrees, other tribes, already formed to delight in brackish or fresh water, will each in their turn monopolize the place[281].

Existing species at a site will not therefore mutate – but will be *replaced* by others following environmental changes. Lyell wrote to a friend,

> When I first came to the notion, which I never expressed elsewhere... of a succession of extinctions of species, and creation of new ones, going on perpetually now and through an indefinite period in the past, and to continue for ages to come, all in accommodation to the

277 Lyell 1853, p. 135.
278 Ibid.
279 Lyell 1853, p. 799.
280 Lyell 1853, p. 700.
281 Ibid.

> changes which must continue in the inanimate earth, the idea struck me as a grandest which I have ever conceived[282].

Lyell faced the problem that Linnaeus had avoided: the disappearance of species from the geological record and the possibility of extinction. Were they replaced by the creation of other species? In a letter to the geologist Adam Sedgwick, he stated:

> I have admitted that we have data only on extinction, and I have left it rather to be inferred ...that the place of the lost species is filled up... from time to time by new species. But I certainly wish it to be inferred from my book that in the oceans, beyond the sphere of human interference, and in the desert, and in the wilderness, and among the infusoria and insects, the extinction has been going on in the last 6000 years, and that the substitution of species to supply the vacancy has also been going on[283].

Creation, Time, and Ecology[284]

Lyell could not accept the 6-day time scale for creation. The Creator must have exercised a lot of pre-planning in the colonization of the earth – and the colonization itself required time. Lyell introduced *time* into Creation:

> Each species may have had its origin in a single pair... and species have been created in succession, at such time and such places to enable them to multiply and endure for an appointed period and occupy an appointed space on the globe[285].

It is very interesting to realize that the need for extended time and planning ahead – arose from *ecological* reasoning: if scientists were to plan the colonization of new land,

> It would be necessary for naturalists to study attentively the climate and other conditions at each spot. It may be no less requisite to introduce the different species in succession so that each animal might have time and opportunity to multiply before the species destined to prey upon it was admitted[286].

The Creator must have proceeded in this way. Many bushes and grasses must have been created and allowed to spread over a large area, before the world could be colonized with sheep, goats and deer – otherwise the first herbivores created would have devoured the first plants and then died of starvation. Time was needed for the herbivores to multiply before they could be able to furnish food for the first pair of lions or wolves. Time was needed for insects to multiply to large enough numbers, before swallows could be allowed to enter the world and destroy thousands of them in one sweep[287].

Lyell was led to almost modern ecological concepts – by the conflict between his deep religious conviction and belief in the immutability of species, on the one hand, and his observation that animal fossils were not the same in different geological periods, on the other. The dismissal of miracles and catastrophes, and the recognition that vast amounts of time had passed, led Lyell to explain the differences in fossil faunas in different epochs as a result of environmental changes[288]. In

---

[282] Lyell, Life and Letters, p. 468.
[283] Lyell to Sedgwick, 20.1.1838 – Life and Letters, p. 36.
[284] The term "ecology" was coined by Ernst Haecked, apparently around 1878.
[285] Lyell 1853, p. 666.
[286] Ibid.
[287] Ibid. p. 665.
[288] See Wool 2001.

his "Principles", first published 180 years ago, Lyell recognized as well as any modern ecologist the importance not only of the physical environment, but also of the interactions of biological species with each other, in determining the fauna of any region:

> The possibility of existence of certain species in a given place, or of thriving more or less therein, is determined not merely by temperature, humidity, soil, elevation and other circumstances of like kind, but also by the existence or non-existence, the abundance or scarcity of a particular assemblage of other plants and animals in the same region[289].

Lyell defined two new theoretical concepts: the "station" and the "habitation", which are in use today under different names:

> *Stations of plants and animals*: stations comprehend all the circumstances, whether relating to the animate or inanimate world, which determine whether a given plant or animal can exist in a given place[290].

The "*station*" is distinguished from the "*habitation*", which is the place on earth in which the species can actually be found. The "*habitation*" is determined not only by the "*station*" but also by factors such as the site where the species was created, the time that passed since then, and its power of dispersal[291].

Lyell was aware of the interdependence of species. For example, suppose a flowering plant grows in a thicket of thorny bushes. Although the growth conditions there are less favorable than in the open meadow, it is possible that the few individuals, protected by the thorns of their "friends" from the grazing cattle, will be the only ones to set seed and propagate the species. Further, a change in the abundance of any one species may have unsuspected consequences at far-away sites. If the number of salmon be reduced, due to predation by seals,

> the necessary consequence must often be that in the course of a few years the otters, at the distance of several hundred miles inland, will be lessened in numbers from the scarcity of fish[292].

Some, but not all, environmental perturbations will cause extinction. Any species of plant or animal that could not endure low temperatures, or a particular disease, or the famine caused by the locusts consuming all green matter, would have been eliminated when the factor was first encountered. But recurrent invasion of locusts, or an epidemic disease, will not cause species extinction, if the species has survived that first encounter and recovered.

Lyell was particularly impressed by the wonderful ability of insects to recover quickly after every destruction: he sees in this proof for the greatness of the Creator, who gave them the ability to be reduced to insignificant numbers "without an obvious violent force" intervening, only later to reproduce to enormous populations – like a machine that requires no fuel when not in use, but immediately springs into action when the need arises!

The Effect of Man on the Environment and Species Diversity

Awareness of Man's negative effects on the environment increased in the 20th century in the affluent societies of America and Europe. Already in the "Principles" however, written 180 years ago, we

[289] Lyell 1853, p. 677.
[290] Ibid. p. 670. This definition is nearly identical to the modern concept of the ecological "niche".
[291] Ibid. The concept is similar to the ecological "habitat".
[292] Ibid. p. 677.

find many statements expressing Lyell's awareness of the negative effects of Man on the environment that would have been approved by any modern ecologist. For example,

> We can only estimate the revolution caused in the animal world by the growth of the human population – even only as consumer of organic matter[293].

> When a powerful European colony lands on the shores of Australia… when it imports a multitude of plants and large animals from the opposite extremity of the earth, and begins rapidly to extirpate many of the indigenous species, a mightier revolution is effected in a brief period than the first entrance of a savage horde, or their continuous occupation of the country for many centuries[294].

Lyell writes that many swamps in Europe now occupy sites where humans destroyed extensive pine and oak forests in historical times. Some forests were cleared by the order of Roman emperors, and others were felled by order of the British Parliament because they were havens for outlaws and wolves[295]. Forest clearing may cause destruction of the land:

> In my travels in Georgia and Alabama, in 1846, I saw in both countries the formation of hundreds of valleys in places where the natural forest was destroyed. Of one such valley there was no trace twenty years earlier, but when the trees were felled, cracks opened to a depth of 3 feet by the heat of the sun, and during the rains, in twenty years a valley was formed no less than 55 feet deep, 300 yards long and 20 to 180 feet deep[296].

Human activities also accelerate the destruction of coastlines by removing from the beaches the blocks of limestone, which fall down when the cliffs disintegrate, to be used in the cement industry:

> If these blocks were left where they fell, they would break the force of the waves before they reach the cliffs and delay the transformation of the peninsula into an island, a process which may bring the destruction of the city of Harwick[297].

The draining of marshes and the felling of forests are useful for Man, but they reduce the diversity of species which exist in these habitats. Some claim that, globally, human agricultural activities improve the productivity of the land, but this is doubtful.

> Man is, in truth, continually thriving to diminish the natural diversity of stations of animals and plants, in every country, and to reduce them all to a small number fitted for species of economic use[298].

> [people] are so much in the habit of regarding of the sterility or productiveness in relation to the wants of man, and not so as regards the organic world generally[299].

However, Lyell had a clear conscience about these harmful effects: Man was destined to rule the world and its inhabitants, and everything that is good for Man must be the will of God.

---

293 Lyell 1853, p. 681.
294 Ibid. p. 150.
295 Ibid. p. 721.
296 Ibid. p. 204.
297 Lyell 1853, p. 320.
298 Ibid. p. 682.
299 Ibid. p. 681.

The Third Circle

Charles Darwin, his Theory, his Supporters and Adversaries

The third circle covers roughly the period from 1830 to 1900. At the center of the third circle is placed, of course, Charles Darwin. Seven years of collecting data and forming the central idea (1837-1844)[300] resulted in a 230-page "abstract". Then followed fifteen years of delay and deliberation, before the publication in 1859 of the book – known by the shortened title "The Origin of Species"[301] – which became a landmark in the history of biology and of science in general.

> He has established a new theory, which reveals to us the natural causes of organic development, the acting causes of organic form production, and of the changes and transformations of animal and vegetable species. This is the theory which we call the Theory of Selection, or more accurately, the Theory of Natural Selection[302].

The philosopher Herbert Spencer summarized the central idea of the book in the expression "The survival of the fittest". Darwin adopted this expression in a later edition of his book.

Darwin's work covered many fields. His books on the descent of man and sexual selection are described in detail in later chapters[303]. His six botanical books, which deserve much more space than can be allotted to them here, are discussed in Chapter 17.

While Darwin pondered over his theory, a book entitled "Vestiges of the Natural History of Creation" was causing a sensation. It was published anonymously, presenting an hypothesis on the development of the organic world (1844). The author later identified himself as Robert Chambers[304].

An important place at the center of the circle is allocated to Alfred Russel Wallace[305], who thought out the theory of natural selection independently while collecting organisms in the Malay islands. Wallace's paper, sent to Darwin in 1858 for evaluation, prompted Darwin to publish his book sooner than he had planned. The works of Darwin's contemporary supporters – Thomas Henry Huxley[306], Ernst Haeckel[307], and August Weismann[308], and his opponent Louis Agassiz[309], are included in the circle. Under the headline "The struggle for existence of the Theory of Evolution" are discussed some of his critics – the Bishop of Oxford, Samuel Wilberforce, the Scottish engineer Robert Jenkin, and the biologist St. George Mivart[310] – who raised many objections to the theory, some of which Darwin could not answer.

---

[300] Chapter 8.
[301] Chapter 9.
[302] Ernst Haeckel, 1876, I: 120.
[303] Chapters 15 and 16.
[304] See Chapter 10.
[305] Chapter 11.
[306] Chapter 12.
[307] Chapter 19.
[308] Chapter 20.
[309] Chapter 13.
[310] Chapter 14.

8

Charles Darwin: The Formative Years

Biographical Notes

Following his grandfather, Erasmus Darwin[311], and his father – both medical doctors – Charles Darwin (1809-1882) was sent to the medical school in Edinburgh, Scotland, but he did not complete his studies: he felt himself unfit to be a doctor, particularly after watching a surgical operation on a young boy, performed without anesthetics[312].

His disappointed father sent him to Cambridge to study for the clergy – the only alternative that provided a general education and the prospect of a decent occupation. Charles admitted that he did not fancy the classical studies (Theology, Latin and Greek) but he enjoyed two courses – Geology, taught by Adam Sedgwick (1785-1873), and Botany, taught by John Stevens Henslow (1796-1861). Both scientists combined their lectures with fieldwork. Young Darwin enjoyed horse-riding, shooting, and collecting beetles, and the fieldwork suited him well. He even accompanied Sedgwick on a research excursion to the Lake District.

At that time, the British naval command (the Admiralty is the term used in Darwin's time) planned a mapping survey of the shores of South America. A ship was equipped for the mission – a 3-mast, 10-gun ship called the "Beagle". The captain of the Beagle, Robert FitzRoy (1805-1865), a young commander, had just returned from a previous voyage and the Beagle was being refitted based on his recommendations. FitzRoy wanted to engage an educated young man as his companion for the planned 3-year voyage (which eventually lasted five). He even offered to share his cabin with a suitable man if found (a real sacrifice in the crowded vessel) and suggested that the companion would be in charge of whatever zoological, botanical and, particularly, geological data would be collected during the voyage.

The Admiralty asked the major universities to recommend a suitable candidate and Darwin's teachers at Cambridge suggested him for the position. He agreed immediately, but his father worried that the voyage would be a waste of time and his son would lose the chance of a decent career. Strong support from his uncle – Josiah Wedgwood II, his mother's brother – was required before he was given his father consent.

[Darwin did not complete a medical degree at Edinburgh, but the two years he spent there were significant for his future as a scientist. The lectures he found mostly "boring", but the scientific atmosphere in the Department of Zoology – headed by Prof. Jamieson (who translated Cuvier's "Introduction" on the fossils into English) was sometimes exciting. In the medical school, dissection of human bodies was not forbidden, unlike in The Anglican universities of Oxford and Cambridge; Professors were in contact with the French Academy of Sciences and familiar with the comparative anatomy studies of Cuvier and Geoffroy Saint-Hilaire.

Moreover, Darwin also met there with Robert Edmond Grant (1793-1874), a former graduate who had remained as a tutor at the university. Grant was investigating marine invertebrates, mainly

---

311 Chapter 2.

312 Darwin's Autobiography. Ed. Nora Barlow, 1958, p. 48. See below, Note added in print.

sponges. In a chance meeting, Grant realized that Charles was the grandson of Erasmus Darwin – whom he held in high esteem. Grant adopted Darwin as his assistant and taught him techniques of dissecting and observing small organisms. Grant was enthusiastic about Geoffroy's anatomical studies. He had visited the Paris laboratory and listened to Lamarck's lectures, which he related to the young Darwin. Although he may not have been aware of it at this early stage, the two years at Edinburg would seem to have left a deep mark on Darwin's mind.]

The Voyage of the "Beagle"

Charles Darwin sailed with the "Beagle" in late December, 1831. During the five years of the voyage, the 23-year-old was transformed from an amateur naturalist into an eager observer and collector of rock samples, fossils, animals, and plants. His books on the Voyage of the Beagle[313], the geology of South America, and the formation of coral islands[314] published in 1842 and 1846, are direct products of that journey. On board the Beagle, in the absence of museums, libraries, and colleagues with whom to consult, he had to resort to his own original thinking about nature.

Darwin was first exposed to the tropical scenery and the consequences of volcanic activity when the Beagle anchored off the island of St. Jago (anchorage at Tenerife in the Canary Islands was denied them because the authorities were worried about the spread of cholera). The world he saw was new to him, and his notes illustrate his inquisitive mind and his interest in everything, small or large. He described in detail, on several pages, the behavior of octopuses that he saw in the shallow water at low tide, and noticed a rock stratum with many seashells – high above the water level, evidence that the land had been elevated.

The Beagle sailed westwards across the Atlantic to the shores of Brazil, then spent three years mapping the shore from Bahia southward to Tierra del Fuego. The survey was a slow and laborious project – small rowboats were sent up the great rivers as far as they could go, to map their course, while the Beagle stayed in the deep water. Darwin was not employed in the mapping project but went ashore each time that the ship stayed long enough in one place, and made long excursions on horseback to the interior of the land. He studied the geological formations, collected plants, and shot and skinned birds and mammals. He observed and recorded everything in detail – from the behavior of worms (*Planaria*) and spiders, to frogs, birds, and mammals. On one excursion from Rio de Janeiro, in the company of a British landowner, he witnessed the cruel treatment of the negro slaves and was deeply moved[315].

From Monte Video, he rode several weeks in the company of local gauchos, and was impressed by their riding skills and their use of the lasso to hunt the Rhea (South-American ostrich). The spoiled young Englishman adjusted well to the rough lives of his companions:

> This is the first night which I passed under the open sky… there is a high enjoyment in the independence of the Gaucho, to be able at any moment to pull up his horse and say, here we shall pass the night[316].

From Bahia Blanca he rode "with the captain's approval" 400 miles across the Argentinian pampas to Buenos Aires, returning by boat on the river. Not far from the shore he found a rock formation with vertebrate fossils of huge size, and collected bones and skeletons of nine species[317]. Darwin gradually became convinced that Lyell's theory of extinction was correct. Species varied in

---

313 Darwin 1912, 1937.

314 Darwin 1890.

315 Frustrated by his inability to do anything about it, he later wrote, "We left the shores of Brazil. I thank God I shall never again visit a slave country" (Darwin 1912, p. 502).

316 Darwin 1912, p. 76.

317 The bones were later examined by Richard Owen in Oxford, who named one of them *Mylodon darwinii* in his honor.

abundance, and when very rare, could become extinct. It was not possible that creatures as big as those he collected as fossils, still existed undetected in some part of the earth.

> Why should we feel such great astonishment at the rarity being carried a step further to extinction? ...[this] appears to me much the same as to admit that sickness in the individual is the prelude to death, but when the sick man dies, to wonder, and to believe that he died through violence[318].

The Beagle was delayed in Patagonia for a private project of Captain FitzRoy. Three years before, on his first command of the Beagle, the captain had taken with him three native youngsters – two boys and a girl. He took care of their needs and their education in England at his own expense. They learned English and were able to help with the communication between the crew of the Beagle and the natives. Now he returned them to their home country. The crew provided them with accommodation and some equipment and endeavored to secure their reacceptance into their tribe. Meanwhile the Captain, with a small crew in two boats, attempted to find a passage westwards: sailing around Cape Horn proved difficult and dangerous.

The Beagle finally sailed through the Magellan Straits into the Pacific Ocean, to map the western shores of Chile. The passage was rough and frightening:

> To the north there were so many breakers that the sea is called "the milky way". One sight of such a coast is enough for a landsman to dream for a week about shipwrecks, peril, and death. With this sight we bade farewell to Tierra del Fuego[319].

In Chile, Darwin disembarked and rode with a few companions, from Valparaiso to the foot of the Andes. He visited gold mines and described the harsh working conditions of the miners. Layers of rock abounding with fossil seashells, miles away from the sea, provided clear and convincing proof that the ground in that part of the world has been elevated. The cause of this elevation Darwin witnessed for himself: an earthquake completely destroyed the city of Conception, while Darwin was staying on the beach in Valdivia, a nearby town.

> The most conspicuous effect of the earthquake is the uplifting of the land... similar shells were found at the altitude of 1300 feet. There is no doubt that this uplifting was caused by repeated cases of upheaval like the one I witnessed in the present earthquake[320].

Darwin spent a few weeks riding with a mule caravan up to the summit of the Andes, crossing to Argentina and returning through another pass to meet up with the Beagle. Paying special attention to the geological strata as he ascended, he writes again his conviction that Lyell was right about the changes in the face of the earth:

> Daily it is forced home on the mind of the geologist, that nothing – not even the wind that blows – is so unstable as the level of the crust of the earth[321].

The Galapagos Islands

Its mission in the South American continent complete, the Beagle sailed in a general westerly direction, on its long return journey to England (circling the globe rather than retracing its steps eastwards). The first stop was at the Galapagos archipelago on the Equator, some 1,000 km west of

---

[318] Darwin 1912, p. 181.
[319] Ibid. p. 247.
[320] Ibid. p. 379.
[321] Ibid. p. 325.

Ecuador[322]. The Beagle visited four of the 40 islands and rocks comprising the archipelago, stopping for a few days at each. The beaches were black and basaltic, with sparse and strange vegetation, and there were no rivers. Higher up on the slopes there was lush vegetation and rainwater accumulated in the craters.

Darwin disembarked and wandered about the islands on foot. He expressed his wonder at the many volcanic craters he saw – estimating their number as 2,000!

> Seeing every height crowned with its crater, and the boundaries of most of the lava streams still distinct, we are led to believe that within a period geologically recent, the unbroken ocean was here spread out. Hence, both in space and time, we seem to be brought near to that mystery of mysteries – the first appearance of new beings on this earth[323].

Already on the first island – (Chatham, now named San Cristobal) – Darwin saw the creatures that have given the islands their name (Galapagos = tortoises in Spanish).

> As I was walking along, I met two large tortoises, each of which must have weighed at least two hundred pounds. These huge reptiles, surrounded by the black lava, the leafless shrubs, and large cacti, seemed to my fancy like ante-deluvial animals[324].

There were no native large mammals on the islands, and the tortoises filled the niche of the large herbivores on the continent. Darwin dedicated several pages in the "Voyage of the Beagle" to descriptions of the habits of these huge reptiles. He saw wide passages made by hundreds of these creatures on their way up to the water sources on the islands – and noted that they were not as abundant as they had previously been: the sailors of whaling boats often stopped at the islands and collected hundreds of tortoises as a source of fresh meat on their boats (he reports being told that over 700 were taken by individual vessels, and that the crew of one ship brought to the shore 200 tortoises in one day). Darwin mentioned that he enjoyed riding on the backs of the tortoises and prodding them with his umbrella – he was not yet 26 years old – and added that he liked the taste of tortoise soup but not the meat[325].

On Charles Island there had previously existed an Ecuadorian penal settlement whose economy depended on pigs, goats, and other introduced animals[326]. Darwin walked to the water source at the summit of the island. The Beagle also briefly visited the large island of Ablemarle (now Isabella), on which there are five active volcanoes.

Darwin was impressed by the abundance of wildlife, and particularly the tameness of the birds: they were not fearful of man and could be captured simply with a stick or a switch. He was surprised to find that many of the species were endemic[327] and that different islands were inhabited by different species. The governor of one island told Darwin that he could tell with confidence from

---

[322] The islands belong to Ecuador. On the first centennial of Darwin's book, in 1960, agreement was reached between UNESCO and the Ecuadorian government to declare the entire Archipelago a national park (due to the efforts of Julian Huxley, the then president of UNESCO). Major efforts have since been made to rid the islands of pigs, goats, and other introduced animals, and the park authorities take strict measures to protect the native birds and animals. A research station – named after Darwin – was established on the island of Santa Cruz as a breeding center for the tortoises, and attempts are made to repatriate them to their natural habitat (tortoises no longer exist on most of the islands), The fees of the visitors – with grants from UNESCO – pay for these operations. The islands have gained fame and, in recent years, the authorities are worried that the great influx of tourists, and the increase in tourist-supporting services, may endanger the wildlife in the park.

[323] Darwin 1912, p. 382.

[324] Ibid. p. 379.

[325] The Beagle took back with it 45 tortoises, most of which were consumed during the voyage. The list of items brought by Darwin to England records "live tortoises" but no number is given.

[326] See Sulloway 2009. These became wild and destroyed the local vegetation and fauna. Introduced mammals are an important cause of the absence of tortoises on many islands today.

[327] Darwin 1912, p. 382.

which island a tortoise had been brought. Darwin, however, did not follow up this line of thought, believing that all the tortoises belonged to one endemic species, and that the differences in carapace shape in different individuals were environmentally caused[328].

Among these endemic species, Darwin listed two species of iguana lizards. He dedicated several pages to their habits – in particular a marine iguana that could dive to feed on algae under water and stay submerged for a long time. Among the 26 species of birds, he mentions that each of the four islands had a different species of mockingbird. He noted that there were differences in the beak size of finches collected from different islands, but did not immediately realize the significance of these differences[329].

From the Galapagos the Beagle, sailing westwards, visited Tahiti and several coral islands in the Pacific and Indian Oceans, which had not been studied in detail. During this activity Darwin worked out his theory of the formation of atolls, of which he reported to Lyell.

> When mountain after mountain, and island after island, sank slowly below the water, fresh bases would be successively afforded for the continued growth of the corals[330].

The Beagle visited the English colonies in New Zealand, Australia (Sydney), and Tasmania and, sailing around India and Africa (Cape Town), arrived again in Brazil. From there it changed course north-eastwards and crossed the Atlantic to return to England.

Darwin summarized his five-year experience:

> In conclusion, nothing can be more important to a young naturalist than travel in remote countries.

After the Voyage

Back in England, Darwin spent two years sorting out his collections and arranging for examination of the material by experts at different museums in Cambridge and London. After weighing up (and listing) the advantages and disadvantages of married life[331], he then married his cousin Emma, daughter of Josiah Wedgwood. After a short stay in London they bought a house in the village of Downe in Kent. Darwin lived in this house until his death 40 years later, rarely leaving England again except for short stays in various health centers.

Darwin soon became a member of the Royal Society, but did not take an active role in its functions. He was reluctant to fight for his ideas and did not appear much in public[332]. He constantly complained of poor health, with periods of nausea and vomiting that prevented him from working – sometimes for weeks or months at a time (historians disagree as to whether the illness was real or psychosomatic). He spent eight years on a detailed study of the anatomy and larval stages of barnacles (Cirripedia), using material he had collected during the voyage and samples sent to him upon request from other investigators – this study was published in four volumes in 1874. He also carried out intensive experimental work with plants, summarized in six volumes[333].

[328] See Sulloway 1984.

[329] Ibid. After Darwin's return to England, when examined by experts, they were shown to be related species of the same genus. This made the finches the prime example of Darwin's principle of "descent with modification" in the Origin of Species.

[330] Ibid. p. 472. see Chapter 7.

[331] Autobiography, ed. Nora Barlow, 1958, pp. 231-233.

[332] Darwin's home and family life are described in Keynes's book, "Annie's Box: Charles Darwin, his Daughter, and Human Evolution", Fourth Estate 2001.

[333] Review in N. Eldredge. 2005.

> I have therefore nothing to record during the rest of my life, except the publication of my various books[334].

Darwin died in 1881, aged 72. His family wanted him to be buried in his village, but pressure from his friends and influential colleagues led to his burial in Westminster Abbey. Seven of his ten children survived him. His son George became a professor of astronomy and mathematics at Cambridge; and his son Francis participated in his father's botanical work, eventually becoming a professor of botany, and edited his father's Life and Letters.

Although Darwin spent many of his working hours in his armchair, he was not an "armchair biologist", unlike many thinkers in the 19th century:

> Darwin sat in his armchair only after becoming an accomplished naturalist, sportsman and global traveler. All his knowledge brought to bear on the origin of species was empirically based. He accumulated the facts, identified the broad patterns in Nature, and formulated mechanistic explanations accounting for the pattern. His empirical observations were accurate, but some of his hypotheses were flawed[335].

Evolution of the Theory of Natural Selection

Darwin opened a folder on "species and varieties" in 1837, and began collecting relevant data on the difference between the two categories. He noted that artificial selection was the most important factor in the development of domesticated varieties bred to possess the particular forms and qualities desirable by man. He carried on an extensive correspondence with farmers, livestock breeders, and plant growers in Europe, initially without any goal for his enquiries. Then, at the end of 1838, he read Thomas Malthus's book "An Essay on the Principle of Population"[336] about the necessary conflict between the geometric population growth and the linear expansion of agriculture (food production), which inevitably leads to poverty and misery in human populations. Malthus had briefly written that the conflict holds true also among animals and plants. Darwin reasoned that under competitive conditions ("struggle for existence") the weaker individuals must have less chance of surviving. As a pigeon-breeder, he was also aware of the variation among individuals within species – even among offspring of the same pair of parents: some are weaker than others and more disease-prone. These observations gave direction to his further data collection.

Darwin, Voyage, Lone Rider

Although he continued to collect data for years, rather early on he changed his mind about the *fixity of species*, and gradually concluded that species can indeed change, and he posited a mechanism for bringing about the change – *natural selection*. Darwin's friends were aware of his views and often urged him to write them down. In 1842 he began to organize his theory in a series of notes, and in

[334] Autobiography, 1958, p. 115.

[335] Peter Price, American Entomologist, 42: 219, 1996.

[336] See Chapter 2.

1844 he wrote an "essay" – a 230-page document that he regarded as an abstract for a much larger "big book"[337]. He let his friends Charles Lyell and Joseph Hooker[338] read the "essay". Both of them recommended that he should publish the book, lest someone else publish similar material before him. But Darwin still hesitated, feeling that he had not yet assembled sufficient supporting data. In a letter to a friend in 1845, he admitted that he was reluctant to publish his newly-reached conviction for fear of being criticized:

> The general conclusion at which I have been slowly driven – from a directly opposite conviction – is that species are mutable, and that allied species are co-descendants from common stocks. I know how much I open myself to reproach for such a conclusion, but I have at least honestly and deliberately came to it. I shall not publish on this subject for several years[339].

While Darwin was still hesitating, the publication – anonymously – in 1844 of a book titled "Vestiges of the Natural History of Creation" caused a stir among the intellectual circles of London[340]. It contained a mixture of ideas about the changes that had taken place in the biological world. Although scientists rejected the book as superficial and unscientific, it nonetheless stimulated thought on the subject and may have prepared public opinion for the evolutionary ideas yet to come. Darwin certainly read the Vestiges, but he continued working on his "big book" and on his own theory for the next 14 years[341].

In 1857, he wrote a long letter to the American botanist, Asa Gray[342]. In it Darwin described his theory in detail (but asked Gray to keep the information to himself[343]). Man has brought about wonderful change in his domesticated breeds by slowly accumulating small changes over time. These changes occurred either from alteration of the environment or because of variation among offspring of the same parents. There is no reason why the same kinds of variation may not occur in nature. Nature has unlimited time for selecting the most suitable from the variety of individuals born every generation: if an environmental change occurs, some individuals will be destroyed, but those which are able to adapt to the new conditions will survive.

> This process I called natural selection... Natural selection, accumulating those slight variations in all parts of its structure which are in any way useful to it during any part of its life. [Natural selection which] selects exclusively for the good of each organic being[344].

---

337 Darwin realized that he had a ground-breaking new theory. Worried that he might die before assembling enough data for publishing the "big book", he left £400 in his will to his wife, asking her to find a suitable editor and publish the book after his death (F. Darwin 1887, I: 16).

338 Joseph Dalton Hooker (1817-1911), botanist and director of the Royal Botanical Gardens at Kew, London. He served as a surgeon-botanist in expeditions to Antarctica on board HMS Erebus, and in botanical expeditions to the Himalayas and India (1849), New Zealand (1864), the Rocky Mountains in America (1877), and Morocco (1878). He found many new species and organized the systematic botanical collections at Kew. Hooker served for five years as President of the Royal Society, and was knighted. He was a close friend of Darwin's and criticized, but supported, his views. He found Darwin's theory helpful for explaining the geographic distribution of plants.

339 Darwin to Jenyns, Darwin, F., Charles Darwin Life and Letters II: 32. See also Quammen 2006.

340 See Chapter 10.

341 The Cambridge historian, J. van Wyhe, who is in charge of Darwin's archives, wrote in 2005 that the 14-year gap between the Essay and the Origin was not due to Darwin's hesitation and fear of possible criticism of the idea of natural selection, but to his work load. Darwin worked simultaneously on several books, and much of his time was spent on the study of barnacles. Since he was a very meticulous writer, the work on each book took years.

342 Asa Gray (1810-1888), Professor of Botany at Harvard, founded the Department of Botany and donated his botanical collection and his library to the university. He was one of Darwin's few confidants and corresponded often with him. Gray accepted the idea of natural selection, but believed that variation was controlled by a divine force. He became a close friend of Joseph Hooker, visited him in England, and traveled with him in the Rocky Mountains.

343 Darwin to Asa Gray. Darwin, F., Charles Darwin's Life and Letters II: 122 . "I ask you not to mention my doctrine. The reason that if any one, like the author of the "vestiges", were to hear of them, he might easily work them in, and then I should have to quote from a work perhaps despised by naturalists".

344 Darwin to Asa Gray. Darwin, F., Charles Darwin Life and Letters II: 122, 123.

The offspring of every species will try to hold as many habitats in nature as they can – and a few will succeed. In this way an hierarchy is formed in nature:

> This I believe to be the origin of the classification or arrangement of all organic beings at all times. These always seem to branch and sub-branch like a tree from a common trunk., the flourishing twigs destroying the less vigorous, the dead and lost branches readily representing extinct genera and families[345].

A.R. Wallace and the Ternate Paper, 1858

In 1858, Darwin received by mail a short manuscript, sent from the island of Ternate in the Malay Archipelago (near Borneo). In the accompanying letter, the author, Alfred Russel Wallace[346], asked Darwin to evaluate the paper and, if possible, to show it to Lyell.

Darwin was astounded to discover that Wallace had independently thought precisely along the same lines as he himself had in the past 20 years, and arrived at the same hypothesis regarding the mechanism of biological change – natural selection. Fearing that he might lose precedence, Darwin let his friends – who had warned him before that this might happen – read the paper; and he added that, as a gentleman, he had doubts as to whether it would be proper to publish his planned book, now that Wallace might deserve the credit.

Lyell and Hooker, aware of their friend's 20 years of work on the question of the origin of species, found a creative solution. They presented Wallace's paper to the next meeting of the Linnaean Society, accompanied by excerpts from the "essay" that Darwin had let them read 14 years earlier, and a copy of Darwin's 1857 letter to Asa Gray. In the absence of both authors, the "joint paper of Darwin and Wallace" was read by Lyell and Hooker and recorded in the Transactions of the Society; Darwin's priority was thereby established. (Wallace, unaware of the proceedings, was in the Malay Archipelago, and Darwin – a member of the Linnaean Society – had chosen not to attend the meeting). The presentation evoked only a mild response by the audience. Darwin set to work and one year later, in 1859, "The Origin of Species" was finally published, with the first edition of 1,250 copies being sold out on the first day.

Darwin, Driving Tortoises

The "Origin" was much smaller than Darwin had planned it to be. Parts of the material for the "big book" were later published in two volumes as Darwin's "Domestication of Plants and Animals" (1865). The remaining material was only published many years after Darwin's death[347].

---

[345] Ibid.

[346] See Chapter 12.

[347] Stauffer 1975.

Darwin's Reaction

Darwin's correspondence from July 1858 contains enthusiastic letters of gratitude to his two friends, thanking them for presenting his theory to the Linnaean Society[348]. In the letter to Hooker he mentions the latter's promise to inform Wallace of the Linnaean Proceedings, and offered to append a "note" to Hooker's letter. It was not until six months later (25.1.1859) that Darwin wrote to Wallace, praising him for his Ternate paper "Which I read a few months ago", and mentioning that other people had also spoken favorably of it. He noted that he had sent Wallace several copies of the "joint" paper, so that he could see that his paper had been printed verbatim and nothing was changed. Darwin's letter delighted Wallace, then a little-known person in the scientific world. Wallace wrote proudly to his mother that important biologists like Darwin and Lyell thought favorably of his work.

In August, 1859, Darwin wrote to Wallace that his own book on the origin of species, on which he had worked for many years, was in press, and that the publisher had promised to send Wallace a copy.

> When it arrives, I would be glad to hear your opinion on it, since you thought so thoroughly on the subject, and in the direction similar to my own[349].

[348] F. Darwin 1887, II: 162; 220.
[349] Darwin, F., Charles Darwin's life and Letters, Aug. 1859.

9

Evolution of the Theory of Natural Selection

Facts are theories we believe in. Facts which we do not believe we call theories[350].

There is something profoundly disturbing about Darwinian evolution: it is cruel, wasteful, and opportunistic[351].

Darwin's early attempts to organize his thoughts on the origin of species – the "notebooks", the "abstract" of 1842, and the essay of 1844, were not intended for publication[352]. The texts were edited by his son, Francis Darwin, and published in a limited edition in 1909. This material is now freely accessible on the internet[353]. The 1844 material provides a glimpse into Darwin's early deliberations.

On Variation and Domestication

> Most organic beings in a state of nature vary exceedingly little. The amount of hereditary variation is very difficult to ascertain, because naturalists… do not all agree whether certain forms are species or races[354].

Natural selection requires unlimited variation among individuals, in all characters – whether morphological, by which species and varieties are recognized, or subtle characters affecting animal fitness. Yet the impression was that each species is homogeneous – almost no data existed in the 19th century on variation in natural populations, except rare aberrant individuals such as albino birds shot by collectors. On the other hand, domestic animals and plants were known to vary greatly, as illustrated by the enormous numbers of strains of domesticated breeds of any plant or animal bred for long enough – for example, more than 1,200 varieties of cabbage.

> Yet considering how many animals and plants, taken by mankind from different quarters of the world for the most diverse purposes, have varied under domestication in every country and every age, I think we may safely conclude that all organic beings… if capable of being domesticated and bred for long periods, would vary[355].

Darwin suggested that conditions during domestication are favorable for variation: when an animal is bred for a particular purpose, the individuals become "plastic" and more liable to change than they are in nature. The domestic varieties differ from each other in almost any character desired by

350 In "Treasure of Citations"; anonymous (an American jurist).
351 D. Hull, Introduction to Lamarck's Zoological Philosophy, 1984, p. 54.
352 Eldredge 2005.
353 Van Wyhe 2005. F. Darwin 1887.
354 Darwin 1844, Chapter II.
355 Darwin 1844, Chapter II.

the breeder. The evidence suggested to Darwin that all characters – whether physical, mental, or behavioral – can vary. Even if the change is small and hard to detect in one generation, such changes accumulate and become stronger with continued breeding.

Darwin noted that some muscles are strengthened by exercise and frequent use, whereas others become weaker through disuse. These effects may be passed on to the offspring. Changes in the food and in climate bring about changes in coat color and texture – but Darwin was not sure that these changes are heritable[356].

Darwin recognized two kinds of variation. The majority were small individual variations ("fluctuating variation") in measurable characters (lengths of limbs, color, weight etc.). Occasionally, individuals differing greatly from all others were found in domestic breeds. These were called "sports" by gardeners and breeders – and sometimes referred to as "heritable monsters". A child might be born with an extra digit on the hand or foot, or a sheep or a dog born without a tail. Such "sports" were known to be heritable, and sometimes deliberately bred as domesticated races. A famous example – which is often mentioned in Darwin's publications and by other authors of the 19th century – is the case of the "ancon sheep", a variety selected in the USA in 1793 from a single aberrant lamb with a long body and short legs: the deformed and almost monstrous form of ancon sheep is transmitted completely to the next generation when they are mated or crossed with other varieties – exactly like a pure breed.

During the process of domestication, the breeder selects for breeding only those individuals which most closely resemble the desired form. Useful variations are preserved and accumulated for human benefit. An important principle in this process is *isolation*, the separation of desirable individuals from the rest of the population. In the absence of isolation their useful characters will be mixed and an homogeneous breed will result. The process requires continuous effort, attention to the smallest detail, and perseverance by the breeder. Even in the case of "sports" which are strongly heritable, it is imperative that uncontrolled mating be strictly avoided[357]. In this way new breeds, adapted to specific needs and desires of man, can be maintained.

"A Natural Means of Selection": Is a process similar to that of selection by the breeder – possible in nature? Darwin starts with a hypothetical case. Suppose that there existed an all-powerful being who controlled the biological world[358]. There is no reason to doubt that such a being could have planned and created a specialized race or species by selecting characters which favored that race in certain environments. If the designer noticed a plant growing only on rotten tree bark (an uncommon habitat), he could give it an advantage by making its fruit conspicuous to attract birds to disperse its seeds, or design the seeds to germinate on the fresh bark of trees, "like our *Verbascum*". Such a being could gradually select an insect's structure to facilitate its obtaining honey or pollen:

> In accordance with the plan by which this universe seems governed by the Creator, let us consider whether there exists any secondary means in the economy of nature by which the process of selection could go on adapting, nicely and wonderfully, organisms, if in ever so small a degree plastic, to diverse ends. I believe such secondary means exist[359].

Under special conditions, a small population of animals can increase extremely rapidly ("Malthusian growth"): When the Spaniards introduced domestic cattle into the Caribbean Islands or the Argentinian pampas, the populations increased rapidly to enormous numbers. Under a severe struggle for existence, a slight change which might give an advantage to one or a few individuals could allow them to reproduce faster than others, while the non-favored individuals would become less common.

---

[356] Ibid. Chapter I.
[357] Ibid. Chapter II: 7.
[358] In modern terms, this would be referred to as an Intelligent Designer, see Chapter 2.
[359] Darwin 1844, II: 4.

> There will be **a natural means of selection**, tending to preserve those individuals with any slight deviation of structure more favorable to the then existing conditions, and tending to destroy any with deviations of an opposite nature. If the above propositions be correct… new races of beings will – perhaps only rarely, and only in some few districts – be formed[360].

"Special Creation" or Common Descent[361]?

At this early stage of building up his theory, in the "essay" of 1844 Darwin listed the reasons that had led him to reject the doctrine of Creation of animal species, and to favor instead the idea of their common descent.

> It is derogatory that the Creator of countless universes should have made, by individual acts of His will, the myriads of creeping parasites and worms[362].

Darwin repeatedly emphasizes phenomena that cannot be reasonably accounted for by Creation, but are readily understandable if all these species have indeed descended from common ancestors:

> These wonderful parts of the hoof, foot, hand, paddle, both in living and extinct animals, being all constructed on the same framework… can by the creationist be viewed only as ultimate facts and incapable of explanation; whilst on our theory of descent these facts all necessarily follow: for by this theory all the beings of any one class, say of the mammalia, are supposed to be descended from one parent stock[363]. On this theory, therefore, all the organisms yet discovered are descendants of probably less than ten parent forms[364].

Supporting evidence of common descent is provided by the embryological stages of different species. The early embryos of given vertebrate species are more similar to the *embryo* of their probable common progenitor, than the *adults* of these species are to the *adult* form of that progenitor[365]. This similarity is not limited to the external limbs, but extends in detail to the internal organs:

> And lastly, that in a still earlier period of life, their arteries should run and branch as in a fish, to carry the blood to gills which do not exist[366]. Hence the similar course of the arteries in the mammal, bird, reptile and fish, must be looked at as a most ancient record of the embryonic structure of the common parent – stock of these four great classes[367].

Darwin admits that it is difficult for people to grasp the meaning of a million years, or one hundred million years, and appreciate the cumulative effects of small changes over such a long time, when the intermediate stages are not observable. This difficulty was overcome by the belief in catastrophes; but there is a more likely explanation:

> The extinction of the larger Quadrupeds… has been thought little less wonderful than the appearance of new species; and has, I think, chiefly led to the belief of universal catastrophes… I believe, however, that very erroneous views are held on the subject. As

360 Ibid. II: 106.

361 A note in Darwin's 1837 notebook "If we choose to let conjecture run wild, then animals – our fellow brethren in pain, disease, death, suffering and famine – our slaves in the most laborious works – our companions in our amusement – may partake our origin in one common ancestor: we may be all melted together" [F. Darwin 1887, II: 6].

362 Ibid. X: 254.

363 Ibid. VIII: 216.

364 Ibid. X: 252.

365 Ibid. X: 230.

366 Ibid. X: 50.

367 Ibid. VIII: 226.

> far as is historically known, the disappearance of species from any one country has been slow – the species becoming rarer and rarer, locally extinct, and finally lost[368].

Darwin then raised an objection to his own theory: if natural species were formed slowly, through continuous and persistent selection, similar in principle to the selection of cultivated breeds by man – it is to be expected that many transitional stages should be found between species of a genus, between genera of the same family, and between families. But such forms – if found – are rare. This remains an enigma. But Darwin emphasizes that even if all the transitional forms will be found as fossils, no clear linear connections should be expected.

On July 5, 1844, Darwin completed his 230-page "sketch", which he let his friends read. He certainly realized the importance of his new theory, writing the following to his wife:

> I have just finished my sketch of my species theory. If, as I believe, my theory is true, & if it be accepted by one competent judge, it will be a considerable step in science. I therefore write in case of my sudden death, as my most solemn and last request... that you devote £400 to its publication[369].

Then he let the matter rest for the next 15 years!

The Final Form

There are two books which everybody knows about, but nobody reads. One is Marks's Das Kapital. The other is Darwin's The Origin of Species[370].

"The Origin of Species by Means of Natural Selection, or the preservation of favoured races in the struggle for life" (1859)

> It may be said that natural selection is daily and hourly scrutinizing, throughout the world, every variation, even the slightest; rejecting that which is bad, preserving and adding up all that is good; silently and insensibly working whenever and wherever the opportunity offers, at the improvement of each organic being in relation to its organic and inorganic conditions of life[371].

Twenty-five years after its publication, and a few years after Darwin's death, Thomas Henry Huxley – his friend and foremost supporter ("Darwin's bulldog"), reflected on the central idea of "The Origin of Species":

> The suggestion that new species may result from the selective action of external conditions upon the variation from their specific type which individuals present – and which we call "spontaneous" because we are ignorant of their causation – ... is the central idea of the Origin, and contains the quintessence of Darwinism[372].

Darwin had based his theory of natural selection on three *"facts of nature"*. The first, as suggested by Malthus, is that every species has the capacity to increase its populations enormously. Darwin calculated that, even taking into account the reproductive limitation in the biology of elephants, "from a single pair, in 740 to 750 years, there will be nearly nineteen million elephants alive." In

---

368 Ibid. V: 147.
369 Darwin, F. 1887, I: 16.
370 Kerkut 1960.
371 Darwin 1859, first edition of the Origin.
372 Huxley to Francis Darwin. Life and Letters II: 195.

Chapter 3 of the book he presents some evidence for this: on his second visit to America, Columbus had released some cattle on the Caribbean island of Santo Domingo. Within a few years, herds of cattle on the island already numbered in the thousands; and in 1587, 55 years after Columbus's visit, the Spaniards exported 35,444 cattle skins from there.

Another recorded case of population increase was the unfortunate consequence of a kind-hearted Spanish captain, who had taken pity on a female rabbit (held on board his ship for food) which gave birth to kits during the voyage, and he released the mother and her litter when anchored at the island of Porto Santo in 1419. Thirty years later, the Spanish colony at Porto Santo had to be abandoned, because the rabbits had multiplied beyond count, consumed all the vegetation on the island, and prevented the growth of sufficient crops to support the human population[373].

The second "fact of nature" was that populations in nature seem to be stable. Other than during short-term resurgence of agricultural pests such as the locust, explosive population growth does not occur. This means that a large proportion of the reproductive output of animals and plants is destroyed – population growth is checked by "loss of seeds, disease, and premature death". A struggle for existence occurs everywhere in nature, albeit in subtle ways:

> [the population] do not increase at all in numbers. Therefore the whole normal increase must be kept down, year by year, by natural or artificial means of destruction. We behold the face of nature bright with gladness; we often see super-abundance of food. We do not see, or we forget, that the birds which are idly singing around us mostly live on insects or seed, and are thus constantly destroying life. Or we forget how largely these songsters, or their eggs, or their nestlings, are destroyed by birds of prey. We do not always bear in mind, that, though food maybe now superabundant, it is not so at all seasons of each recurring year[374].

The third "fact of nature" is *variation*. The entire second chapter of the "Origin" is dedicated to the description of individual variation – and the existence of distinct varieties within species. Darwin's knowledge of variation, however, was almost entirely based on domesticated breeds (the information on variation is discussed in much more detail in Darwin's "Domestication").

Darwin was aware of individual differences even among siblings: a pigeon-fancier and breeder himself, he was familiar with variation from experience. From correspondence with other breeders he knew that individuals differed from each other not only in morphology, but also in behavior, strength, and resistance to disease. In his 1844 "essay" he considered that the environmental conditions prevailing under domestication may be favorable for animals and plants to vary. He inferred that such variation could be produced in nature, but was not sure whether or not these variations are inherited:

> The most favorable conditions for variation seem to be when organic beings are bred for many generations under domestication... The size and vigor of the body, fatness, periods of maturity, habits of body or consensual movement, habits of mind and temper, are modified or acquired during the life of the individual, and become hereditary… Food and climate will occasionally produce change in colour and texture of the external covering of animals… but whether these peculiarities, thus acquired during individual lives, are inherited – I do not know[375].

Darwin was aware of two kinds of variation in domesticated breeds, and assumed that such existed also in nature: ordinary inter-individual differences; and,

[373] Darwin (1898) 1868, "Domestication" II: 117-120.

[374] Darwin, "Origin", 6th edition (1872) 1898, p. 46.

[375] Darwin 1844, Chapter 1.

> Besides these slight variations, single individuals are occasionally born considerably unlike in certain parts, or in their whole structure, to their parents. These are called by horticulturists and breeders as "sports"... such sports are known in some cases to have been the parents of some of our domestic races... especially of those which in some senses may be called hereditary monsters. For example, where there is an additional limb, or where all the limbs are stunted – as in the Ancon sheep[376].

> At long intervals of time, out of millions of individuals reared in the same country and fed on nearly the same food, deviations of structure so strongly pronounced to be called monstrosities arise. But monstrosities cannot be separated by any distinct line from slighter variations. Even strongly marked differences occasionally appear in the young of the same litter, and in seedlings from the same seed capsule[377].

When the struggle for existence is severe, Darwin reasoned, the stronger and healthier will have a better chance of surviving than the weaker. Darwin called this process "*natural selection*", by analogy with the artificial selection exercised by the breeder on his stock. This process is the cause of the beautiful adaptations we see in the biological world:

> We see beautiful adaptations everywhere and in every part of the organic world... All these results... follow from the struggle for life. Owing to this struggle, variations however slight and from whatever course proceeding, if they be in any degree profitable to the individual of a species, will tend to the preservation of such individual, and will generally be inherited by its offspring. I have called this principle, by which each slight variation, if useful, is preserved, by the term Natural Selection[378].

The struggle occurs not only among individuals within species. In a paragraph written for the "big book" – perhaps around 1856 – but not included in the "Origin", Darwin illustrated the process of introduction and competition among species in nature:

> Nature may be compared to a surface covered with ten thousand sharp wedges, many of the same shape & many of different shapes, representing different species, all closely packed together & driven in by incessant blows, the blows being far severer at one time than at another. Sometimes a wedge of one form & sometimes another being struck, the one driven in forcing others out[379].

Speciation

> As natural selection acts solely by accumulating slight, successive favourable variations, it can produce no great or sudden modification... hence the cannon of "natura non facit saltum"... is on this theory intelligible[380].

Natural selection will have two results. First, by favoring the individuals best fit to the current environmental conditions, and selecting out the less fit, it should make the populations better adapted. And second, it should encourage the formation of new species.

Individuals of the same species compete most strongly for the same resources in the struggle for existence. Competition will be lessened if they diverge somewhat from each other in characters,

376 Ibid.
377 Darwin 1898 (1872), p. 6.
378 Ibid. p. 45.
379 Stauffer 1975, p. 208.
380 Darwin 1898 (1872) , Conclusions, p. 361.

especially if the environment differs somewhat at different sites[381]. The differences among individuals will gradually expand into variation among varieties, and the varieties will then tend to become different enough to be called different species.

> As the modified descendants of each species will be enabled to increase by as much as they become diversified in habits and structure, so as to be able to seize on as many and widely different places in the economy of Nature, there will be a constant tendency in natural selection to preserve the most diverging offspring of any one species. Hence, during a long continued course of modification, the slight differences characteristic of varieties of the same species, tend to be augmented into the greater differences characteristic of species of the same genus[382].

"Descent with Modification"

While in the Galapagos Islands, Darwin had collected some birds – especially mockingbirds – that differed in appearance on the different islands. After his return to England the birds were examined by experts at the British Museum. A group of finches, varying in bill dimensions, were shown to be closely related. Summarizing his impressions, Darwin wrote in "The Voyage of the Beagle" (1845):

> The most curious fact is the perfect gradation of the beaks in the different species of *Geospiza*. There are no less than six species, with insensibly graduated beaks. Seeing this gradation, and diversity of structure in one small, intimately related group of birds, one might really fancy that from an original paucity of birds in this archipelago, one species has been taken and modified for different ends[383].

However different the species thus produced may be, their common origin can be detected (for example, in vertebrates), by comparisons of skeletal characters. This is a general principle, true for the entire natural world:

> The similar framework of bones in the hand of a man, wing of a bat, fin of a porpoise and leg of a horse – the same number of vertebrae forming the neck of the giraffe and of the elephant – and innumerable other such facts at once explain themselves on the theory of descent with slow and slight modifications[384].

> The real affinities of all organic beings, in contra-distinction to adaptive resemblances, are due to inheritance or community of descent. The natural system is a genealogical arrangement, with the acquired grades of differences marked by the terms varieties, species, genera, families etc. And we have to discover line of descent by the most permanent characters, whatever they maybe and of however small importance[385].

---

381 This is referred to as the "Principle of divergence". Darwin wrote to Asa Gray in September 1857: "The same spot will support more life if occupied by many diverse forms".

382 Darwin 1898 (1872), Conclusions, p. 360.

383 Darwin 1845, p. 384. The phenomenon that Darwin "fancied" is today referred to as *adaptive radiation*. The term was suggested by Osborn in 1902 (Osborn 1902,J. Huxley 1942, p. 486) . The Galapagos finches are the most celebrated example of adaptive radiation, thanks to the work of David Lack (1947), who named them "Darwin's Finches".
Another famous example of adaptive radiation is the evolution of a group of colorful birds – the "sickle bills", Drepanididae – inhabiting the Hawaii archipelago. Taxonomists divide the "sickle bills" into 18 different genera - among them seed eaters, insect eaters, and nectar-feeding species with curved beaks which gave the group its common name. Like the Galapagos, the Hawaiian islands are volcanic, thousands of kilometers from the nearest continent.

384 Darwin 1898 (1872), p. 306. The anatomical facts were known to the great comparative anatomists of the 19th century at the Museum of Natural History in Paris - Georges Cuvier [Chapter 4] and Geoffroy Saint-Hilaire [Chapter 5]; but they treated the skeletal similarities as taxonomic tools, not as indications of a common descent.

385 Ibid. p. 366.

Support from Embryology

Darwin was delighted to find supporting evidence for his theory of descent in the embryological studies of K.E. von Baer[386]. Von Baer had observed that the earlier stages of the embryos of different mammalian species are almost indistinguishable, although their adult forms are very different. Contrary to the prevalent opinion[387], von Baer wrote that an embryo of a higher animal never passes the *adult* stage of a lower animal: the gill slits in the mammalian embryo resemble fish embryos, not adult fish – the adult fish or reptile cannot live in the embryological environment. Similarities among embryos of different vertebrates are greatest when the embryos are undifferentiated in the early stages, and they diverge from each other as the embryological development proceeds.

Darwin recognized that von Baer's work lent strong support to his own theory of descent: the similarity of embryos may indicate a closeness of common ancestors. Phylogenetically-closer organisms will be more similar to each other as embryos. He quoted von Baer at length:

> [as von Baer wrote] the feet of lizards and mammals, the wings and feet of birds, no less than the hands and feet of man, all arise from the same fundamental form... Generally, the embryos of the most distinct species belonging to the same class are closely similar, but become – when fully developed – widely dissimilar[388].

Darwin found support for this conclusion in his own work. The fledglings of very different varieties of pigeons were quite indistinguishable when very young. He found an excellent case of this in his taxonomic work with the Balanidae (Cirripedia). Taxonomists were confused over the affinities of the Balanidae, and Cuvier had included them in the Mollusca. Darwin examined their larval stages and immediately saw where they really belonged:

> Even Cuvier did not perceive that a barnacle is a crustacean, but a glance at the larva shows this in an unmistakable manner[389].

On the Evolution of Instincts

> An action, which we ourselves require experience to enable us to perform, when performed by an animal – more especially by a very young one – without experience – and when performed by many individuals in the same way, is actually said to be instinctive[390].

Darwin saw no difficulty in explaining the evolution of "mental" faculties by natural selection. If the level of expression of an advantageous character – like an instinct – is variable among individuals, it is not impossible that natural selection will improve this character.

Darwin sought to trace the stages in the evolution in cuckoos, of the instinct of laying eggs in the nests of other birds: it may have originated as an accident, which became an established habit if it had a selective advantage in some particular circumstances. This behavior was accompanied by the instinct of the fledgling to throw the eggs of its foster parent out of the nest[391].

---

[386] Karl Ernst von Baer (1792-1876) was born in Germany but did most of his work in Russia, where he was elected a member of the Academy of Sciences. He proved that mammals developed from eggs, which he seems to have been the first to observe under a microscope. Examination of the development of chick embryos led him to comparative embryology. Von Baer discovered the early development of the nervous system and the five primary vesicles in the vertebrate brain.

[387] Fifty years later, the German biologist Ernst Haeckel (Chapter 19) interpreted the similarity of early embryos to the adult forms of more primitive animals as proof that the ontogenetic differentiation of each individual is "a short and concise recapitulation of phylogeny" (Haeckel's "biogenetic law"). This erroneous interpretation persisted for many years.

[388] Darwin 1898 (1872), p. 364.

[389] Darwin 1898 (1872), p. 365.

[390] Ibid. p. 191.

[391] Ibid. pp. 197-200.

Among examples of instinctive behavior in insects, Darwin lists the behavior of slave-making ants[392] and the "voluntary contribution" by aphids of their sweet secretions to ants, which he describes as "one of the strongest instances of an animal apparently performing an action for the sole good of another"[393].

Darwin deals at length with the instincts of comb-building in bees[394].

> He must be a dull man who can examine the exquisite structure of a comb, so beautifully adapted to its end, without enthusiastic admiration[395].

He lists cases of variation among bees in comb-building, and wonders how – due to the neuters (workers) in bee and ant colonies being sterile – could natural selection operating on variation among sterile individuals, improve social instincts and morphological differentiation? Darwin suggests that selection also works among groups:

> The fertile males and females... transmitted to their fertile offspring a tendency to produce sterile members with the same modifications... selection may be applied to the family, as well as the individual, and may thus gain the derived end.
>
> How the workers have been rendered sterile is a difficult question. But if such insects had been social, and if it had been profitable to the community that a number have been ...capable of work, but incapable of procreation... I can see no special difficulty in this having been affected through natural selection[396].

"The Origin of Species" – especially the latest (6th) edition – in fact covers much more than the establishment of natural selection as the moving force in evolution. Chapters VI and X deal with the "difficulties" that Darwin himself felt in his theory, and Chapter VII is an attempt to reply to the many objections raised against the theory – in particular by the biologist St. George Jackson Mivart[397]. Chapter VIII deals with "Instinct"; Chapter IX with hybrid sterility as a mechanism for the stability of species; and Chapters XII and XIV with the geographical distribution and classification of animals.

It is noteworthy that the word "*evolution*" nowhere appears in the first edition of "The Origin of Species" (1859). The term was introduced by the British philosopher Herbert Spencer (1820-1903). In a book entitled "First Principles"[398], Spencer described "The Law of Evolution" as a general process of change from the simple to the more complex, from the diffused to the more organized, and from the homogenous to the more heterogeneous. Evolution is applicable to everything from galactic bodies to human society. As for evolution in biology, Spencer wrote:

> That, in the course of time, species have become more sharply marked off from other species, genera from genera, and orders from orders, is a conclusion not admitting of a more positive establishment than the foregoing... If, however, species and genera and orders have arisen by"Natural Selection", then, as Mr Darwin shows, there must have been a tendency to divergence, causing the contrasts between groups to become greater[399].

[392] Ibid. pp. 202-205.
[393] Ibid. p. 193.
[394] Ibid. pp. 207-212.
[395] Ibid. p. 205.
[396] Ibid. pp. 214-215. The question of group selection separate from individual selection was controversial in the late 20th century (e.g. papers by Wade 1977, 1978).
[397] See Chapter 14 below.
[398] Spencer 1862. First Principles. W.J. Johnson, printer, 121 Fleet Street, London.
[399] Ibid. p. 371.

10

On the Same Track: Ideas about the History of Life before 1859

Lamarck and Darwin: similarities and differences of two theories:

> Heaven forefend me from Lamarck's nonsense of "a tendency to progression", adaptation from the slow willing of animals &c. But the conclusions I am led to are not widely different from his, although the means are wholly so[400].

> You often allude to Lamarck's work. I do not know what you think of it, but it appears to me extremely poor. I got not a fact nor an idea from it[401].

By the early 20th century, Darwin's theory had been accepted by the majority of biologists and become the basis of biological thinking, while Lamarck's views on evolution[402] were rejected as false, non-scientific, or ridiculous. However, in the 19th century there had been no doubt in almost everybody's mind that characters acquired during the lifetime of an individual were transmissible to the next generation, and that the characters of the parents were "blended" in the offspring. Although Darwin was aware that some characters "refused to blend"[403], he often referred to the cumulative effect of "use and disuse" – Lamarck's mechanism for adaptation and transmutation in animals – to assist natural selection in producing change [in the "Origin" as well as in his later books, "The Descent of Man" and "Sexual Selection"].

Scientists in the 19th century frequently found it difficult to tell Darwin's theory apart from that of Lamarck. Charles Lyell regarded Darwin's theory as an extension and improvement on Lamarck's – not as an alternative theory:

> I cannot go to Huxley's length in thinking that Natural Selection and variation account for so much, and not so far as you, if I take some passages in your book separately. I think the old Creation is almost as much required as ever, but of course it takes a new form **if Lamarck's views improved by yours** are adopted[404].

Darwin denied any connection to Lamarck's theory (see quotations above). The confusion between the two theories infuriated him:

> Plato, Buffon, my grandfather before Lamarck, and others, all propounded the view that if species were not created separately, they must have descended from other species, and I can see nothing else in common between the Origin and Lamarck. I believe this way of

400 Darwin to Hooker. 1844. F. Darwin (ed.) 1887: Darwin's Life and Letters II: 23.
401 Darwin to Lyell. Ibid. II: 215.
402 See Chapter 6.
403 See Chapter 18 below.
404 Lyell to Darwin, 11 March 1863, Lyell's Life and Letters II: 363.

> putting the case is injurious to its acceptance, and closely connects Wallace's and my views with what I consider, after two deliberate readings, as a wretched book, and one from which (I still remember my surprise) I gained nothing[405].

In Darwin's theory, although natural selection had replaced the concepts of "desire" and "needs" of the animals, there is nevertheless, a fundamental similarity between the two theories. Darwin, like Lamarck fifty years before him, believed that all life had descended from a common origin, and that species are not of fixed form and could "transmutate" to become newer, more complex organisms. Both theories allowed the inheritance of acquired characters (expressed in the form of Use and Disuse). Darwin, like Lamarck, had no knowledge of the mechanisms of heredity. That acquired characters are inherited was at the time a plausible hypothesis. For example,

> Embryological resemblances of all kinds can be accounted for... by the progenitors of our existing species having varied after early youth... and having transmitted their newly acquired characters to their offspring at corresponding age[406].

> Disuse, aided sometimes by natural selection, will often have reduced organs when rendered useless under changed habits of life. And we can understand on this view the meaning of rudimentary organs[407].

Until, at the end of the 19th century, August Weismann denied the possibility of inheritance of acquired characters[408], no blame was put on Darwin for using this explanation.

Forerunners of Darwin?

> The complex of ideas which later went to make up Darwinism was widely enough diffused in the eighteenth century, that finding an unknown or forgotten evolutionist has about it something of the fascination of collecting rare butterflies[409].

Patrick Matthew: [410]Matthew wrote a short note in a book on naval timber, in which he described a process of natural selection: individuals who are not strong, fast, or cunning enough are eliminated without reproduction, as prey to predators or due to disease, making room for "more perfect" ones. Darwin, who was unaware of this note in 1859, acknowledged Matthew's contribution in the introduction to the 6th edition of the "Origin". This acknowledgement, however, did not satisfy Matthew, who insisted that he had had priority.

William Charles Wells: (1757-1817). Wells, a Scottish physician, discovered the mechanism of formation of dew, published a paper on human vision, and became a member of the Royal Society. In 1818 he wrote a paper[411] (a case study of a patient) in which he suggested that the black skin color in Africans was the result of natural selection – but not, however, selection of individuals *within* populations: he believed that this was impossible because of blending inheritance – but rather selection *among* populations. Wells considered that ancient humans had been scattered in very

---

405 Darwin to Lyell. F. Darwin 1887, Darwin's Life and Letters, II: 198-199.
406 Darwin 1898 (1872), Origin, p. 182.
407 Darwin, Origin, Conclusions, p. 366.
408 See Chapter 20 below.
409 Loren Eiseley, 1961, p. 119. See also R. Stott, "Darwin's Ghosts", 2012.
410 Ibid.
411 Wells, W.C. "An account of a female of the white race of mankind, part of whose skin resembles that of a negro". 1818, Hurst, Robinson and Co., London

small family groups, and that dark-skinned families had some advantage (unrelated to their color; perhaps some resistance to disease) which allowed them to spread in tropical Africa[412].

Edward Blyth: (1810-1873). Blyth was a well-known zoologist whom Darwin consulted more than once [413], and the author of two papers that seem to describe natural selection. The American historian, Loren Eiseley, suggested that Darwin had deliberately ignored the two papers[414]. Eiseley appended the two papers, published in 1835 and 1837 respectively, to his 1981 book. The more important of the two is entitled "The variety of animals" (1835). Blyth describes the results of the struggle for resources, and suggests that this result was intended by Providence for improving the species:

> And as …the stronger must always prevail over the weaker, the latter, in a state of nature, is allowed but few opportunities of continuing its race. The one best organized must always obtain the greatest quantity [of resources]… and thus be enabled, by routing its opponents, to transmit its superior qualities to a greater number of offspring[415].
>
> The same law which was **intended by providence to keep up the typical qualities of a species** can be easily converted by man into a means of raising different varieties. The original form of a species is unquestionably better adapted to its natural habitat than any modification of that form[416].

This view is the opposite of what Darwin had in mind. Darwin, therefore, seems to have been justified in not considering these papers as suggestive of his theory.

Robert Chambers and "The Vestiges of Creation"

The Mysterious Book

A book bearing the title "Vestiges of the Natural History of Creation" was published in 1844. The combination of "creation" and "natural history" in the title attracted the attention of secular as well as church-oriented readers. The author's name was not revealed and the publisher kept it secret – contributing to the public interest in the book.

The book was a synthesis, or mixture, of ideas borrowed from known authors like Lyell, Lamarck, and even Darwin ("Voyage of the Beagle"), not always acknowledged. The book was very popular – 12 editions were printed within a short period[417]. It was discussed in tea parlors and clubs of the middle and upper classes of English society, not only in scientific circles. Speculations about the author reached as far as the Royal Family.

It was only in the 12th edition that the name of the author was disclosed: Robert Chambers (1802-1871), a Scottish journalist who ran a publishing house in Edinburgh together with his brother[418]. It is claimed that the popular Vestiges was a thought-stimulant among British intellectuals, preparing the way for Darwin's theory.

---

412 See also Gould, The Flamingo's Smile, pp. 335-346.

413 Edward Blyth lived and worked in Calcutta, India, and organized the natural history museum there. He participated in scientific expeditions in India and China.

414 Eiseley 1981, "The Mysterious Mr. X".

415 Blyth 1835, in Eiseley 1981, p. 103.

416 Ibid.

417 An anonymous commentary in the press of the period states: "The fascinating and revolutionary book 'Vestiges of the natural history of creation', published anonymously in 1844, was simply an international sensation of its time".

418 Darwin reported in 1847 that he had met Chambers and had the impression that he was the mystery author.

The History of the World

Chambers composed what seemed to him to be a unifying theory of the biological world and its evolution. He suggested that there is order and direction in nature, beginning with the mineral and cosmic system. In the living world there is a slow and gradual progression according to natural laws, but these laws themselves are God's creation:

> That God created animated beings… is a fact so powerfully evidenced, and so universally received, that I at once take it for granted… Organic creation is… the result of natural laws, which are, in a manner, expressions of His will[419].
>
> [there is no doubt] that the almighty author produced the progenitors of all existing species by some sort of personal or immediate exertion. But how does this notion comfort with what we have seen of the gradual advance of species, from the humblest to the highest? Taking care Himself of all the particular minute creations would surely take a very mean view of the creative power, in short, to anthropomorphize it[420].

The first two chapters of the "Vestiges" deal with the creation of the universe. Since this singular event, the processes of erosion and volcanic activity have been in operation as they are today (the "Uniformitarian" view as expressed by Lyell is cited on p. 51 of "Vestiges"). The crystallization of minerals in the laboratory, and the formation of ice crystals on windows (in winter) in forms reminiscent of plants, suggested to Chambers that the shapes of plants may have been determined by physical or electrical forces[421]. Animals and plants are composed of the same elements – carbon, oxygen, hydrogen and nitrogen – which combine to form various proteins, and combinations of proteins constitute cells. Cell division is common to every living organism, suggesting that they all are descended from a common stock.

The living world progresses from single cells to complex organisms[422]. The basic living unit is spherical, within which new units are formed and released to become independent organisms – as in the *Volvox*.

There are no remains of plants and animals in the oldest, granitic rocks – this is because these rocks were formed before life was created. The geological strata above the granite are calciferous, which testifies to the presence of life: plants and corals absorb CO_2 from the air and the water and deposit calcium carbonate in their skeletons.

The first evidence for the existence of dry land is in the Carboniferous era, where the remains of terrestrial plants are found, with evidence of a progression from simple ferns to advanced trees. From the sizes of the trees, Chambers infers that the climate during this period must have been tropical[423] and that the atmosphere was rich in CO_2. Such an atmosphere could encourage the growth of vegetation, but not the proliferation of animal life[424]: this only became possible when oxygen was released by plants into the atmosphere.

The first animals to appear were marine – their skeletons and shells form the calciferous strata. Early organisms vanished and their place was occupied by more advanced ones, as can be observed in the more recent geological formations: fish as well as invertebrates, which are still unlike the extant species. A progressive trend is noticeable "from simple to more complex forms", as suggested by Lamarck[425].

---

419 Chambers 1994 (1844), p. 154.
420 Ibid. pp. 152-153.
421 Ibid. p. 165.
422 Ibid. p. 172; these ideas are probably borrowed from Lamarck.
423 Ibid. p. 84.
424 Ibid. p. 172.
425 Ibid. p. 72.

Following the Carboniferous era, the "red sandstone period" supported giant reptiles and terrestrial plants (this statement is based on Darwin's notes from the Voyage of the Beagle). Above the sandstone, the (Cretaceous) geological strata are rich in insect and mammalian remains.

The Law of Development

Chambers adopted Lamarck's principle that life progressed from simple to complex forms. But he also adopted Cuvier's taxonomic structure of four independent branches based on the general animal form: Radiata, Mollusca, Articulata, and Vertebrata[426]. The progression from the simple to the complex occurs within each of Cuvier's four "embranchements"– and even in subunits within the vertebrates – fish, reptiles, birds, and mammals. A similar progression also occurs in plants[427]. Above all, this was planned and supervised by the Almighty:

> It has pleased Providence to arrange that one species should give birth to another, [that the simplest and most primitive type …gave rise to the type next above it, that this again produced the next highest], until the second – highest gave birth to man, who is the very highest. Be it so, it is our part to admire and to submit[428] (ibid. p. 234).

Further, all animals are built as variations on the same basic plan – altered as adaptations for different purposes[429]. Chambers lists some examples: the nose, originally an organ of smell, became adapted as an organ of work in the elephant. The ribs became an organ of locomotion in snakes. On the other hands, different organs in different organisms may serve the same function: the lungs are adapted for respiration on land and the gills in the water. Rudimentary organs may indicate the primitive and common origin, in particular in animals which represent intermediate stages: the ostrich and the platypus are intermediates between birds and mammals[430].

> The whole train of animated beings, from the simplest and oldest up to the highest and most recent, are then to be regarded as a series of advances of the principle of development which have depended upon external physical circumstances, to which the resulting animals are appropriate. I contemplate the whole phenomena in the first place arranged in the councils of divine wisdom[431].

Chambers found evidence supporting the Law of Development in embryology. Every individual embryo passes in its ontogeny developmental stages which resemble animals below it in the scale of life. Even Man develops from a single cell ("animalcule") and passes through stages resembling a fish, a reptile, a bird, and a primitive mammal before showing the features of a human being (Chambers explains that the resemblance is not with the *adult* animals of these lower classes, but with equivalent stages in their embryonic development)[432] A progressive pattern can be seen in the development of the heart from a simple tube ("insect heart"):

> In mammals, the heart has four chambers; In reptiles, only three, and in fish, only two. Let us trace this law also in the production of certain classes of [human] monstrosities… the heart, for instance, goes no further than the three-chambered form, so that it is the heart

[426] See Chapter 4.
[427] Chambers 1994 (1844) pp. 196-197.
[428] Ibid. p. 234.
[429] This idea is borrowed from Geoffroy Saint-Hilaire [Chapter 5].
[430] Ibid. p. 194.
[431] Ibid. p. 203.
[432] Ibid. p. 212. This interpretation follows Von Baer.

> of a reptile. There are even instances of this organ being left in the two-chambered fish form[433].

Chambers illustrates the Law of Development in a simple diagram: a main line of embryonic progression leads to the highest mammal – Man – with the embryos of fish, reptiles, and birds branching off from it at early stages.

As regards transmutation, Chambers suggests the platypus and the ostrich as possible intermediates between birds and mammals; and that such transformations may in fact occur, but are not reported when observed, due to the strong belief in the fixity of species[434].

> ...How easy it is to imagine an excess of favourable conditions, sufficient... to make a fish mother develop a reptile heart, or a reptile mother develop a mammal one... This under-adequacy would suffice in a goose to give its progeny the body of a rat, and produce the Ornithorhynchus [= platypus] Or might give the Ornithorhynchus the mouth and feet of a true rodent, and thus complete at two stages the passage from Aves to the Mammalia[435].

Reaction to the Book

A year after publication of the "Vestiges", Chambers published – again anonymously – a second volume entitled "Explanations". He explained that he was not interested in publicizing his own views – that was why he had published the "Vestiges" anonymously – but in distributing the knowledge and thus contributing to the happiness of mankind:

> [the book is] the first attempt to connect the natural sciences into the history of Creation, with as little disturbance as possible with existing beliefs.

Chambers hoped to bridge the gap between science and religion, and that the book would be accepted by both parties. He was greatly disappointed. The scientists objected to the book as shallow and confused, and as offering nothing new. The religious establishment objected that it did not follow the Scriptures. T.H. Huxley curtly expressed his objections in a letter to Darwin:

> As for the Vestiges, I confess that the book simply irritates me by the prodigious ignorance and thoroughly unscientific habit of mind manifested by the writer[436].

The geologist Adam Sedgwick, in Cambridge, wrote a detailed criticism of both of Chambers's books[437]. Sedgwick found serious errors in the latter's interpretation of the geological and paleontological evidence. He commended Chambers for recognizing the control of nature by a "First Cause" – unlike the atheist authors[438] – but rejected his theories for lack of supporting evidence:

> The author is neither well acquainted with the first principles of physics, nor well read in any sound work on physiology.
>
> Spontaneous generation (in the Author's sense) has not one good unambiguous fact to rest upon The Theory of Development has no firmer support in nature, and the only pretended

433 Ibid. p. 219.
434 Ibid. p. 220.
435 Ibid. p. 219.
436 Huxley to Darwin, Darwin's Life and Letters II: 189.
437 Adam Sedgwick, 1850; A discourse on the studies of the University of Cambridge, preface to the fifth edition, pp. 20 ff.
438 Like many of his colleagues in British universities, Sedgwick was a Creationist. His criticism is directed not only at Chambers, but against all who believed in a gradual change and transmutation in the biological world.

> fact [of transmutation] turns out to be nothing better than a misconception and a blunder[439]. No anatomist has ever observed... changes bringing the nascent [vertebrate] embryo into a true similitude with the Radiata, Arthropoda, or Mollusca[440].

In particular, the theory of progressive development in the embryonic state, with transformation in an ascending direction from fish to man, is not supported by evidence: the embryonic development of the Mammalian heart does not go to the stages of amphibian or reptilian hearts, and – after all the transformations – each species retains its constant adult form. Sedgwick held that the apparent gradual ascending trend in the fossil record in the geological strata was a divine arrangement:

> What beings may be produced out of the abortive imagination of the human brain is not a question worth a moment's pause. So far as nature is concerned, philosophy has nothing to do with what may be, but with what is[441].
>
> But the elevation of the fauna of successive periods was not made by transmutation but by creative addition. There was a time when Cephalopoda were the highest type of animal life, the primates of the world. Fishes then took the lead, then reptiles. And during the Secondary period they were anatomically raised above any form of reptile class now living in the world. Mammals were added next, until Nature became what she is now, by the addition of Man[442].

Despite this criticism, it seems that Chambers' popular book played a positive role in preparing the public for the views that Charles Darwin was to publish 15 years later.

439 Ibid. p. 43.
440 Ibid. p. 29.
441 Ibid. p. 44.
442 Sedgwick 1850 (1831), cited by Lyell 1873, p. 442.

11

Alfred Russel Wallace

His Life

> Once in a generation, a Wallace may be found, physically, mentally and morally qualified to wander unscathed through the tropic wilds of America and Asia; to form magnificent collections as he wandered; and withal to think out sagaciously the conclusions suggested by his collections[443].

Alfred Russel Wallace (1823-1913) was born in Usk, Wales, the eighth of nine children (only four of them survived childhood). His father, a lawyer, enjoyed literature, and before he was 13, Alfred had heard his father read aloud the works of Shakespeare, Cervantes' Don Quixote, and Dante's *Inferno.* When his father then lost his fortune and his estate, the family had to leave their home. Alfred also had to leave school and was sent to live with his older brother William, a surveyor, who did not have a permanent address. The brothers changed lodgings frequently as they moved from place to place with the surveying jobs (Alfred later became a qualified surveyor himself and worked in this profession for some years). As a boy he became interested in the plants that he saw while helping his brother. He had only seven years of formal education and acquired his knowledge by reading. Aged 20 and unemployed, he found a teaching job in a private school in Leicester. He taught draughtsmanship, English, and arithmetic for two years. The town had a good public library and Alfred read in particular books on biology – Darwin's "Voyage of the Beagle" and Chambers' "Vestiges" apparently captured his imagination. At the library he met Henry Walter Bates[444] and they became close friends. Bates was younger than Wallace but already an accomplished entomologist, and Alfred became infected with the fever of insect collection.

The stay at Leicester was cut short by the death of his brother William, and Wallace spent the next year or two taking care of his brother's finances and helping his mother and sister. The friendship with Bates was renewed when they met up again in London. Both unemployed, they worked out a plan: to travel to the Amazon forests in Brazil, collect insects, birds, and other organisms, and sell them to British collectors. They consulted with experts at the British Museum, learned the necessary techniques, arranged with an agent in London to sell the collections for them, and in 1848 set out for Brazil.

In his diary, Wallace expressed his surprise at the lack of color and of animal life in the dense forest:

> It is only over the outside of the great dome of verdure exposed to the vertical rays of the sun that flowers are produced, and on many of these trees there is not a single blossom to be found at a less height than a hundred feet. The whole glory of these forests could only be

---

[443] T.H. Huxley, Man's place in Nature, 1900.

[444] Henry Walter Bates (1825-1892) was in his youth a clerk in a brewery, but an enthusiastic collector of insects. With his friend Wallace he traveled to the Amazon in Brazil and stayed there for 11 years. Back in England, he published his important book "Naturalist on the River Amazon" and other important papers. Among other topics he described and explained the phenomenon of *mimicry.* Bates served as President of the Royal Geographical Society.

> seen by sailing gently in a balloon over the undulating flowery surface above. Such a treat is perhaps reserved for a traveler of a future age[445].

After two years of collecting together, they parted company. Wallace remained for two more years in the Amazon, collecting while traveling slowly up the Rio Negro to its source. His younger brother, who came to assist him with his collection, died of yellow fever. Then misfortune struck again, when the ship Wallace boarded to return to England caught fire and sank, and all his most recent collections were lost. After a few days in life boats, the crew and passengers were rescued by a passing ship. Fortunately the collections had been insured by his agent, who provided him with enough money for a temporary stay in London.

Wallace attracted the attention of the scientific circles by lecturing on his travels to interested audiences – at the time the Amazon was little known biologically[446]. He had managed to save from the sunken ship the notes and drawings of many species of palm trees, which he published as a book.

A few years later, Wallace set out again for the tropics, this time to the Malay Archipelago (Indonesia of today). He spent eight years collecting on these islands (see below). Wallace was a first-rate field biologist and observer. He became familiar with the lives and ecology of the creatures he collected, and observed the fine adjustments in form, coloration, and behavior between the animals – insects in particular – and the plants they fed or alighted on. He sent his collections periodically to his agent in London, who sold them to interested buyers, providing him with enough money for a continued stay.

During his travels in the tropical islands he thought about animal life in general, and expressed his ideas in two articles. The second of the two he sent to Charles Darwin for evaluation. This article clearly described the process of natural selection, unaware of Darwin's work on the subject. Wallace did not know of the frantic activity his paper caused in Darwin's circle of friends (see below). When Wallace finally returned to England, his collections comprised around 125,000 items, and he was recognized as co-author of Darwin's theory of natural selection and evolution.

Wallace married late and had three children. He lived to the age of 90 and devoted his life to the support of Darwinism. After the death of both Darwin and Huxley, Wallace remained the "grand old man" of Evolution, and was invited to give a series of lectures in the USA, which he expanded in order to see more of the country (in California he met his brother John, whom he had not seen in 40 years). Wallace wrote hundreds of articles and some 20 books on various subjects: noteworthy are "Darwinism", "Natural Selection" (a collection of essays on the subject), "The Malay Archipelago", and "The Geographical Distribution of Animals". He seems to have been a restless individual, changing his lodgings frequently[447]. Financially, he remained unsuccessful. He invested his savings in his brother-in-law's failing photographic business; and became involved in a lawsuit against a believer in "flat earth" who dragged him through the courts for years at great expense, although Wallace eventually won the case. He vigorously supported several liberal movements, such as the Land Reform[448] and the Anti-vaccination movements (he claimed that enforced infant vaccination was based on improper data and statistical analysis[449]). He tried to convince his scientific friends that spiritualism was worth investigating scientifically. This was one reason why his applications for a permanent job as manager of some nature reserve, which would have suited him well, were never approved. Even a government pension of £200 did not relieve his difficulties[450].

---

[445] Wallace 1858, p. 36.

[446] Wallace, 1858: A narrative of travels on the Amazon and Rio Negro; Reeve & Co., London.

[447] See his biography: Raby 2001.

[448] Ibid.

[449] Fichman, M., Keelan, J.M. 2007: Registrar's logic: the anti-vaccination argument of Alfred Russel Wallace and their role in the debates over compulsory vaccination in England, 1870-1907. Studies in History and Philosophy of Biological and Biomedical Sciences 38: 585-607.

[450] Raby 2001.

The Malay Archipelago

> I myself had hoped, rather than expected, ever to reach the Ultima Thule of the east. And when I found that I really could do it now, had I but the courage to trust myself for a thousand miles in a Bugis prau, and for six or seven months among lawless traders and ferocious savages – I felt somewhat as I did when, as a schoolboy, I was for the first time allowed to travel outside the stage-coach to visit that scene of all that is strange and new and wonderful to a young imagination: London[451]!

Wallace's first book, published in 1869 and dedicated to Charles Darwin, is a review his of eight years of life in the Malayan islands – of traveling 14,000 miles in "60 or 70" sea crossings in local boats. He describes the topography, the vegetation, the fauna he collected, and his observations of the native tribes around whom he had lived.

The book opens with a summary of Wallace's achievement as a collector (lamenting that he was actually collecting only 75% of the time): a list of the 125,000 specimens of animals he had hunted – including 17 orangutans of different ages. These apes – called Mias by the locals – were much in demand in European museums and their skins and skeletons were sold at high prices. Wallace hired local hunters to increase his collections of rare birds and mammals, and shot some of the Mias himself – as the natives did not have firearms[452]. His account of the habits of these apes – building nests, throwing sticks to ward off attackers – was the first source of data on these animals. Wallace even attempted to nurse an orphan orang, after killing its mother, but failed to bring it back alive to England.

The book gives some idea of the problems this lonely Englishman, accompanied by one or two trustworthy Malay assistants, had to overcome. He needed to plan his visits to the different islands according to the irregular availability of boats traveling between the islands. He needed to obtain letters of introduction to prominent businessmen or to the local village chief or Rajah. He needed to negotiate and rent a place to stay in every island and village, or occasionally build himself a bamboo hut when no accommodation could be found. He needed to take care of his own health and that of his assistants, nursing them when they became sick (as they nursed him when he became ill or disabled by injury). Travel from one village to the next – on foot – required the hiring of up to 20 porters. The villagers, who had never seen a European before, were in some cases frightened by the tall Englishman with the ragged beard, and his appearance caused panic! And there was a language barrier: Wallace learned to speak Malayan, but in some places the help of his Malay assistants was needed for negotiation (with the Dutch governors he spoke French). All that plus obtaining provisions and cooking food – on top of the main purpose for his stay: hunting for numerous vertebrates and preparing their skins and skeletons, and preparing the insect collection.

Wallace describes the indigenous Malayan tribes in the islands where he had stayed, sometimes several weeks in one place – especially in a trading-village called Dobbo, at the far eastern end of the archipelago. These observations later earned Wallace membership in the British Anthropological Society. He describes with approval the intensive native agricultural techniques on the islands – sponsored, or imposed, by the controlling Dutch governors. He thought that firm control was advantageous for the indigenous people: there was always a stage of oppression between barbarism and civilization. The Dutch form of control through local Rajahs was superior to the way the British

[451] Wallace A.R. 1962 Malay Archipelago, p. 309.

[452] Animal protection societies and animal-lovers would probably condemn Wallace for the massacre of so many creatures, many of them rare species and some that have since become extinct. But they should remember that hunting and collecting was the only way of documenting the variety of the biological world before the 20th century. The museum collections were the basis for the future studies of biology and evolution. True, nonetheless, that scientific travelers and collectors contributed their share to the extinction of rare species. The American Ornithological Society is named after Audubon, who left us with color pictures of the American avifauna which he hunted – some of them do not exist today (he discarded the corpses).

administered their colonies via slavery: the British colonies never succeeded in making the natives more civilized[453].

Spiritualism

The belief in the reality of spirits and communication with the dead are dismissed today as nonsense, but this was not the case in the 19th century.

Wallace lists the paranormal phenomena that were reported by well-educated professionals – judges, lawyers, diplomats, clergy, physicians, chemists and engineers. These phenomena included moving tables supposedly without human touch, names written on paper by an invisible hand, and verbal communication with the deceased. Some of the sessions (seances) with famous "mediums" were set up deliberately as tests to in order to uncover them as fraudulent – even using photography[454].

In a 40-page paper Wallace argued that the phenomena are not necessarily true, but that the subject is worth a thorough scientific investigation rather than an out-of-hand dismissal as fraud. Scientists like Huxley dismissed these reports, Wallace claimed, because they could not suggest alternative explanations for these phenomena, not because they were proved wrong:

> Men "with heavy scientific appendages to their names" refuse to examine these reports when invited... My position therefore is, that the phenomena of spiritualism in their entirety do **not** require further proof. They are proved quite as well as any facts are proved in other sciences. And it is not denial or quibbling that can disprove any of them[455].

Parenthetically, Wallace suggested that the belief in afterlife and in messages from the dead may be far more effective in shaping the moral conduct of people than long lectures from the pulpit. "When the bodily sensations and pleasures do not exist, Man may devote his thoughts to intellectual and moral issues"[456].

Wallace and the Theory of Evolution

Wallace's first scientific contribution was a short paper sent from Sarawak (in Borneo) in 1855. The Sarawak paper, dealing with the origin of new species from older, related forms, did not attract too much attention.

Wallace noted that while representatives of large taxonomic groups, like mammals or birds, may be found on all continents, smaller divisions are more limited in distribution: Marsupials abound in particular in Australia, and Monotremata are found exclusively there. In particular, related families and genera are often geographically close, and closely-related (= morphologically similar) species are often found in the same geographical region. From these observations Wallace deduced that

> Every species has come into existence coincident both in space and time with a pre-existing closely-allied species. The great law... is that every change shall be gradual, that no new creature shall be formed widely different from anything before existing[457].

Wallace expanded this deduced law into a few rules, which he described in his paper: 1) each species had originated from a previously-existing species at the same location. The extent of similarity between species reflects the order of their appearance. [He noted that the same phenomenon can be demonstrated in fossil faunas: closely-related species of fossils are found in adjacent strata or

---

453 Wallace 1962. The Malay Archipelago, pp. 196-197.
454 Wallace 1874, pp. 27-29.
455 Ibid. pp. 30, 34.
456 Ibid. pp. 33-34.
457 Sarawak paper, p. 18.

periods]. 2) If from the same ancestral species (antitypes, in Wallace's terminology) there emerged two or more species, a complex branching sequence can be expected. Even if their common original founder species no longer exists, the similarity of the populations will reveal their common origin. This is the situation often observed in nature. 3) Geographically isolated locations contain different species. 4) More new species will be formed in periods of geological stability then in periods of geological upheaval.

Wallace seems to have been much impressed by Lyell's "The Principles of Geology", but objected to Lyell's idea of "special creation": If species were created separately and independently, then it is not possible to account for the presence, in different species, of rudimentary limbs and organs – like rudimentary legs in snakes and whales. But if each species was formed from a pre-existing species, the presence of rudimentary organs is understandable[458].

The Sarawak paper contained little new factual data, apart from some observations on the parrots he had seen during his stay in the Amazon forests. The paper did not attract much notice when it was published in 1855. Lyell mentioned it to Darwin, who read it and remarked "There is nothing new in it".

The Ternate Paper

Wallace had read Malthus's "Essay on Population" and the idea of a struggle for existence must have been on his mind. The idea of natural selection occurred to him during a two – hour struggle with a bout of malaria, while he was staying on the island of Ternate. This is how he recalls the event 40 years later:

> It occurred to me to ask the question, Why would some die and some live? And the answer was clearly, that on the whole the best fitted live. From the effects of disease the most healthy escaped; from enemies, the strongest, the swiftest, or the most cunning... Then it suddenly flashed upon me that this self – acting process would necessarily improve the race, because in every generation the inferior would inevitably be killed off and the superior would remain – that is, the fittest would survive[459].

During the bout of malaria he thought of all the details, and when the symptoms eased he wrote up the paper and sent it, by the next available ship, to Darwin for evaluation, with a request to also show it to Lyell; it took several months to arrive in England. The receipt of this paper was the trigger for the publication of "The Origin of Species"[460]. This is how Wallace reflects on the event:

> The immediate result of my paper was that Darwin was induced to at once prepare for publication his book on the origin of species, in the condensed form in which it appeared, instead of waiting an indefinite number of years to complete a work on a much grander scale. I feel much satisfaction in having thus aided in bringing about the publication of this celebrated book, with ample recognition by Darwin himself of my independent discovery of Natural Selection[461].

In the Ternate paper Wallace describes, in clear and concise language, the principles of natural selection – the same theory that Darwin had worked out in 20 years of painstaking data collection.

---

458 Interestingly, in 1855, Wallace explained the rudimentary organs as preliminary primitive forms of organs, to be developed further in the future. The wings of penguins he considered an early stage in the development of wings in flying birds; the rudimentary legs in snakes he considered the beginnings of reptilian walking legs. [In a later edition of the Sarawak paper (1870) he added a note, that according to Darwin's theory, these rudiments are the result of disuse.]

459 Cited in P. Raby, A.R. Wallace: A Life, p. 131. This is one of several ways in which Wallace retrospectively narrated the event.

460 See Chapter 8.

461 Wallace 1891, Natural Selection, p. 21.

The Struggle for Existence: Wallace was a field naturalist and observed the organisms in their natural environment. He noticed that animal population sizes are stable in the long run, although every species can increase indefinitely if unchecked. Populations are limited by the shortage of food.

> The greater or less fecundity of an animal is often considered to be one of the chief causes of its abundance or scarcity. But a consideration of the facts will show that it really has nothing to do with the matter. Even the least prolific of animals would increase rapidly if unchecked. Wild cats are prolific and have few enemies. Why then are they never as abundant as rabbits? The only intelligible answer is that their supply of food is more precarious.
>
> Large animals cannot be so abundant as small animals. The Carnivora must be less numerous than the Herbivora. Eagles and lions can never be so prolific as pigeons and antelopes[462]... **The life of animals is a struggle for existence**... Those that die must be the weakest – the very young, the aged, and the diseased[463].

On Variation: In the absence of data on natural populations, many people believe that species are fixed entities, but this belief is supported only by observations on domesticated species. This may not be true in nature. No inference about species in Nature should be based on observations on domestic varieties:

> Among the former, their well-being and very existence depend upon the full exercise and healthy condition of their senses and physical powers, whereas among the latter these are only partly exercised, and in some cases absolutely unused. Now when a variety of such an animal occurs [in a domesticated animal], having increased power or capacity in any organ or sense, such increase is totally useless, is never called into action. In the wild animal on the contrary... any increase ...creates as it were a new animal, one of superior powers and which will necessarily increase in numbers and outlive those inferior to it.
>
> Again, in the domestic animals, all variations have an equal chance of continuance, and those which would decidedly render a wild animal unable to compete with its fellows and continue its existence are no disadvantage whatever in the state of domesticity. Our quickly fattening pigs, short-legged sheep, pouter pigeons and poodle dogs could never have come into existence in a state of nature[464].

Natural populations are variable, and some variants may survive better and diverge further away from the original type. There is no need to suppose, like Lamarck, that the "will" and the "needs" of the organisms are required to bring about adaptation of organisms to their environment: selection of the most adapted individuals does just that. A selective advantage to individuals with a longer neck in procuring food will bring about the elongation of the neck. This is how Wallace described the evolution of new species:

> If... any species should produce a variety having slightly increased powers of preserving existence, that variety must inevitably in time acquire superiority in numbers. Now let some alteration in physical conditions occur in the district – a long period of drought... the least numerous and most feebly organized variety would suffer first, and were the pressure severe, must soon become extinct. The superior variety would then alone remain, and on

[462] Ibid.
[463] Ibid. p. 23.
[464] Ibid.

> return of favorable circumstances would rapidly increase in numbers and occupy the place of extinct species and variety. This new, improved and populous race might, in the course of time, give rise to new varieties... exhibiting several diverging modifications of form... Here then we have progression and continued divergence[465].

Wallace as a Darwinist: Further Contributions

Upon his return from the Malay Archipelago, Wallace became a prominent defender of the Theory of Evolution. His book, "Darwinism"[466], is a collection of essays in reply to criticisms of the theory. To refute claims that species in nature lack the variation needed for selection, Wallace patiently measured different traits on samples of individual birds and mammals to show that much measurable variation did exist within the same species. His book "Natural Selection"[467] is a collection of Wallace's own original papers on the subject. The central paper, which gave the book its title, was published in 1867.

Wallace disagreed with Darwin on three important issues. One was the interpretation of variation in domestic animals as a model for natural processes: for the reasons cited above,

> Domestic animals are abnormal, irregular, artificial. They are subject to varieties which never occur and never can occur in a state of nature... The two [domestic and natural animals] are so opposed to each other in every circumstance of their existence, that what applies to the one is almost sure not to apply to the other.

The second disagreement was over the interpretation of sexual color dimorphism in birds. The males are generally more colorful than the females. Darwin suggested that color and pattern characters do not seem to affect fitness, and could not be improved by natural selection: a separate mechanism, *sexual selection*[468], favoring the more outstanding males, is therefore needed to explain the dimorphism in these characters. Wallace insisted that the color in animals is useful for protection from predators (see below), and that the dimorphism is the result of natural selection – favoring the cryptic colors of the females and young, while the males are free to vary.

The third disagreement concerned the evolution of Man. Wallace admitted, like Darwin, that Man's origin from the primates was uncontestable, and the morphological and anatomical characters of Man evolved by natural selection. Man's intellectual abilities, however – the mathematical, musical, and artistic talents – could not have evolved from the primate line by natural selection, and another factor must have been involved – the Divine hand. Wallace's arguments are discussed below.

The Function of Animal Colors: Camouflage: Creationists interpreted the colors and patterns of animals as designed to please the Creator and/or Man, as part of the beauty and harmony in nature. Wallace, the field observer, believed that the role of color is to create a correspondence between the animal and its surroundings and background. For example, many Arctic animals are white and blend with the color of the ice and snow; desert animals are gray and brown, blending with the color of their environment; and jungle birds are predominantly green. This correspondence gives the animal better protection from predators (if it is hunted) or better hiding while stealthily approaching its prey (if it is a hunter). Wallace brings many examples from his field observations of insects[469]. In diurnal butterflies, the bright and attractive colors on the upper surfaces of the wings

---

465 Ibid. p. 23.
466 Wallace 1889; Macmillan, London.
467 Wallace 1891; Macmillan, London.
468 See Chapter 16.
469 Wallace, Natural Selection, p. 41 ff.

disappear from view when the butterfly alights with its wings held vertically together, exposing the dull-patterned underside. In nocturnal moths that rest exposed to view in the daytime, the dull-patterned forewings form a cover to conceal the colorful hind wings. To fully appreciate the value of camouflage, Wallace adds, one should observe the behavior of certain insects that resemble twigs or other plant parts, like stick insects (Phasmidae). The combination of color, form, and behavior creates a stunning illusion[470].

Variation in color among individuals of the same species certainly occurs in nature, although rarely observed. White individuals are seldom found in natural populations – except in Arctic habitats – although they abound in domesticated animals, from mice and rabbits to horses and cattle. The reason for their rarity in nature is that white is highly conspicuous, and white individuals are hunted and selected against[471].

Warning Colors: Darwin approached Wallace with the question: Why are the caterpillars of some butterflies very brightly colored, despite this making them easy prey for birds? Wallace suggested that they must have other means of protecting themselves – they must be toxic or repellent to predators, and their colors warn the potential enemies not to attack them (even if the caterpillar not ultimately consumed, the injury from an attack could be fatal to it). Similarly, insects that can protect themselves with stings, such as wasps – often have bright and conspicuous colors. Wallace suggested this idea in a lecture to the Royal Society and asked the Fellows to test this suggestion. He received many confirming reports[472]. This explanation is extended to the phenomenon of *mimicry*, which was described and elaborated by Wallace's fellow-traveler in the Amazon, Henry Walter Bates.

> [mimicry is]… a resemblance in those parts only that catch the eye – a resemblance that deceives[473].

In tropical South America, there are whole groups of defenseless Pierid butterflies (mimics) which resemble toxic ones (models), (*Heliconius* spp.). There are species of beetles which resemble wasps. Some harmless snakes have red, yellow, and black colors mimicking dangerous species.
This type of mimicry ("Batesian mimicry") can be maintained when the mimic and the model coexist in the same area, and the mimic is rare compared with the model[474].

The Geographical Distribution of Animals

Returning from his travels in the Malay Archipelago, Wallace collected material for a comprehensive book on the geographical distribution of animals, with the ambitious purpose declared in the full title of the book:

> The geographic distribution of animals, with a study of the relation of living and extinct faunas elucidating the past changes of the earth's surface[475].

The geographic distribution of animals raised many questions in the 19th century. The material assembled in museums since the time of Linnaeus, and the information collected by world travelers

[470] Ibid. p. 43.
[471] Ibid. p. 49.
[472] Ibid. p. 86.
[473] Ibid. p. 54.
[474] In other cases, a number of mimics – of different species – may resemble the same model and share similar warning colors, thus enhancing the protection of all of them ("Mullerian" mimicry). See also R.A. Fisher's analysis of the phenomenon in Chapter 27.
[475] Wallace 2003 (1876).

like Alexander von Humboldt[476], had revealed that the faunas of different continents and islands were not the same despite similar climatic conditions. This was difficult to reconcile with the biblical model of dispersal from Mount Ararat[477]. In particular, the peculiar faunas of oceanic islands – as reported by Darwin in the Voyage of the Beagle[478] – required explanation. Darwin attempted to explain these facts as a result of the phenomenon of *dispersal* of organisms: either actively by flying or swimming, or borne passively by winds and ocean currents. Floating tree trunks could serve as rafts carrying organisms such as tortoises and snakes from continents to isolated islands. Darwin tried to prove experimentally that seeds and fruits like coconuts could float in the sea, retaining their power of germination for months[479].

Wallace considered that dispersal was insufficient to explain the faunal similarities between continents and islands. He suggested the existence in the past of *land bridges* between continents. Changes in sea level ("not very large changes") could have expose land areas now submerged, or alternatively inundated land areas now exposed, thus enabling free exchanges and relocation of faunas that are now separated by oceans[480].

> The hypothetical view as to the more recent of the great geographical changes of the earth's surface here set forth, is not the result of any pre-conceived theory, but has grown out of a careful study of the facts accumulated, and has led to a considerable alteration of the author's previous views.

Wallace divided the globe into six *Zoographical Regions*[481]: the Palearctic (northern Asia and Europe); the Ethiopian (Africa); the Oriental (Asia and the Pacific islands); the Australian; the Nearctic (North America); and the Neotropical (South America). He supplemented his personal observations on the geographic distribution of animals by examining summaries and lists of records in museums, tabled in the book. In his tables Wallace listed families and genera, but not species: he explained that unlike the larger taxonomic units, the same species could be found – and could have been formed – in different regions if the environmental conditions were suitable. He proposes to accomplish the following:

> All that we can in general hope to do is, to trace out – more or less hypothetically, some of the largest changes in physical geography that have occurred during the ages immediately preceding our own, and to estimate the effects they will probably have produced on animal distribution. We may then… be able to determine the probable place of birth and subsequent migrations of the more important genera and families[482].

Paleontological evidence for great changes in the fauna in Europe and North America in the relatively recent past – between the Miocene and Pliocene – excited Wallace's imagination:

---

[476] Alexander von Humboldt, (1769-1859) German-born traveler; in 1800, he travelled widely in western South America and Mexico and recorded the flora and fauna; he stayed in Paris for 23 years and published there his 30 volumes of field work; He founded the science of biogeography.

[477] Haeckel 1876.

[478] Darwin 1845.

[479] Darwin 1898 (1872), Chapters XII and XIV.

[480] Wallace's explanation was erroneous. The only land bridge recognized today is the now-submerged northern connection between North America and Asia – the Behring Strait – which seems to explain the distribution of fossil equidae (horses).

[481] This division was suggested by the British ornithologist Philip L. Sclater (1829-1913), for the distribution of birds. Wallace preferred it to an alternative suggestion of T.H. Huxley, because it seemed suited for the distribution of mammals and reptiles and even for insects. (Wallace 2003, Geographic Distribution, I: 64-69; 82-86). These major divisions are still largely used today.

[482] Wallace 2003 (1876) I: 49.

> It is surely a marvelous fact …this sudden dying out of so many large Mammalia, not in one place only but over half the land surface of our globe. It is clear therefore, that we are now in an altogether exceptional period in the earth's history[483].

Wallace postulated that the animal world – and mammals in particular – had originated in the massive northern land mass of the globe. The southern land masses – Australia, South America, and southern Africa – were colonized later by migrants from the north, when temporary contact between them was established:

> All the palaeontological, no less than the geological and physical, evidence points to the great land masses of the Northern Hemisphere as being of immense antiquity, and as the area in which all higher forms of life developed[484]. Into these [southern land masses] flowed successive waves of life, as they each in turn became temporarily united with some part of the northern land[485].

Wallace did not accept the idea that a submerged continent existed between the Americas and Africa:

> To those who would create a continent to account for the migrations of a beetle, nothing would seem more probable than that a South Atlantic continent then united parts of what are now Africa and South America. There is, however, no such evidence for a general permanence of what are now the great continents and deep oceans[486].

In contrast to the Atlantic and Pacific oceans, the Mediterranean Sea does not form a boundary between zoogeographical regions. In Wallace's lists, the boundary between the Palearctic and the Ethiopian regions is south of the Sahara. In the Mediterranean there are at least two areas of shallow water: between Gibraltar and Morocco and between Malta and the African shore. A decrease in sea level "by a few hundred meters" could expose these areas as dry land bridges connecting Europe and Africa. Paleontological evidence shows that large mammals, which are today restricted to Africa, had existed in the past in Europe. Wallace offers the following scenario as an explanation of why they no longer exist in Europe today:

> During the height of the glacial epoch, these large animals would probably retire to the Mediterranean land and North Africa, making annual migrations northwards during the summer. But as the connecting land sank, and became narrower and narrower, the migrating herds would diminish and at last cease altogether. And when the glacial cold had passed away, would be altogether prevented from returning to their former haunts[487].

Wallace suggested that South America was until recently a large island separated from North America by a wide sea. In this island flourished a unique fauna of primitive mammals, the Edentata, while North America had a widely different and diverse fauna, including ungulates and carnivores[488]. One of the amazing zoogeographical phenomena, which supported the idea that South America had once been isolated, was the extraordinary diversification of the single family of hummingbirds in that continent:

---

[483] Ibid. I: 150.
[484] Ibid. II: 154.
[485] Ibid. II: 163.
[486] Ibid. I: 156.
[487] Ibid. I: 114.
[488] Ibid. II: 57.

> How vast must have been the time required to develop those highly specialized forms out of ancestral swift-like type. How complete and long – continued the isolation of their birthplace to have allowed of their modification and adaptation to such divergent climates and conditions, yet never to have permitted them to establish themselves in the other continents! No naturalist can study in detail this single family of birds, without being profoundly impressed with the vast antiquity of the South American continent, its long isolation from the rest of the land surface of the globe[489].

It was more difficult to trace the zoogeographical origins of the unique faunas of Madagascar, the Malay Archipelago, and in particular, the fauna of Australia[490]. Marsupials are abundant in Australia, but some species also exist in South America and even in North America. Could this be proof of a former land bridge between these continents? Wallace did not think so:

> Primeval forms of marsupials we know abounded in Europe during much of the Secondary period, and no doubt supplied Australia with the ancestors of the present fauna. It is clear, therefore, that in this case there is not a particle of evidence for any former union between Australia and South America[491].

The diversity of the fauna in the Hawaiian Archipelago (known at the time as the Sandwich Islands) seemed to Wallace to be too great to have originated in such small areas of land. He suggested that the chain of islands had once been a part of a wide land area, now partly submerged, which continued for 1,000 miles north-west of the existing archipelago[492]. A similar speculation was offered to explain the ancient fauna of New Zealand, which had once included many species of flightless birds, Struthiones [the only surviving representative of the group in New Zealand is the kiwi]: Wallace postulated a former wide, fringing land mass around each island, and believed that the fossil birds must be related to the existing Asian, Australian, African, and South American flightless birds (the Emu, the Ostrich, and the Rhea), and must have colonized the islands from somewhere:

> The wide distribution of the Struthiones may… be best explained by supposing them to represent a very ancient type of bird, developed at a time when the more specialized carnivorous Mammalia had not come into existence. This points to the conclusion that New Zealand not long since had formed a much more extensive land, and that the diminution of its area by subsidence has been one of the causes – and perhaps the main one – in bringing about the extinction of the larger species of these flightless birds[493].

The most speculative hypothesis – by his own admission – relates to the explanation of the faunal diversity in the Malay Archipelago, and particularly on the island of Celebes [now Sulawesi]:

> It is true that there we reach the extremest limit of speculation. But when we have before us such singular phenomena as are presented by the fauna of the island of Celebes, we can hardly help endeavoring to picture to our imagination by what past changes of land and

---

489 Ibid. II: 8-9.

490 Ibid. I: 286; I: 315; I: 388, respectively.

491 Ibid. I: 399.

492 Wallace's speculations about the Hawaiian Islands were erroneous. These are volcanic islands that emerged individually from the ocean floor and were never part of a continent. His speculations about New Zealand were also considered wrong. But in 2016, the American Geological Society reported that a "hidden continent", named Zelandia, was in fact discovered sunken east of Australia, and that the two islands of New Zealand are part of that continent. A research expedition was planned in 2017 to investigate this sunken continent (GSA Today, 21 Dec. 2016). Wallace's imagination seem to have carried him the right way.

493 Wallace 2003 II: 370.

> sea (in themselves not improbable) the actual conditions of things may have been brought about[494].

Finally, Wallace expressed some thoughts on the past faunal changes from the point of view of Man.

> This is certainly not a great while ago, that the great organic revolution, implying physical changes… has taken place since Man lived on the earth… It is clear, therefore, that we are now in an altogether exceptional period of the earth's history. We live in a zoologically impoverished world, in which all the hugest and fiercest, and strangest forms have recently disappeared. And it is, no doubt, a much better world for us now that they have gone[495].

Wallace chose to end his two-volume book on the geographical distribution of animals with an appeal to the younger generation of biologists to realize the advantages of studying this new field:

> In concluding this task, the author ventures to suggest that naturalists who are disposed to turn aside from the beaten track of research, may find in the line of study here suggested a new and interesting pursuit, not inferior to the lofty heights of transcendental anatomy, or the bewildering mazes of modern classification. And it is a study that will surely lead them to an increased appreciation of the beauty and the harmony of nature, and to a fuller comprehension of the complex relations and mutual interdependence which link together every animal and vegetable form, with the ever-changing earth that supports them, into one grand organic whole[496].

Epilogue: The Modern Explanation

Wallace's six zoogeographical regions are still acceptable descriptions of the distribution of animals on the planet. But the explanation of the processes that brought about the current distribution has changed since Wallace's time. Wallace's theory of land bridges was replaced in the early 20th century by Alfred Wegener's (1880-1930) theory of continental drift, published in 1923-1929. Wegener, a German meteorologist, noticed the similarity of the Atlantic shoreline of western Africa to the shoreline of eastern South America. Aligning paper-cutouts of the contours of the two continents, he suggested that they may once have formed a single land mass. He further suggested that the terrestrial continents are drifting apart on the ocean floor. The original land mass – called *Pangea* – split about 250 million years ago (mya): the northern part is roughly equivalent to the palearctic zoogeographical region; and the southern part – called *Gondwanaland* after a region in central India – composed of Antarctica, Australia, South America, Africa, and India (hence the similarities in the flora of these continents) – separated and drifted apart slowly about 150 mya. Wegener used Wallace's zoogeographical data to support his argument. Scientists contested the continental drift theory because it was not clear how the heavy land masses could overcome friction and move about on the ocean floor.

In the 1960s, this theory was replaced by the theory of *plate tectonics*, suggested by the Canadian Tuzo Wilson. The theory describes the upper crust of the earth – land and sea – as comprising of several plates that slide over the molten core of the planet. When the slowly-moving plates collide with each other, earthquakes result. Mountain-chain elevation and volcanic eruptions are associated with the edges of these plates. Deep-sea research has revealed the origins of plate movements, as new molten lava emerges from cracks in the earth mantle along the edges of the plates, driving them apart.

---

494 Ibid. I: 438.

495 Ibid II: 149-150.

496 Ibid. II: 553.

Wallace's Line

In the eight years of his travels in the Malay Archipelago, assembling vast collections of animals, Wallace noticed that he could draw a clear zoogeographical line among the islands. This line passes from Mindanao in the Philippines, southwards to the islands of Bali and Lombok. To the west of this line, including the major islands of Borneo and Java, the fauna is very similar to that of India and Asia, with tigers, elephants, rhinoceroses, wolves, monkeys (apes), and tapirs. East of this line, the islands have an Australian fauna – kangaroos and other marsupials, emu, and distinctive families of parrots. This was a surprising phenomenon in view of the short distances between the islands (Bali and Lombok are only 25 km apart). Wallace described this line in 1859, and again in 1863, to the Linnaean Society. The phenomenon is mentioned in his book on the Malay Archipelago.

One hundred years later, Wallace's line acquired strong support from the theory of plate tectonics. It turns out that the line corresponds, quite accurately, to the borderline between the Asian and Australian tectonic plates. The frequent and intense volcanic activities in that area are the result of the movements of the plates[497]. The difference in the fauna on either side of Wallace's line thus resulted from an ancient evolutionary change on the face of the earth!

Wallace, whom some authors regarded as "Darwin's moon", thus earned fame as an independent thinker. "Wallace's line" is a permanent zoogeographical monument to him on the globe.

Wallace on The Evolution of Man

Huxley and Darwin did not doubt that Man, like his primate ancestors, had evolved by natural selection[498]. Even the anti-Darwinian anatomists recognized that the human body is very similar to that of the primates:

> I cannot shut my eyes to the significance of that all-pervading similitude of structure – every tooth, every bone, strictly homologous – which makes the determination of the difference between Homo and Pithecus the anatomist's dilemma[499].

At the end of a review of Lyell's books on geology as related to the origin of species[500], Wallace stated his objections to Darwin's ideas on the origin of Man:

> [But] if the researches of geologists and the investigations of anatomists should ever demonstrate that he [man] was derived from the lower animals in the same way they have been derived from each other, we shall [still] not be debarred from believing, or from proving, that his intellectual capacities and his moral sense were not wholly developed by the same process. Neither natural selection nor the more general Theory of Evolution can give any account whatever of the origin of sensational or conscious life. We may even go further, and maintain that there are certain purely physical characteristics of the human race which are not explicable on the theory of variation and survival of the fittest. The brain, the organs of speech, the hand, and the external form of Man offer some difficulties in this respect[501]. The supreme beauty of our form and countenance has probably been the source of all our aesthetic ideas and emotions, which could hardly have arisen had we maintained the shape and features of an erect gorilla[502].

[497] One such event was the famous eruption of the volcanic island of Krakatoa in 1883 [Winchester 2003].
[498] See Chapter 15.
[499] Richard Owen, cited by Wallace,1889, p. 450.
[500] Wallace, April 1869 Quarterly Review.
[501] Ibid.
[502] Ibid. p. 393.

Certainly the human moral sense did not evolve by natural selection. Wallace considered that a supreme power was involved in all aspects of human evolution:

> The structure of the human foot and hand seem unnecessarily perfect for the needs of savage man... Again, what a wonderful organ is the hand of man! Of what marvels of delicacy is it capable, and how greatly it assists in his education and mental development! The hand is equally perfect in the savage, but he has no need for so fine an instrument[503].

> While admitting to the full extent the agency of the great laws of organic development in the origin of the human race, as in the origins of all organized beings, yet there seems to be evidence of a power which has guided the action of these laws [of variation multiplication and survival] in definite directions and for special ends[504].

Darwin regretted this turn in Wallace's approach and wrote to him: "I differ grievously from you, and I am very sorry for it. I can see no necessity for calling in an additional and proximate cause in regard to man"[505].

In his later books, Darwinism (1889) and Natural Selection (1891), Wallace included a deeper discussion of Man, with some emphasis on the anatomical differences between man and ape: smaller brains, longer canine teeth, arms being longer in apes than in humans (longer than the legs), human erect posture, opposable thumbs and toes, and bare versus hairy skin. Wallace continues:

> An enormously remote epoch, when the race that was ultimately to develop into man, [it] diverged from the other stock which continued the animal type, and ultimately produced the existing varieties of anthropoid apes[506]. The erect posture and free hands were acquired at a comparatively early period... and were the characteristics that gave him superiority over other animals[507].

Wallace agreed with Darwin about the origin of Man from the primates and the evolution of the structure of the human body by natural selection. But there is nonetheless something distinct in humans.

> [man is] not only the head and culminating point of the grand series of organic nature, but is in some degree a new and distinct order of being – a being who is in some degree superior to nature, inasmuch as he knew how to control and regulate her actions[508].

Wallace thought that once man had acquired the intellectual and moral characteristics that distinguished from the other animals, he was far less affected by natural selection. Because he could build shelters to protect himself from the climatic hardships, invent tools and weapons for hunting food and defense against predators, light fires and prepare food in a more edible and digestible form than animals could, man lay beyond the reach of many of the selective forces that operate on animals. Moreover, primitive man lived in family groups and societies in which there was a division of labor, the weak and disabled were protected, and human intelligence solved adaptational difficulties. Those groups or tribes in which mutual assistance was stronger had an added advantage in the struggle for existence. Such groups could deal more successfully with environmental hardships without the need for morphological or anatomical changes via natural selection.

---

503 Ibid. p. 392, p. 202.
504 Ibid. p. 394.
505 F. Darwin 1887, III: 116.
506 Ibid. p. 454.
507 Ibid. p. 458.
508 Wallace, Natural Selection, 1891, p. 181.

In a short, early paper, Wallace suggested that the differences observed among different human races had evolved by natural selection. But at a very early stage of human evolution, once Man's mental faculties were fully developed, his physical form ceased to be affected by selection:

> From the time... when the social and sympathetic feelings came into active operation, and the intellectual and moral faculties became fairly developed, man ceased to be influenced by "natural selection" in his physical form and structure... From the moment his body became stationary, his mind would become subject to the very influences... The better and higher specimens of our race would therefore increase and spread, the lower and more brutal would give way and successively die out[509].

Wallacc was swept by admiration for the power of Man over Nature, and offered the following prediction:

> The development of perfect beauty... ennobled by the highest intellectual faculties and sympathetic emotions... till the world is again inhabited by a single, nearly homogeneous race, no individual of which will be inferior to the noblest specimen of existing humanity[510]. We can anticipate the time when the earth will produce only cultivated plants and domestic animals[511].

Wallace was impressed by the publications of the Philadelphia physician, Samuel Morton[512]. He dedicated many pages in "Natural Selection" to a discussion of the differences in brain size between ape and man, and between different human races[513]. He commented that Europeans are more successful than the "mentally underdeveloped" natives in Australia and America due to their greater intelligence, and it was only natural that they had replaced the natives when they came into conflict with them in the struggle for existence "Just as the weeds of Europe overrun North America and Australia, extinguishing local products[514].

The brain size of primates, Wallace wrote, is smaller than in humans ("even that of an Indian") although the body sizes are similar. This indicated an abrupt, "saltational" jump between ape and man. A higher "controlling intelligence" had planned ahead the structure of modern man. The savages, lowest on the human brain scale, have larger brains than the primates but they do not have the intellectual (mathematical and moral) faculties of Europeans – although they do have artistic talents (ancient, beautiful paintings were known from caves in Spain and France). Wallace deduced that the savages possessed a great intellectual potential, which was not realized because it was not required for their way of life. The transition of Man's progenitors to bipedal walking and the freeing of the hands for the use of tools also manifests pre-planning by the Higher designer: the reduction and loss of the ability to hold on to branches must have been deleterious to the arboreal primitive man, but it was important in preparation for the needs of modern man. There is no proof that the origin of human mental faculties – Wallace wrote – is traceable to the animal world. They must be attributed to a spiritual, invisible world:

> [The Darwinian theory] shows us how Man's body may have been developed from that of a lower animal form under the law of natural selection. But it also teaches us that we possess intellectual and moral faculties which could not have been so developed, but must have

509 Wallace 1864: J. Anthrop. Soc. London, p. 6.
510 Natural selection, 1891, p. 185.
511 Natural Selection, p. 182.
512 See above, Chapter 13.
513 Natural Selection, pp. 187-204.
514 Ibid. p. 178.

> had another origin. And for this origin we can only find an adequate cause in the unseen world of spirit[515].

Women's ability to sing could not have developed by natural or sexual selection: bird and mammal females do not sing, and the ability to sing is not used as a criterion for selecting wives by the primitive savages[516].

Wallace states his final conclusion:

> We are …driven to the conclusion that in his large head and well – developed brain, he possesses an organ quite disproportionate to his actual requirements – an organ that seems prepared in advance, to be fully utilized as he progresses in civilization[517]. A controlling intelligence has directed the laws of variation, multiplication and survival for His own purposes[518]. We should infer the action of mind, foreseeing the future, and preparing for it[519].

515 Darwinism, p. 478.
516 Ibid. p. 198.
517 Ibid. p. 193.
518 Ibid. p. 208.
519 Ibid. p. 188.

12

Thomas Henry Huxley

Biographical Notes

Thomas Henry Huxley (1825-1895), like Wallace, did not come from an affluent family. One of six children of a rural schoolteacher living in Ealing, he was sent to live with his married sister after only two years of formal school education. He accompanied his brother-in-law – a young medical practitioner in the outskirts of London – on his rounds in the poor neighborhoods of the city as an apprentice, and was appalled by the living conditions of the people[520].

Young Thomas was an avid reader. He read all the books in his brother-in-law's library. He taught himself German (later also French) and early on he marked for himself those fields of science he wanted to learn more about: "mathematics, logic, theology, physiology, histology and physics"[521]. When he was 17 he was encouraged by his sister and brother-in-law to enter an open public competition in botany, arranged by the Charing Cross School of Medicine. He sat down to read the relevant books in the public library, entered the competition and won second prize. Consequently he received a scholarship to the medical school and enrolled in formal studies. He excelled in anatomy and organic chemistry, and at age 19 discovered a previously unknown membrane at the base of a human hair – a discovery published in a medical journal.

When he graduated from medical school he was still too young to enter the College of Surgeons, a necessary step before full qualification as a doctor, and he did not have the means to start his own medical practice. In order to earn a living and still be in the profession, he enlisted in the Royal Navy as an assistant surgeon on board H.M.S. "Rattlesnake". The ship – bigger than the "Beagle", as it carried 28 guns – set sail under the command of Captain Stanley to survey the eastern shores of Australia and map the passes between the Great Barrier Reef and the mainland. The mission lasted five years (1846-1850), during which the ship was based in Sydney. When in port, the officers of the ship were entertained by local businessmen, and Huxley got to know Henrietta, the sister of his host's wife. They fell deeply in love, but the Rattlesnake had to move on and Henrietta could not accompany him. They vowed to wait for one another until Huxley could return to England and get a position that could support a family. Eight years passed before that promise was fulfilled. Henrietta arrived in England in 1855 and they married two months later. The marriage lasted 40 years, until Huxley's death from heart failure in 1895. They had eight children. Two of Huxley's grandchildren earned fame in their own right – the writer Aldous Huxley and the biologist Julian Huxley, who became President of UNESCO.

Huxley's Early Years as a Scientist

During the voyage of the "Rattlesnake", Huxley occupied himself with research. He worked on the anatomy of Coelenterata which were caught by net – medusas, sea anemones, and the 'Portugese man o'war' *Physalia*. These were delicate organisms that could not be preserved, and the histological work had to be carried out on board the moving ship "with the microscope tied to the table in the

[520] Huxley, L., T.H. Huxley's life and letters, I: 16.

[521] Ibid. I: 13.

map room". Huxley demonstrated that these organisms shared a simple structure, with only two cellular layers – ectoderm and endoderm – surrounding a central digestive cavity. His two papers on the anatomy of the Coelenterates, and one on the blood of the lancelet (Amphioxus), were sent by Captain Stanley to his father, the Bishop of Norwich, who arranged to get them published by the Linnaean Society before the ship returned from the voyage.

Back in England, Huxley began looking for a suitable job, while continuing to publish research papers. He discovered a structural similarity between the tail of the amphibian tadpole and the larva of the Tunicata, which later supported placement of the tunicates at the origin of the vertebrate phylogeny. His papers gained him recognition by the scientific community. A year after his return, aged 26, he became a Fellow of the Royal Society and two years later already served in various posts in the Society (in 1853 he was awarded the highest medal of honor of the Society). But these were non-paying activities, and he badly needed a job. Contrary to Huxley's expectations, the Navy – which owned the rights to the results of research carried out on board its ships – was not interested in publishing his papers. His request to extend his appointment while the "Rattlesnake" was in port – so that he could continue to be paid – was only partly approved. His applications for jobs in academic institutions in England, and even in Toronto, Canada, were unsuccessful. Almost five years passed before he found a reasonably paying position as a lecturer in paleontology at the School of Mines in central London. Disappointed and frustrated, Huxley wrote in 1851:

> To attempt to live by any scientific pursuit is a farce... A man of science may earn great distinction but not bread... He will get invitations to all sorts of dinners, but not enough income to pay his cab fare.

Years later he was still complaining about the unfair treatment of scientists in matters of income:

> What men of science want is only a fair day's wages for more than a fair day's work. And most of us, I suspect, would be well content if, for our days and nights of unremitting toil, we would secure the pay which a first-class treasury clerk earns without any obviously trying strain upon his faculties[522].

Contributions to Biology

Within a few years after his return to England, Huxley had become a prominent scientist. His public lectures always attracted large audiences due to his ability to explain complex phenomena in terms that could be understood by laymen (at least nine volumes of his lectures and addresses have been published): his lectures "On a piece of chalk", which discusses Geology and Paleontology[523], "On yeast ", which deals with fermentation – a subject familiar to all via the production of beer and bread [524] – and "on the formation of coal", a product familiar to all because it was used for heating and cooking in the homes[525], are examples. One important lecture traced the history of the theories of the beginnings of life ("Biogenesis and Abiogenesis"[526]. Another lecture, with a strong political and anti-clerical edge, was dedicated to Joseph Priestley, the priest and scientist who was persecuted and had to flee his home in Birmingham when both his laboratory and his house were burned by an agitated mob in 1791[527].

---

[522] Administrative Nihilism, Critiques and Addresses, 1871, p. 30.
[523] Huxley 1890e.
[524] Huxley 1890d.
[525] Huxley 1890c.
[526] Huxley 1890. See Chapter 1.
[527] Huxley 1910. See Chapter 2.

Huxley's main field of biological research was vertebrate anatomy. He published books and laboratory manuals on the skeletons of all vertebrate orders. A series of eight lectures, with many illustrations, on the comparative structure of the vertebrate skull, began with a detailed description of the human skull, and continued to fish, amphibian, reptile, bird and mammalian skulls, from the viewpoint of the theory of common descent[528]. He explained in simple language the anatomical evidence for the evolution of birds from small, predatory dinosaurs, with a comparison of the pelvis of a bird and a crocodile, and illustrations of two fossil species that could be intermediates between the two classes – birdlike creatures that had reptilian teeth in their beaks[529]. Among his scientific achievements was the placement of Amphioxus at the base of the vertebrate phylogeny.

Huxley emphasized the anatomical similarity between man and the primates, and in 1863 published a paper – and later a book – entitled "Man's Place in Nature" (see below)[530]. He also recognized the physiological and ecological differences between animals and plants:

> It is not so much that plants are deoxidizers and animals oxidizers, as that plants are manufacturers and animals, consumers. It is true that plants manufacture a good deal of non-nitrogenous produce… but it is this which is chiefly useful to the animal consumer[531].

In 1864, a few young scientists joined Huxley and formed the "X-club", which aimed at the advancement of science. Among them were the botanist Joseph Hooker, the philosopher Herbert Spencer, and the physicist John Tyndall. The members of the X-club became the central nucleus of the Royal Society and together founded the first scientific journal, *Nature***,** in 1869.

Huxley served, over the years, as President of the Royal Society, the Geological Society, and the Society for the Advancement of Science. He became a member of many royal commissions and committees, which dealt with a variety of subjects from the control of fisheries to the construction of the school curriculum and the teaching of life sciences in schools and universities. For three years he served as the Rector of King's College in Aberdeen, Scotland, a temporary administrative job. He was invited to lecture all across Britain, as evident from the numerous letters he wrote to his wife, who remained at home with the children.

In 1876, already a famous scientist, he was invited to the USA for a series of lectures. This time he traveled with his wife – the first time since their marriage – and they regarded this tour as a second honeymoon, staying for a week in Niagara Falls[532]. At Yale University he examined the large fossil collection assembled by the American paleontologist, Othniel C. Walsh (1831-1899). This collection led him to reconstruct the phylogenetic tree of the horse, a well-known example of phylogeny which was later incorporated in many textbooks of evolution[533]. Extrapolating backwards from the fossil evidence then available, he predicted that an even older horse should be found in Eocene formations. A small horse-ancestor, with four toes on each foreleg, was in fact discovered a few months later.

Despite his fame and prominent scientific position, Huxley was never able to secure an academic position at a university. It is not impossible that this was due to his anti-religious stance. Huxley categorized himself as "agnostic", at a time when the British universities were institutions of the Anglican Church. Huxley did not try to bridge the gap between science and religion – on the contrary, he did his best to show that the two represent totally separate approaches to life.

---

[528] Huxley 1938b.
[529] Huxley 1897, pp. 94-114.
[530] Huxley 1900.
[531] Huxley, L., T.H. Huxley's life and letters, p. 230.
[532] Huxley, L., T.H. Huxley's Life and Letters, p. 460.
[533] E.g. Morgan 1925; 1916; Evolution and Genetics, Princeton University Press, Princeton.

Huxley and the Theory of Evolution: "Darwin's Bulldog"

Soon after reading "The Origin of Species" in November, 1859, and anticipating that the book and its author would be severely criticized, Huxley wrote to Darwin:

> Depend upon it, you have earned the lasting gratitude of all thoughtful men... You must recollect that some of your friends, at any rate, are endowed with an amount of combativeness which... may stand you in good stead. I am sharpening my claws and beak in readiness[534].

And a year later he wrote to Hooker: "We have a devil of a lot to do, in the way of smiting the Amalekites"[535].

Huxley declared his support for Darwin's theory because it did not contradict any known biological fact – moreover, the known facts of [embryonic] development, comparative anatomy, and paleontology complemented each other and are fully explained by the theory.

However, Huxley noted, there was one "missing link": there was no convincing proof that natural selection could bring about the evolution of new species.

> [Darwin's theory] is at the moment the only theory of the origin of species that has any scientific value"[536]. So long as all the animals and plants, which were selected by man from the same progenitor, are interfertile, and their offspring are interfertile, this link is still missing. I therefore accept Darwin's hypothesis, provided that a proof be found that physiological species can be produced by selection[537].

Huxley maintained close contact with Darwin, reported new finds to him, and consulted him on difficult issues. For his unrelenting support of the Theory of Evolution and his fighting spirit, Huxley earned himself the title of "Darwin's Bulldog".

Huxley and Religion: "Science and the Hebrew Tradition"

In a series of lectures, prepared in 1876 for his American tour, Huxley expressed his criticism of the biblical stories about the origin of life as taught by the Church. He endeavored to convince his readers that the report of creation of the world in six days could not be a description of a real historical event. The fact that organisms found as fossils in older strata are less and less similar to those that live today, the older is the stratum they are found in, supports the alternative theory of gradual evolution[538].

> My belief ...is, and has long been, that the pentateuchial story of creation is simply a myth. I suppose it to be a hypothesis respecting the origin of the universe, which some ancient thinker found himself able to reconcile with his knowledge... of the nature of things, and therefore assumed to be true[539].

---

534 Huxley to Darwin, 23 Nov. 1859.
535 Huxley to Hooker. Huxley's Life & Letters I: 215.
536 Huxley 1900, p. 147.
537 Ibid. p. 149.
538 Huxley 1897, lectures III-V. p. 89.
539 Huxley 1897, lecture V, p. 180.

Huxley argued that he did not oppose religion, but only the pagan interpretation of the first chapter in the biblical text of Genesis, as transmitted from generation to generation by the Church[540]. He considered that the principles of true religion do not require a belief in the biblical Genesis[541]:

> Surely the prophets would have made swift acquaintance with the head of the scholar who had asked Micah whether per-adventure, the Lord further required of him an implicit belief in the accuracy of the cosmogony of Genesis!
>
> The antagonism of science is not to religion, but to the heathen survivals and the bad philosophy under which religion itself is often well-nigh crushed[542].

Huxley dedicated two long lectures to the biblical story of Creation. In particular he strongly rejected the publications of William Gladstone[543], who claimed that paleontological evidence supported the biblical story as narrated in Genesis. Gladstone had argued that the evidence shows a parallelism with the order of Creation: on the fifth day "the water animals" were created, then the "air animals", and then on the sixth day "the land animals", and finally Man. Huxley presents contradictory evidence to show that the order of creation is in no way supported by science.

> Mr Gladstone was totally misled by assuming that his interpretation of creation is in any way supported by natural sciences... Even if the expression "...evening ...morning" will be extended to mean not a single day but a period that be millions of years, and even if it is agreed that "creation" does not indicate a single act of a Superior Power but a continuous process, even then there is no scientific support to the description that "Water animals", "Air animals", and "Land animals" were created as groups, each separate from the others[544].

For examples, indicating that the scientific evidence was different from the order of creation in Genesis, Huxley referred to the book of Leviticus, which lists the "creeping things" created on the 6th day as "land animals". He points out that although the zoological identity of the "She-retz" and "Reh-mes" is not at all clear, there is abundant fossil evidence that many species of "creeping things" existed in different epochs. Among the "creeping things" listed in Leviticus are reptiles: the lizard, the crocodile, and the chameleon, definitely land animals – but they are found in much older strata than the mammals, even before some of the fishes! The Cetacea – dolphins and whales, which Gladstone listed as "water animals" – were derived from land animals, according to their anatomy, and therefore could not have existed before them. "Air animals" such as birds and bats appeared later than some of the land animals. Many more examples can be found[545]. Huxley concludes that no reconciliation of the scientific finds with the church beliefs is likely:

> And, for my part, I trust that this antagonism will never cease, but that to the end of time true science will continue... retrieving men from the burden of false science, which is imposed on them in the name of religion[546].

540 Huxley 1897, lecture IV, p. 139-190.

541 The true religion Huxley found in the book of Micah, Chapter 6:8: And what doth the Lord require of thee, but to do justice and to love mercy, and to walk humbly with thy God.

542 Huxley 1897, p. 162.

543 William E. Gladstone (1809-1898), then the British Prime Minister.

544 Huxley 1897, pp. 154, 155.

545 Ibid. p. 115-132.

546 Ibid. p. 162.

The Biblical [Noah's] Deluge

The deluge was a pivotal event in the history of the world, according to the biblical tradition. The belief that pairs of representatives of every animal species had emerged from Noah's Ark when it landed on Mount Ararat, and spread from there to every corner of the earth, accounted for the distribution of plants and animals on the planet. The German evolutionist Ernst Haeckel[547] explained why Mount Ararat was so well suited to be a temporary abode for all the animals before their dispersal:

> Mount Ararat, in Armenia, being situated in a warm climate, and rising over 10,000 feet in height, combines in itself conditions for a temporary abode of such animals as live in different zones. Accordingly, animals accustomed to polar regions could climb up the cold mountain ridges, those accustomed to warm climate could go down to the foot of the mountain, and the inhabitants of temperate zone could remain midway up the mountain. From this point it was possible for them to spread north and south over the earth[548].

Huxley seems to have closely studied the geography and topography of the "fertile crescent", which included the valley of the Euphrates – present-day Iraq – to Armenia, where Mount Ararat is located, as well as the Jordan valley from Mount Hermon to the Arava valley. In two long lectures (VI, VII), Huxley argues convincingly that a flood of such dimensions that covered the area with standing water for 40 days "until the dove found some resting place" was an impossibility. He claims that a column of water deep enough to cover the earth to the top of Mount Ararat could not be maintained, because the area slopes down to the Indian Ocean and the water would have rushed there. In the Jordan valley there is no evidence that the area had been flooded since the interglacial period. In addition, Huxley criticizes the described dimensions and structure of the Ark, and claims that it would have been unable to float to the top of Mount Ararat[549].

> Thus… in the face of the plainest and most commonplace of ascertained physical facts, the story of the Noachian deluge… is utterly devoid of historical truth[550].

Contributions to the British Educational System

Huxley was a member of a governmental commission on public school education. His recommendations[551], to which today few would object, highlight the deficiencies in the school curriculum of his time. He strongly advocated the inclusion of physical education in the curriculum of all elementary schools. Sports arenas and playgrounds should be part of every school. Rules of conduct should be based on moral, rather than religious principles. The Bible should be taught by a secular teacher, with geographical and historical explanations and with no theological emphasis. The curriculum should include not only reading, writing, and arithmetic, but also music, poetry, and art. Despite opposition, Huxley demanded that the teaching of science should be mandatory in all schools:

---

547 See Chapter 19.

548 Haeckel, The History of Creation 1876, p. 44.

549 It is surprising that Huxley did not use his sarcasm and ability to ridicule nonsensical ideas in this case, and chose instead to assemble evidence, like a lawyer, in many pages of his lectures, in order to convince his readers and prove the stories wrong by scientific arguments.

550 Ibid. p. 226.

551 Huxley, T.H. 1890f: The School Boards.

> I am strongly inclined to agree with some learned schoolmasters, who say that, in their experience, the teaching of science is all waste of time. As they teach it, I have no doubt it is[552].

In a sarcastic paragraph characteristic of many of his addresses, Huxley advocated the teaching of basic home economics to pupils – particularly to girls – in the elementary schools:

> Why Englishmen, who are so notoriously fond of good living, should be so helplessly incompetent in the art of cookery, is one of the greatest mysteries of nature. But from the varied abominations of railroad refreshment rooms to monotonous dinners of the poor, English feeding is either wasteful or nasty, or both[553].

Teaching Natural Science in the Universities

The teaching of science – in particular biological science (physiology, in Huxley's terminology) in universities was one of Huxley's main concerns. He held strong principles about how the biological sciences should be taught: unlike the Humanities, studying from books alone is not sufficient.

> Practical work in the laboratory is absolutely indispensable, and that practical work must be guided and superintended by a sufficient staff of demonstrators, who are for science what tutors are for other branches of study.
>
> Ask the most practiced and widely informed anatomist what is the difference between his knowledge of a structure which he has read about, and his knowledge of the same structure when he had seen it himself, and he will tell you that the two things are not comparable[554]!
>
> Let those who want to study books devote themselves to literature, in which we have the perfection, both as to substance and as to form.

Huxley published a book on the teaching of animal physiology in schools, with laboratory manuals, advocating the dissection of representatives of all animal groups for demonstration. He had to defend his principle of mandatory laboratory instruction against a public anti-vivisectionist movement, culminating in a law banning animal maltreatment, enacted in Parliament in 1875. As a member of the government committee appointed to discuss the proposal, Huxley fought hard against the law –

> Observation and experiment alone can give us the real foundation for any kind of natural knowledge... Unless the fanaticism of philozoic sentiment overpowers the voice of humanity, and the love of cats and dogs supersedes that of one's neighbor, the progress of experimental physiology and pathology will, indubitably, in the course of time, place medicine and hygiene upon a rational basis[555].
>
> I did my best to prevent the infliction of needless pain for any purpose – expressing my regret at the condition of the law which permits a boy to troll for pike or set lines with live bait, for idle amusement, and at the same time lays the teacher of that boy open to penalty of fine and imprisonment if he uses the same animal for the purpose of exhibiting... the circulation of the blood in the web of the foot[556].

---

552 1890e Universities, Actual and Ideal, p. 54.
553 The School Boards. Critiques and Addresses, p. 47.
554 1890e Universities – Actual and Ideal.
555 Huxley, L., T.H. Huxley; Huxley's Life and Letters, p. 434.
556 Ibid. p. 441.

Huxley served for three years as Rector of the University of Aberdeen, Scotland. In a lecture to the faculty and students of King's College, he recommended changes in the medical curriculum, adjusting the required material to include greater emphasis on subjects that would help the young doctors actually at patients' bedsides – and less on subjects like general zoology and botany. Students just did not have enough time to learn everything:

> Methuselah might, with much propriety, have taken half a century to get his Doctor's degree. And, very fairly, have been required to pass a practical examination on the contents of the British Museum, before commencing practice as a promising young fellow of two hundred, or thereabouts. But you have four years to do your work in, and are turned loose, to save or slay, at two- or- three and twenty[557].

Many university professors might agree with Huxley's pessimistic note, based on his many years of experience as an examiner of dissertations: Students work to pass, not to know. And outraged science takes her revenge: they do pass, and they don't know[558].

One subject on which Huxley expressed, clearly and unambiguously, his opposition to the prevailing practice was that of the education of women. Girls were not at that time admitted to universities or into scientific clubs. In the higher circles of British society, women were educated at home by private teachers. Huxley was furious:

> I don't see how we are to make any permanent advance while one-half of the race is sunk – as nine-tenth of the women are – in mere ignorant parsonese superstition... My own plans are... to give my daughters the same training in physical science as their brothers will get... they at any rate shall not be got up as man-traps for the matrimonial market[559].

"Man's Place in Nature" (1864; 1900)

Huxley's book, summarizing six public lectures on primate anatomy, came out in the midst of the public controversy over Darwin's theory. The subject of the origin of Man from the primates was not discussed in the "Origin", but erupted after the 1860 debate and the clash between Huxley and the Bishop of Oxford, Samuel Wilberforce[560]. Huxley recognized the importance of the subject,

> For it will be admitted that some knowledge of man's position in the animate world is an indispensable preliminary to the proper understanding of his relation to the universe[561].

The book begins with the history of knowledge of man-like creatures in Europe. Travelers in Africa and the Far East in the 16th and 17th centuries had reported the existence of terrible, dangerous, man-like monsters, based on imagination and on stories heard from the natives. Based on such reports, Linnaeus listed four species of "Anthropomorpha" without having seen any physical evidence of any of them.

Two "monsters" were described from Africa: a small monster (called Pongo, later – Chimpanzee) and a big monster (later, Gorilla). Huxley relied on the description reported by the traveler Thomas Savage[562]. The first reliable description of a specimen of an African primate was that of a young animal brought alive from Angola and described by the British zoologist and physician, Edward Tyson (1650-1708). He named the creature pygmie, observed it while alive and described its skeleton

557 1890e Universities - Actual and Ideal, p. 60.
558 Ibid. p. 67.
559 Huxley to Lyell; T.H. Huxley's Life and Letters, p. 212.
560 See Chapter 14.
561 Huxley 1900, p. 80.
562 Huxley 1900, pp. 59-63.

after it died, and reported it in 1699 as an intermediate between Man and monkey[563]. Years later, the skeleton was examined and described in detail by Huxley, who confirmed that it was a young chimpanzee[564].

Dutch travelers from the Far East mentioned an animal called an orangutan, and also gave some details of the biology of the gibbon: it could walk upright on its hind legs, but moved more quickly in the tree canopy using its hands. A young orangutan was brought alive to Paris in 1758 (it died shortly afterwards), together with some skeletons and skins of adult specimens. The animal was described by Geoffroy Saint-Hilaire[565] as a primate restricted to the islands of Borneo and Sumatra, and unrelated to the African fauna. First-hand reports on the orangutan in nature were provided by Wallace when he returned to London[566] (Wallace failed in his attempt to bring a young animal back alive with him).

Skeletal Similarity: Huxley examined primate skeletons in the collection of the College of Surgeons. He measured the dimensions of the limbs, the vertebral column, the pelvis and the skull, and compared them with human skeletons. The comparisons of skeletal characters, and especially the denture, led Huxley to the definite conclusion: Man's closest relatives are the apes.

Huxley studied in the detail the embryological development of vertebrates. He used the embryo of the dog as a standard for comparison with the other vertebrates. He confirmed that the early embryos of all vertebrates, including Man, were very similar, not only in general form but also in details of the limbs. Again the apes were the closest to human embryos:

> Without question, the mode of origin and early development of man are identical with those of the animals immediately below him in the scale. Without doubt, in these respects, he is nearer the apes, than the apes are to the dog[567].

Huxley concluded that Linnaeus was fully justified when he included man with the primates:

> And thus the sagacious foresight of the great lawgiver of systematic zoology, Linnaeus, becomes justified, and a century of anatomical research brings us back to his conclusion that Man is a member of the same order – for which the Linnaean term Primates should be retained – as the Apes and Lemurs[568].

The Neanderthals. A fragment of a human skull was discovered in 1829 in the Engis cave in Belgium, together with bones of extinct mammals. Another fossil fragment was discovered in the Neander valley in Germany in 1856, and is listed as the "Type" specimen of the Neanderthal humans. These discoveries raised great interest among biologists – in an article entitled "On some fossil remains of man", Huxley described in detail the fossil skulls, their dimensions, and the opinions of different contemporary anatomists over more than 60 pages. The verdict was that "the Neanderthal cranium has most extraordinary characters… it belonged to one of the wild races of Northern Europe"[569]. However:

---

563 S.J. Gould, "To show an ape", in The Flamingo Smile, p. 263-280. Tyson, a believer in the Chain of Being, considered the "pygmie" a missing link between monkey and man.
564 Reported in the Transactions of the Royal Society in 1869.
565 Chapter 5.
566 Huxley 1900, pp. 46-58.
567 Ibid. p. 90.
568 Ibid. p. 145.
569 Huxley 1900, p. 168.

In no sense can the Neanderthal bones be regarded as the remains of a human being intermediate between man and the apes[570].

Humans and Apes – Cranial Capacity and Mental Ability

Many arguments were raised against the origin of Man from the apes: "We are men, not improved monkeys with shorter or longer limbs. How about our intellectual powers, knowledge, morals, the ability to tell right from wrong, human compassion – are these not sufficient to separate us from the brutes?"

Huxley was familiar with Morton's[571] measurements of cranial capacity in humans. There was a wide difference in brain size between the gorilla and Man – despite the similarity of body size: a four-year-old European child has a brain twice as large as that of an adult gorilla[572].

Nevertheless, the difference in cranial capacity between Man and gorilla was no greater than between two human skulls, and the difference between Man and chimpanzee was negligible compared with the difference between chimpanzee and lemur. Huxley did not believe that the difference in brain size was the cause of the great difference between Man and the apes.

> I by no means believe that it was any original difference of cerebral quality or quantity which caused that divergence between the human and the pithcoid strips which has ended in the present enormous gulf between them…
>
> Objectors… plausibly argue that the vast intellectual gap between Ape and Man imply a corresponding structural chasm between them in the organ of intellectual function; so that, it is said, the non-discovery of such vast difference proves, not that they are absent, but that science is incompetent to detect them[573]. A man born dumb, notwithstanding his great cerebral mass and his inheritance of strong intellectual instincts, would be capable of few higher intellectual manifestations than an Orang or a Chimpanzee, if he were confined to the society of dumb associates.
>
> And believing, with Cuvier, that the possession of articulate speech is the grand distinctive character of Man… I find it very easy to comprehend, that some equally inconspicuous structural difference may have been the primary cause of the immeasurable and practically infinite divergence of human from the simian strips[574].

To those objectors who argued that Man's origin from the apes is degrading for the human species, Huxley replied at length:

> Is it indeed true that the Poet, or the Philosopher, or the Artist, whose genius is the glory of his age, is degraded from his high estate by the undoubted historical probability, not to say certainty, that he is the direct descendant of some naked and bestial savage, whose intelligence was just sufficient to make him a little more cunning than the Fox, and by so much more dangerous than the Tiger? Nay more, thoughtful men once escaped from the

[570] Ibid. p. 205. Many more Neanderthal fossils were since discovered in Europe (and in Israel, Rak and Erensburg, 1987). They are considered a parallel species, not in the line of descent of Homo sapiens. Interest in the Neanderthals was greatly renewed in the late 20th century, when molecular studies discovered some mtDNA sequence similarity between Neanderthal and Homo genomes (Krings et al. 1997). The overlap in the ranges of the two species raised the possibility – still debated – of some interbreeding. A complete Neanderthal mtDNA sequence [taken from a single specimen] indicated that it is outside the range of modern human variation (Clark 2008).

[571] Gould 1981; see Chapter 15.

[572] Huxley 1900, note on p. 107.

[573] Ibid. note on p. 142.

[574] Ibid. note on p. 143.

blinding influence of traditional prejudice, will find in the humble stock whence Man has sprung, the best evidence of splendour of his capacities[575].

Evolution and Ethics

Late in life, Huxley adopted the pessimistic Malthusian attitude that the struggle for existence in human populations is inevitable and prevents any form of social cooperation:

> One of the most essential conditions, if not the chief cause, of the struggle for existence is the tendency to multiply without limit, which man shares with all living things. It is notable that "increase and multiply" is a commandment much older than the Ten, and …perhaps the only one …obeyed by the great majority of thc human race. But, in civilized society, the inevitable result of such obedience is the re-establishment, in all its intensity, of that struggle for existence – the war of each against all – the mitigation or abolition of which was the chief end of social organization[576].

In his last public lecture in Oxford, Huxley suggested that the struggle for existence in the animal form should not prevail in human society: society flourishes by selection for the morally best, not the physically strongest individuals[577]. He began his lecture with a metaphor. Human society resembles a cultivated garden. The gardener selects a suitable piece of land in "a state of Nature" and encloses it with a wall (Huxley emphasized that the "state of Nature" is not constant, but varies naturally in space and time). In the state of nature, species compete for limited resources, and some of them fail and become extinct. To maintain his garden, the gardener must restrict the tendency of every individual to acquire as much as possible of the resources and to multiply to maximum capacity. The gardener thins the population, removes the weak and the sick, and supplies the remaining individuals with the best conditions for growth.

Huxley argued that the expression "survival of the fittest" is misleading, because the term "fittest" carries the meaning of "best". Survival in nature is not a function of quality, but depends on the environment. If the world cools down drastically – as had happened in the past – the "fittest" will probably not be the most highly developed, but the simplest – the lichens and unicellular organisms.

The human population, Huxley argued, is variable not only in physical characteristics but also in intelligence and ability. Every individual has the ambition to reach a higher standing; hence there is competition among humans. But this competition is not a "struggle for existence" but a "struggle for means of enjoyment". This competition is the reason for suffering and pain in the world[578].

In human society there is no room for the struggle for existence in its animal form. Huxley emphasizes the irreconcilable conflict between the natural law ("Law of the Cosmos") and law in the human society ("Law of Ethics"). Life in human society requires a restriction on individual desires. This is why the savage law of "eye for an eye" was already replaced in ancient times by considerations of "crime and punishment". Huxley argues that the "Law of Ethics" must prevail in society:

> Let us understand, once and for all, that the progress of society depends, not on imitating the cosmic process, still less in running away from it, but in combating it. (The ethical law)

---

575 Ibid. p. 154.

576 Huxley 1989. The Struggle for Existence in Human Society. In response to this paper, the anarchist Peter Kropotkin – and others – came out with the contrary opinion, that cooperation, rather than conflict and struggle for existence, are the dominant forces in Nature. (Kropotkin 1902; Drummond 1896).

577 The lecture was published, together with a comprehensive preliminary essay ("Prolegomena"), as a book entitled "Evolution and Ethics" (Huxley 1989).

578 Huxley 1989, p. 100.

> is directed, not as much to the survival of the fittest, as to the fitting of as many as possible to survive[579].

The "Strangulation Age"

Huxley recorded an exchange of letters on the length of time a scientist should be allowed to live: scientists should be strangled when 60 years old[580]. Darwin[581] recalls a conversation with Lyell on the subject of the "strangulation age". When Lyell finally accepted the Theory of Evolution by natural selection, he reminded Darwin of this joke, and added that he hoped to be permitted to live, now that he accepted the theory.

> Sixty was the age which he had long declared that men of science aught to be strangled, lest they should harden them against the reception of new truths, and make them into clogs upon progress, the worse in proportion to the influence they had deservedly won[582].

In 1881, a short while after the death of Darwin, Huxley's daughter fell ill and died after a long period of hospitalization. This personal disaster together with his grief over Darwin's death caused Huxley to withdraw from all public involvement and retire (the then Prime Minister, Gladstone, awarded him a government pension).

Huxley died of a heart attack in 1895. He was 70, ten years older than the "strangulation age" he had proposed for scientists.

---

579 Ibid. pp. 140, 141.
580 Huxley's Life and Letters, II: 109.
581 Darwin's Autobiography, p. 100.
582 Huxley's Life and Letters, II: 109.

13

Louis Agassiz: Classification and the Plan of Nature

Biographical Notes

Louis Agassiz (1807-1873) was born in Switzerland, studied medicine in Germany, and worked for several years at the Museum of Natural History in Paris. His important studies on the glacial period in Europe were published in 1847 and attracted much attention. He acquired fame in the United States for his contributions to science – and his firm rejection of Darwin's theory.

Agassiz immigrated to the USA in 1846, and two years later was awarded a professorship at Harvard University. He advocated the need to expand the knowledge of zoology and paleontology in America and raised funds on a large scale for the purpose. His efforts led to the establishment of the Harvard Museum of Comparative Zoology (1859)[583] and the American Academy of Sciences (1863). Later, he led a scientific expedition to Brazil, which enriched the Museum collections by thousands of specimens.

Agassiz considered himself the successor to Cuvier, with whom he had worked on the classification of fishes. He published Cuvier's work on the subject years after Cuvier's death[584]. He adopted enthusiastically Cuvier's system of classification and his theory of catastrophes (see below).

Zoology According to Agassiz

Fifty years after Paley's book "Natural theology"[585], Agassiz – a famous paleontologist and zoologist – returned the glory of Creation and the control of nature to the Divinity:

> In one word, all these facts in their natural connection proclaim aloud the One God, whom man may know, adore and love. And natural history must in good time become the analysis of the thoughts of the Creator of the universe, as manifested in the animal and vegetable kingdoms, as well as in the inorganic world[586].

Shortly after his immigration to America, Agassiz published a textbook on zoology[587] – apparently because no textbook was available in his new country and he felt the need to expand knowledge in this field. From the Introduction to the final chapters, the description of the anatomical and physiological systems in living animals, from the simplest to the most complex – as they were known at the time – reflected Agassiz's belief in Creation.

---

[583] Agassiz's son, Alexander Agassiz (1873-1910), invested his fortune in the building and activities of the Museum and became its second director. Unlike his father, he supported Darwin's Theory of Evolution.

[584] Cuvier 1995 (1850), see Chapter 4.

[585] See Chapter 2.

[586] Agassiz, Essay on Classification, 1857, p. 137.

[587] Agassiz 1875 (1847).

Agassiz believed – like Cuvier – that the species had remained unchanged since their creation. Every organism was created perfect, possessing all the necessary qualities for its survival and reproduction – therefore there is no need for further modification[588]. It is possible to arrange the living species according to a progressive scale of perfection, similar to the Chain of Being[589]: An organism is more perfect, when it possesses more organs and therefore more faculties for interacting with the environment. Placing each organism on this scale requires consideration not only of anatomical and physiological characters, but also unmeasurable characters such as intelligence, judgement, and foresight. The scale is an illustration of God's wise planning:

> [Natural history] indicating to us, in Creation, the execution of a plan, fully matured, the work of a God infinitely wise, regulating nature according to immutable plan which He imposed on her[590].

Observation of the embryonic and larval development of different creatures reveals that they are more similar to each other at their early than at their late stages. At the earliest stages, even the great differences between plants and animals are hard to detect (differences in the egg stage must exist, although our senses are unable to detect them). Agassiz concluded that, in the classification of the animal world, only adult forms should be used[591]!

Geographical Distribution

Agassiz argues that only Man is truly cosmopolitan: all other organisms are restricted to certain areas on the globe. In each part of the world are found some species that are found nowhere else (endemic). A study of the present geographic distribution of organisms leads to the following conclusions: there is a direct relationship between the climate and the richness of the *flora* in a region. A weaker relationship is found between the richness of the flora and the *fauna* of the region. It is not unusual that species that occupy regions far apart are very similar to each other, while closely related species within the same region are often very different[592].

The geographic distribution of animals is affected by climate: the three great climatic regions – the Arctic, the Tropical, and the Temperate – are detectable on all continents and host very different faunas. Local factors also limit the ranges of species – like mountain ranges or seas[593]. Further restrictions are due to the activities of Man. When Man extended his range, he drove many species to extinction – like the Dodo and the Irish Elk – and introduced other species, whether intentionally or inadvertently, into different regions. But all these changes in species distribution are small and insignificant: most species still occupy their assigned place in Nature since Creation.

Agassiz emphasizes repeatedly that the shape and structure of animals do not result from the effects of external environmental conditions, but are a reflection of the pre-conceived, clever plan of the Creator, with each species being created in the intended shape at a particular site.

> There is only one way to account for the distribution of animals as we find them, namely… that they… originated, like plants, on the soil where their remains are found[594].

[588] Ibid. p. 30.
[589] See Chapter 2.
[590] Ibid. p. 34.
[591] Ibid. p. 170. This is contrary to Darwin's suggestion – since then accepted by all evolutionists – that larval stages reveal the affinity by descent of different species.
[592] Ibid. p. 109.
[593] Ibid. pp. 107, 203.
[594] Agassiz 1875, p. 212.

The Geological Past

Agassiz was a paleontologist and was aware of the evidence that the surface of the earth had been populated, at different ages, with different species from those now living, and which appeared periodically, separated by wide gaps in time.

> [the fossils] are of the greatest importance, since they furnish us with the means of ascertaining the changes and modifications the Animal Kingdom has undergone in successive creations since the first appearance of living things[595].

Agassiz acknowledged that four distinct eras could be recognized: The Paleozoic ("the Reign of Fishes"), the Mesozoic ("The Reign of Reptiles"), The Tertiary ("The Reign of Mammals"), and the Recent ("The Reign of Man"). The faunas of these eras are very different, but the fauna of any one stratum is not continuous with that of the strata above or below it: Paleozoic fishes were not the progenitors of the Mesozoic reptiles. There is a complete hiatus between the faunas of successive strata. Even if two adjacent strata contain very similar fossil organisms, and even if they so resemble extant animals that it is impossible to tell them apart, it is not possible that these fossils were the ancestors of the extant ones.

> The present epoch succeeds, but is not a continuation of, the Tertiary age. The two epochs are separated by a great geological event, traces of which we can see everywhere around us. There is nothing like parental descent connecting them. The fishes of the Palaeozoic are in no respect the ancestors of the reptiles of the Secondary age, nor does Man descend from the mammals who preceded him in the Tertiary age… and their connection is to be sought in the view of the Creator Himself, whose aim was to introduce Man on the surface of the globe[596].

Agassiz adopted the Theory of Catastrophes propounded by Cuvier[597]. The one catastrophic event that Agassiz had in mind was the Ice Age, the subject of his early studies. The subsequent melting of the ice must have caused the flooding of Europe and North America by the sea, except for the highest mountain peaks, which remained as isolated islands of dry land. No other sites were available, he claimed, for terrestrial and freshwater species. Since they could not have re-migrated from elsewhere, they must have been re-created when the water receded[598]!

Essay on Classification (1857)

> I am daily more satisfied that the primary divisions of Cuvier are true to nature, and that never did a naturalist exhibit a clearer and deeper insight into the most general relations of animals than Cuvier[599].

Agassiz's book "Essay on Classification" was planned as an introductory chapter to a 10-volume series on the fauna of North America, but only one volume – on Chelonia (turtles) – was eventually published. The "Essay" was published in Boston in 1857. Two years later – almost simultaneously with Darwin's "Origin of Species" – it was republished in England, as requested by two of Darwin's scientific opponents, Richard Owen and Adam Sedgwick.

---

595 Ibid. p. 27.

596 Agassiz 1875, p. 236. That the fossil evidence does not mean a common descent of species from earlier ones was also expressed by Cambridge geologist Adam Sedgwick (Introduction to the fifth edition, Cambridge research notes, 1850).

597 See Chapter 4.

598 Ibid. p. 236.

599 Agassiz 2004 (1857), pp. 146-147.

Agassiz's concept of classification was based on his belief in Creation and divine planning. Geology tells us that there was a primeval period when no life existed. But once created, every species was created together with the environment suitable for its existence and survival, according to the plan of nature.

> When created, the plants and animals were adapted to the home assigned for them, with all the characteristics of the Kingdom, Class, Genus and species to which each of them belongs. These characteristics were not formed by the environment of the place nor any physical factor[600].

There are four models (types) of animal structure, as described by the four "embranchements" (circles) of Cuvier: Radiata – mainly marine forms; Articulata, including insects, spiders, myriapoda, and crustaceans; Mollusca, including snails clams, and cephalopods; and Vertebrata – fishes, reptiles, birds, and mammals. Agassiz argues that the fact that the entire animal world is divisible into these four types only – cannot be accidental or due to natural forces: it must be the result of the deep thinking of a designing mind[601]. There is no sense in comparing animals that belong to different types: similarities should be looked for within each type, because their limbs are homologous – like the skeletons of different Vertebrata. Across types, the similarities are only by analogy (like insect brain and vertebrate brain). Useless organs, like the teeth in a whale or the mammary glands in male mammals, exist because they are parts of the mammalian body model, regardless of their function.

According to Agassiz, the systematic units – classes, orders, families, genera and species – are natural units, planned in a hierarchical order by divine intelligence. Species, genus, etc. are ideal concepts, since the *individuals* constituting a species die and disappear, but the *species* as a natural unit persists and can never change. The job of the taxonomist, who tries to describe the order of nature, is to approach as closely as he can the ideal structure of creation. Agassiz criticizes his predecessors who classified animals and described species, genera, and families arbitrarily without establishing fixed criteria for assigning individuals to such groups.

> ...And after all, what does it matter to science that thousands of species more or less should be described and entered in our system, if we know nothing about them[602]?

Agassiz considered it very important in classification to include the habits and complex adaptations to its habitat among the features that characterize a species[603]. The inclusion of the habitat will enable a true, natural classification, because when a species was created it was designed to occupy a specific niche in nature. For the same reason, the habits (behavior?) of the animals should be noted when assigning individuals to the same or to different species.

> [This] subject is neglected by zoologists, because they tend to concentrate on comparative anatomy and physiology[604].

600 Ibid. p. 36.
601 Agassiz 2004, pp. 24-25.
602 Agassiz 2004 (1857) p. 66.
603 [NOTE: This is in contrast to Darwin, who advocated that classification should reflect common descent, therefore adaptational characteristics should not be used!
604 Ibid. p. 6d.

Geographic Distribution and Centers of Creation

Every geographical region has a characteristic fauna and flora. Every island in the ocean has its own endemic species[605]. These facts were interpreted by Agassiz as an indication that each species had a single site of creation. Individuals of a given species, located in different regions – were created independently.

> [As] facts now point distinctly to the independent origin of individuals of the same species in remote regions, or of closely – allied species representing one another in distant parts of the world[606].

Specific sites of creation existed also for the ancient, now fossil, species. Although they are sometimes very similar to extant species, they must have been very different organisms. "Every modern research", wrote Agassiz, "can only confirm Cuvier's conclusion that species are permanent and their shapes are their constant properties. Between geological epochs, species were *replaced* by others, but did not change!"

> It is not because species have lasted for longer or shorter time that naturalists consider them immutable, but because in the whole series of geological ages… not the slightest evidence has yet been produces that species are actually transformed one into the other. We only know that they are different at different periods, as are works of art of different periods and of different schools[607].

Agassiz wrote that previously people had believed that species were created in sequence according to their level of perfection, and that the oldest strata should contain only the remains of the simplest organisms. But this opinion is not supported by the facts: remains of organisms belonging to all classes occur in all strata.

> I remember hearing [it said] that corals were the first to colonize the earth, followed by molluscs and arthropods, and that vertebrates only arrived long after them. But every zoological museum contains the evidence to the contrary. It is well known that representatives of all four divisions of animal life were present simultaneously in the geological strata, even the most ancient. Fish were present in all strata together with Radiata, Mollusca and Arthropoda[608].

Nevertheless, there seems to be some relationship between the structural complexity of animals and geological time – sufficient to affirm that in deeper, more ancient strata we find fossils of simpler, less complex animals, and that more complex ("higher and higher") forms are found in the more recent strata. This relation holds within each of the "embranchements" of Cuvier: for example, in the Echinodermata we find first the Crinoidea, then sea stars, then sea urchins, and finally the sea cucumbers. Among the vertebrates we first find fishes, then reptiles, birds, and mammals in that order, until at the head of the series we find man, the highest of them all. "But we should not", wrote Agassiz, "be misled to think that this order indicates gradual evolution: this is only proof of the planning genius of the Creator":

605 Ibid. p. 38.
606 Ibid. p. 39.
607 Ibid. note on pp. 59-60.
608 Ibid. p. 26.

> Who can look upon such a series... and not read in them the successive manifestations of a thought, expressive at different times, in ever new forms, and yet tending to the same end, onward to the coming of Man[609].

The variation among individuals, even within a single lineage or family, makes the job of the taxonomist very difficult. "I examined", wrote Agassiz, "hundreds of individual tortoises and could not find even two of them identical". How can we explain this variation? Could it be the result of environmental effects? The environment interacts with organisms in various ways – they are born into the environment, grow in it, feed in it, and contribute their bodies to it when they die. But the environment does not determine their shapes and structures:

> If the shapes of animals were determined by the environment, the respiratory system of a fish, a frog, a crab, and a clam – which live in the same place and respire in the same medium – should have been similar[610].

What are Species?

Agassiz considered the term *species* to be an ideal, philosophical concept[611]. The individuals comprising a species are temporary, and are replaced over time, while the species is constant and remains unchanged since its creation.

> What really exist are individuals, not species. We may at the utmost consider individuals are representatives of species, temporary representatives, inasmuch as each species exists longer in nature than any of its individuals[612].

Agassiz declared that there is no justification for using reproductive isolation, or the absence of live and fertile offspring, as the criterion for recognizing species.

> Until it is proven that all the varieties of dogs... originated from a single source (and not mixed), and as long as the one and only origin of all races of Man can be doubted, it is not logical to accept the principle of reproductive isolation – even when the offspring are fertile – as a criterion for belonging to the same species[613].

He rejected another common belief, that each species had originated from a single pair of individuals, survivors of the deluge in Noah's Ark. Agassiz proposed that *many* individuals of every species must have been created simultaneously.

> It becomes more apparent that each species did not originate in single pairs, but were created in large numbers, in these numerical proportions which constitute the natural harmonies between organized beings[614].

This strange proposition, which was unacceptable even to dedicated Creationists, stems from the abhorrence – in the civilized world – of incest. If the origin of each species was a single pair of individuals, their offspring would necessarily be brothers and sisters, who would have had to mate with each other to procreate. This would be especially repulsive for the first human family. Agassiz could not accept that the Bible would advocate such immoral behavior.

609 Ibid. p. 109.
610 Ibid. p. 71.
611 Ibid. p. 177.
612 Ibid. p. 175.
613 Ibid. p. 172.
614 Ibid.

> The necessity which it involves of a sexual intercourse between the nearest blood relatives of that assumed first and unique human family, when such a connection is revolting even to the savage[615].
>
> The fact that all animals were created in large numbers becomes clearer, so that the idea that each species was represented by a single pair is gradually discarded by naturalists. The high probability that individuals of the same species were created independently in remote and disconnected geographical regions removes the need to include the assumption of a common descent in the definition of a species[616].

On Man and Human Races

Agassiz expressed the opinion common to all religious people, that Man is a different kind of organism and stands above all other creatures.

"I believe that it is possible to show, by anatomical facts, that Man is not only the last and the highest of all living creatures, in the present era, but he is the last of the series, after which no further improvement is possible in the model, according to which the entire world was planned"[617].

"Yet there is not enough evidence", wrote Agassiz, "for a clear separation of the soul [inner characters, distinct from physical] of animals and Man, for example in the matter of instincts". A detailed comparison shows similarity between the maternal behavior of mothers in Man and beasts. The caring for and protection of offspring in beasts is very similar to human behavior. When animals fight for a common cause, warn one another of dangers, and help each other, they act according to the same impulses which motivate humans in similar situations "As every sportsman [= hunter], zoo keeper, or every experienced shepherd or farmer can testify".

The question of whether human races originated from the same source, and should be treated as brothers – or that each had a different origin and represents a different biological species, was controversial in the 19th century. In the discussions of slavery at the time of the American Civil War of the 1860s, Agassiz supported the common opinion that the colored races – blacks in particular – do not share a common origin with, and are inferior to, the white races. He insisted that his opinion was based exclusively on scientific analysis of the facts, removed from all political opinions[618].

In America, the question of the origin of Mankind took on a special emphasis. The settlers in the early colonies were deeply religious, and strongly held to the biblical account of the creation of Man: all mankind issued from Adam and Eve, as taught by the Church. Thus all men were related and equal in the eyes of the Lord. This attitude contrasted with the facts of life: the black slaves in the southern states were certainly not treated as equal to the white masters. If the blacks and whites are of the same origin (monogeny), then the treatment of slaves constitutes a crime! But this is not so if the black race is of a different origin (polygeny). The concept of polygeny enabled a ranking of races, with the black race being lowest in the scale and thus justifying their being treated differently.

Prominent among American polygenists were two important scientists: Samuel Morton and Louis Agassiz.

---

615 Ibid. p. 174.
616 Ibid. p. 175.
617 Ibid. p. 28.
618 Gould, 1982; The Panda's Thumb, Norton, New York, Chapter 16. Gould suggested that Agassiz's attitude towards the blacks had a deep emotional basis. Before his arrival in the USA, Agassiz had never seen a black person: he first met them in his Philadelphia hotel where some were employed. He must have been shocked. Gould quoted a letter that Agassiz wrote to his mother a short time after his arrival in the USA. His letter expresses both disgust and repellence [Gould, 1982, p. 173].

Samuel George Morton (1799-1851)

Morton, a Philadelphia physician (who also held a degree in medicine from Edinburgh, Scotland) was interested in the differences among human races. In 1835, in an album-form publication entitled "Crania Americana", he illustrated these differences.

Morton assembled a collection of several hundred human skulls from all over the world. He measured 13 morphological characters on each skull, then grouped the skulls according to their geographical origin and race, and calculated the group means of the variables. Of major interest for Morton was the cranial capacity of the skull – reflecting the size of the brain, which was considered a measure of intelligence. This he measured as accurately as he could by pouring mustard seeds [later, lead pellets] through the foramen magnum until no more could enter, then emptying the contents into a graduated cylinder. He perceived this as an objective, scientific measure of intelligence[619].

The group means illustrated that "Caucasians"[620] (white English, German and other Europeans) had larger brains than other races, notably American Indians and especially blacks (negroes, in the terminology then prevalent), which he considered the lowest and least intelligent race.

In a later publication, entitled "Crania Aegyptiaca" (1844), Morton analyzed more than 100 skulls from ancient Egyptian tombs, sent to him by the American consul in Cairo (he claimed that he was able to tell from the morphology which of these skulls belonged to a negro). The races were "as perfectly distinct in Egypt 3000 years ago as they are today". These data led to the conclusion that the differences between the white and the black races were ancient, and they could not have differentiated from a common stock [for one reason, there was not enough time for that: Morton dated Creation at 4179 years before his own time, and the Egyptian skulls 1000 years later):

> It seems implausible that an omnipotent Creator would leave his creations to fend for themselves in unsuitable environments until they adapted: each race was adapted from the beginning to its local destination[621].

Louis Agassiz

Agassiz was a Creationist and a strict monogenist before his arrival in America in 1846. He had never seen a black person before. His first encounter with the black waiters in his Philadelphia hotel must have been a shock. In a long letter to his mother, translated by Gould, he expressed his disgust:

> Contrary to all ideas on the con-fraternity of the human type, I expressed pity at the sight of this degraded and degenerate race. None the less, it is impossible to repress the feeling that they are not of the same blood as us. In seeing their black faces with their thick lips and grimacing teeth, the wool on their head, their bent knees, and the livid color of the palm of their hands, I could not take my eyes off their face in order to tell them to stay away.
>
> What unhappiness for the white race to have tied their existence so closely with that of negroes in certain countries! God preserve us from such a contact![622]

Agassiz believed that the geographic distribution of animals was established at Creation, around "centers of creation" in different parts of the globe. The human races, similarly, were created separately in different countries, and were not of the same origin. Basing his polygenic theory of human races on the Scriptures, Agassiz wrote:

---

[619] Gould (1951) checked Morton's means and discovered a built–in bias in the calculation and interpretation.

[620] The term "Caucasians" for the white race was introduced in 1775 by J.F. Blumenbach, who assumed that the inhabitants of the Caucasus – today's Georgia – were the most beautiful humans (Gould 1994).

[621] Morton 1839, cited in Dewbury 2007, p. 125.

[622] Gould 1951. See also Dewbury 2007.

> The unity of mankind and the diversity of the origin of the human races are two distinct questions, having almost no connection with each other[623].
>
> The unity of species does not involve a unity of origin, and that a diversity of origins does not involve a plurality of species[624]. We maintain, therefore, that the unity of mankind does not imply a community of origin for men. We believe on the contrary, that a higher view of this unity of mankind can be taken than that which is derived from a mere sensual connection[625].

A search of the Scriptures supported the diversity of human origins:

> It is not for us to inquire further into the full meaning of the statements of Moses. But we are satisfied that he never meant that **all** men originated from a single pair, Adam and Eve[626]. We would particularly insist upon the propriety of considering Genesis as chiefly relating to the history of the white race, with special reference to the history of the Jews[627]!
>
> We maintain that the Mosaic record there is not a single passage asserting that the differences [among races] have been derived from changes introduced in a primitively more uniform stock of men. We challenge those who maintain that mankind originated from a single pair, to quote a single passage in the whole Scriptures pointing to those physical differences which we notice between the white race and the Chinese, the New Hollanders, the Malays, the American Indians and the Negroes, as having been introduced in the course of time among the children of Adam and Eve[628].
>
> The racial characteristics are permanent from their creation and cannot be changed: "have we not …the distinct assertion that the Ethiopian cannot change his skin, nor the leopard his spots"[629]?

Race and Racism

> It seems to me to be mock-philanthropy and mock-philosophy to assume that all races have the same abilities, enjoy the same powers, and show the same dispositions. History speaks for itself. There has never been a regulated society of black men on that [African] continent… Does not this indicate in this race a peculiar apathy, a peculiar indifference to the advantages afforded by civilized society[630]? If the United States should hereafter be inhabited by the effeminate progeny of mixed races, half Indian, half Negro, sprinkled with white blood… I shudder from the consequences[631].

Agassiz later expressed extreme ideas about the type of education black children should receive: the blacks should be trained in manual work, while the white children should be prepared for "brain work". He strongly criticized the penetration of "white genes" into the black population, and saw the resulting ("hybrid") mulattos a crime against nature:

---

623 Agassiz 1850, p. 3. Note: page numbers are for the copy I was provided by the University Library.
624 Ibid. p. 5.
625 Ibid. p. 10.
626 Agassiz 1850, p. 185.
627 Ibid. p. 139.
628 Agassiz 1850, p. 137.
629 Agassiz 1850, p. 27 on my copy.
630 Cited in Gould 1951, p. 34.
631 Ibid. p. 49.

> The production of half-breeds is as much a sin against nature, as incest is in a civilized community a sin against the purity of character. I hold it to be a perversion of every natural sentiment (Gould 1980, p. 175).

A similar position regarding black people was not rare in the USA at the time, and laws and regulations in the same spirit were accepted in many of the states[632].

Responses

Agassiz's contemporary scientists could not accept his extreme ideas about species. Even Charles Lyell, who had advocated "special creation" twenty years before Agassiz, felt that the latter had brought the matter of creation to an extreme, absurd position:

> I admit that Agassiz's last publication moved me far into Darwin's camp, or the Lamarckian point of view, because when he suggests the origin of every human race to a separate, independent act of creation – and when this is not enough, he creates whole nations simultaneously, each individual separately "from earth, air and water" as Hooker put it – these miracles do not appeal to me... So that I cannot avoid thinking that Lamarck was right[633].

And in another letter he wrote,

> As a matter of fact, the freedom he [Agassiz] allowed himself in his work on "classification", to multiply the miracle of Creation every time that he is faced with the difficulty in understanding how some bird or fish could migrate to a place from their original habitat – or any habitat – has prepared many to accept Darwin's and Lamarck's hypothesis[634].

Strong rejection of the idea of "special creation" was expressed, years later, by T.H. Huxley. In a lecture to the Paleontological Society, he wrote sarcastically:

> It is indeed a conceivable supposition that every species of Rhinoceros and every species of Hyaena, in the long succession of forms between the Miocene and the present species, was separately constructed out of dust, or out of nothing, by supernatural power. But until I receive distinct evidence of the fact, I refuse to run the risk of insulting any sane man by supposing that he seriously believes such a notion[635].

---

632 [see below, Chapter 22 on Eugenics].

633 Lyell to Ticknor, 9 January 1860; Lyell 1881, II: 331.

634 Lyell to Dawson, 15 May 1860; Lyell 1881, II: 332.

635 Huxley 1870, "Palaeontology and Evolution"; in Critiques and Addresses, 1890, p. 207.

14

The Struggle for Existence of Darwin's Theory

The Dilemma

William Whewell, the Master of Trinity College, Cambridge, described in 1847 the dilemma of interpreting the fossil fauna discovered by geologists:

> The study of Geology opens to us the spectacle of many groups of species which have, in the course of the earth's history, succeeded each other at vast intervals of time. One set of animals and plants disappearing, as it would seem, from the face of our planet, and others, which did not before exist, becoming the only occupants of the globe. And the dilemma then presents itself anew: either we must accept the doctrine of the transmutation of species, and must suppose that the organized species of one geological epoch were transmuted into those of another by some long-continued agency of natural causes; Or else, we must believe in many successive acts of creation and extinction of species, out of the common course of nature – acts which, therefore, we may properly call miraculous[636].

The first alternative was strongly rejected by many scientists in the major universities [which belonged to the Church]. A typical example is an article by the Cambridge geologist, Adam Sedgwick, in 1850[637]. Sedgwick wrote a devastating criticism of "The Vestiges of Creation" and its author, Robert Chambers[638] – and the supporters of like ideas – for suggesting the possibility of gradual change and transmutation of species – which, he claimed, is not supported by geological evidence – and for denying the role of God in the creation and regulation of the universe.

Whewell had doubts about the latter alternative as well:

> Mr Lyell, indeed, has spoken of a hypothesis that the successive creations of species may constitute a regular part of the economy of nature. Are these species created by the production, at long intervals, of an offspring different in species from the parents? Or are the species so created produced without parents? Are they gradually evolved from some embryo substance? Or do they suddenly start from the ground, as in the creation of the poet[639]?

Although the first edition of "The Origin of Species" was fully sold out on the first day, the idea of gradual evolution by natural selection was not welcome in England in 1859. The strongest opposition came from the religious establishment – which included the universities (Cambridge, Oxford,

---

636 Whewell, "History of the Inductive Science" (1847) 1866, II: History of Geology, p. 564; Cited in Darwin, F., Charles Darwin's Life and Letters, II: 194.
637 Sedgwick 1850, Preface to the fifth edition of Cambridge research notes.
638 Chapter 10.
639 Whewell, (1847) 1866, II: History of Geology p. 573; Cited in Darwin's Life and Letters II: 192.

and other institutions of advanced studies that were run by the Church). The objections raised by Sedgwick against Chambers were equally well directed against Darwin. Particularly strong was the opposition to the idea that Man is related to other primates and shared a common ancestor with the apes. Darwin, foreseeing the reaction to this idea, said nothing about human evolution in the "Origin", other than a statement at the end of the "Conclusions" that the theory will shed light on the origin of Man; but judging from the public reaction, his views must have been well known when the "Origin" came out (Darwin's views on the evolution of Man were only clearly expressed 12 years later, in his book "The Descent of Man"[640].

The Debate

A plaque on the wall near the main lecture hall, at the Natural History Museum in Oxford, commemorates a well-publicized event in June, 1860, at the annual meeting of the British Zoological Society. One of the sessions was scheduled for a discussion of Darwin's theory, which had appeared in print just a few months earlier. The keynote speaker at the session was the Bishop of Oxford, Samuel Wilberforce. Wilberforce was known as a very good speaker, and there were rumors that he had promised to "smite" Darwin's theory at the meeting.

Darwin himself did not attend the meeting. Two of his friends did most of the fighting for the new theory – the botanist Joseph Hooker and the zoologist Thomas Henry Huxley, whose work on vertebrate anatomy was well known. Huxley recalled later that he was reluctant to go to the meeting, because he suspected that a majority of the audience would be clergy attracted by the presence of the Bishop and hostile to Darwin – but a chance meeting with Robert Chambers made him change his mind.

No protocol was kept of the proceedings of the meeting. The available record was gathered twenty years later, by Francis Darwin, who as editor of his father's "Life and Letters", asked surviving witnesses of the meeting to write their recollections of the event[641]. All the witnesses agreed that a large audience – 700 persons – had gathered for the session, which had to be moved to a larger hall. The discussion began with a lecture by an American visitor, who spoke against Darwin's theory. A few more people spoke briefly, and then Wilberforce took the podium. According to the record, he talked for half an hour, saying as expected that Darwin's theory contradicted Genesis and therefore must be wrong. He added that Man could not have descended from the apes, because there was a fundamental difference in the structure between the brains of Man and the Gorilla[642]. At this point Wilberforce turned to Huxley, who was seated in the audience, and asked him sarcastically – as a supporter of the idea that Man had descended from the apes – "on which side of his family he descended from the apes: his grandfather's or his grandmother's side?" This was a direct insult, and the chairman of the session – the Cambridge botanist Henslow – called upon Huxley to respond.

Regrettably, Huxley was unprepared and did not write down his response, and when asked to recollect the event twenty years later he did not remember his exact words. There are different accounts of what happened next. One account has it that Huxley explained patiently that the honorable bishop did not understand the essence of Darwin's theory of natural selection, which does not deal with changes in one or two generations, but with a long, gradual process over thousands of generations. He then declared that he had just finished a detailed study of the brains of man and the primates, and the bishop's "evidence"of a structural difference was simply contrary to fact. Finally he responded to the personal insult. According to that account he said,

[640] See Chapter 15.

[641] Darwin, F.: Charles Darwin's Life and Letters II: 320-323. The same information is contained in Huxley's "Life and Letters", edited by his own son Leonard Huxley. There was no journal dedicated to science in 1860 – as mentioned in Chapter 12, "Nature" was founded in 1869 – but articles of the event appeared in the local press.

[642] Wilberforce was no biologist, and he had obtained this evidence from his fellow, Richard Owen, a famous professor of anatomy at Oxford. Huxley, L. 1900; T.H. Huxley's Life and Letters", p. 183.

> But if the question is treated, not as a matter for calm investigation of science, but as a matter of sentiment, and if I am asked whether I would choose to be descended from the poor animal of low intelligence and stooping gait, who grins and chatters as we pass, or from a man endowed with a great ability and a splendid position, who should use these gifts [the end of the sentence was drowned out by a great outburst of laughter as the point became clear] to discredit and crush humble seekers after truth, I hesitate what answer to make[643].

A second eye-witness wrote that Huxley had said,

> I asserted and I repeat, that a man has no reason to be ashamed of having an ape for a grandfather. If there were an ancestor whom I would shame in recalling, it would rather be a man – a man of restless and versatile intellect, who, not content with an equivocal success in his own sphere of activity, plunges into scientific questions with which he has no real acquaintance, only to obscure them by an aimless rhetoric and… skilled appeal to religious prejudices[644].

According to the same sources, the response by Huxley did not the end the session. Other people took the stand. Darwin's friend Joseph Hooker explained at length the value of the new theory for the understanding of the development and variation of the vegetable kingdom. On the other side, the former captain of Darwin's ship "Beagle", now Admiral FitzRoy, walked down the aisle carrying a Bible and explained that he had argued with Darwin during the voyage about his ideas being contrary to the Scriptures. The event was greatly publicized in the press, and Darwin's theory certainly gained popularity.

The records of the meeting give the impression that Huxley has won the debate and that Natural Selection had carried the floor – Huxley writes that for the next 24 hours he had the feeling that he was the most important figure in Oxford.

"Knight Takes Bishop?"

The American evolutionist Steven J. Gould (1941-2002) spent a sabbatical year in Oxford in 1970, and surveyed the newspaper articles of the period. In a short article entitled "Knight Takes Bishop?" Gould suggested that the records of the meeting, as given by Francis Darwin and Leonard Huxley, were biased by the editors and that Wilberforce's side was not given a fair chance. Gould claimed that Wilberforce, indeed, lost in the debate, but not because evolution was proven right, but because he had failed to behave as a gentleman should, turning to personal insults[645].

Gould found that two weeks *before* the public debate, Wilberforce had published a critical review of "The Origin of Species", in which he had delivered "lethal blows" at the theory –as he had promised to do. Darwin was aware of this review and commented on it in a letter to Hooker:

> It is uncommonly clever. It picks up with skill all the most conjectural parts and brings up all the difficulties[646].

Darwin's comment is sufficient justification for describing Wilberforce's 1860 review in detail. Modern readers should consider how relevant his objections might still be today, more than 150 years later.

---

643 Vernon-Harcourt to Leonard Huxley. Huxley, L. 1900: T.H. Huxley's Life and Letters, I: 185.
644 J.R. Green to Leonard Huxley. Ibid.
645 Gould 1994, pp. 397-414. The title of the article refers to the figures in the game of chess.
646 Darwin to Hooker, July 1860.

Bishop Wilberforce's Criticism

In his review, Wilberforce presented himself as a reasonable critic of the new theory. He started his article by discussing at length Darwin's reference to the artificial selection of domesticated animal varieties – especially pigeons – as a model of the operation of natural selection[647]. He cites several amusing stories from the "Origin", like the causative relationship between house cats kept by British spinsters in the villages – and the abundance of field mice, bumble bees, and the pollination of clover, or the story of slave-driving ants. But he cuts short his humorous approach:

> Now all this is, we think, really charming writing... but we here are bound to say our pleasure terminates. For when we turn with Mr. Darwin to his "Argument", we are almost immediately at variance with him[648].

Wilberforce objected to two aspects of Darwin's theory: first, the descent of all creatures from common ancestors – and second, the idea that Natural Selection is the primary force driving the changes in the animal world:

> Man, beast, creeping thing, and plant of the earth, are all Linnaean descendants of some one ens, whose various progeny have been simply modified by the action of natural and ascertainable conditions into the multiform aspects of life which we see around us. Mr. Darwin finds then the disseminating and improving power, which he needs to account for the development of new forms in Nature, in the principle of natural selection[649].

Darwin's suggestion that varieties of domestic pigeons, selected by breeders, are a model for the evolution of new species by natural processes, is unacceptable: no proof is brought forward that these varieties ever become new species:

> Now all this is very pleasant writing, especially for pigeon-fanciers. But what step do we really gain in it at all toward establishing the alleged fact that variations are but species in the act of formation, or in establishing Mr. Darwin's position that a well-marked variety may be called an incipient species? We affirm positively that no single fact tending even in that direction is brought forward[650].

The great differences in morphology and other characteristics between the various breeds of dogs are extreme products of artificial selection by breeders, but there is not a shred of evidence that they are becoming separate species. The great Newfoundland dog recognizes the tiny domestic varieties as fellow dogs and treats them quite differently from other carnivores like the fox, the wolf or the jackal:

> The dumb animal might teach the philosopher that unity of kind or of species is discoverable under the strangest mask of variation[651].

Darwin's central assumption is that natural selection will pick up and select any variation, however small, that improves the fitness of an individual in the struggle for existence. Wilberforce agreed that this could be true in theory, but declares that the practice of breeders cannot be used as a model for this theory: Not one of the variations picked up by breeders of domestic animals is of any benefit for the animal itself:

---

[647] Darwin 1952, pp. 23 ff
[648] Wilberforce 1860. p. 230.
[649] Ibid. p. 232.
[650] Ibid. p. 235.
[651] Ibid. p. 237.

> Every variation introduced by man is for man's advantage, not for the advantage of the animal... there is not a shadow of ground for saying that man's variations ever improve the typical character of the animal as an animal. They do but by some monstrous development make it more useful to [Man] himself... Hence it is that nature, according to her universal law with monstrosities, is ever tending to obliterate these deviations and return to type[652].

Darwin's use of an assumption not supported by facts as the foundation of a theory, makes the theory no better than a fairy tale. Wilberforce delivered what seems to be a devastating blow to the theory, expressing his objections very strongly:

> In the name of all true philosophy we protest equally against such a mode of dealing with nature, as utterly dishonourable to all natural sciences, as reducing it from its present lofty level... to being a mere idle play of the fancy, without the basis of the discipline of observation. In the Arabian Nights, we are not offended when Amina sprinkles her husband with water and transforms him into a dog, but we cannot open the august doors of the reasonable temple of scientific truth to the genii and magicians of romance[653].

Finally, Wilberforce turned to the origin of Man. He strongly objects, of course, to the suggestion that Man is the product of the same processes that directed the development of the apes:

> First, then, he [Darwin] not obscurely declares that he applies his scheme of the action of the principle of natural selection to man himself, as well as to the animals around him. Now we must say at once, that such a notion is absolutely incompatible not only with single expressions in the word of God on that subject of natural science... but with the whole representation of the moral and spiritual condition of man.
>
> Man is not, and cannot be, an improved ape. Man is the sole species of his genus, the sole representative of his order and subclass. Thus I trust (says Owen) has been furnished the confutation of the notion of a transformation of an ape into the Man[654].

At the end of his 40-page review, Wilberforce declares that he has rejected Darwin's theory not only because it is contrary to the word of the Lord, but also on scientific grounds:

> Our readers will not have failed to notice that we have objected to the views with which we have been dealing, solely on scientific grounds. We have done so from our conviction that it is thus the truth or falsehood of such arguments should be tried. We have no sympathy with those who object to any facts, or alleged facts, in Nature, or to any inference deduced from them, because they believe them to contradict what it appears to them is taught by Revelation. We cannot, therefore, consent to test the truth of natural sciences by the word of Revelation. But this does not make it less important to point out on scientific grounds, scientific errors, when these errors tend to limit God's glory in Creation, or to gainsay the revealed relation of that creation to Himself. To these classes of error – though, we doubt not, quite unintentionally on his part – we think Mr. Darwin's speculations directly lead[655].

Contemporary Criticisms of Darwin's Theory

Wilberforce was not the only one who rejected the new Theory of Evolution by natural selection. In 1864, when the Royal Society awarded Darwin its highest medal of distinction – the Copley Medal

---

[652] Ibid. p. 238.
[653] Ibid. p. 250.
[654] Ibid. p. 261.
[655] Wilberforce 1860, p. 256.

– the reviewers wrote that the medal was awarded for achievements in zoological and botanical research, and completely ignored the controversial new theory.

Disagreements were raised by members of the scientific community, including some of Darwin's friends. Charles Lyell believed in "special creation"[656]. He accepted Darwin's new ideas on the formation of coral reefs, but was reluctant to accept evolution by natural selection. In 1863 he wrote to Darwin:

The Debate (Knight takes bishop)

> I cannot go to Huxley's length in thinking that natural selection and variation account for so much, and not so far as you, if I take some passages in your book separately. I think the old Creation is almost as much required as ever, but of course it takes a new form if Lamarck's views improved by yours are adopted[657].

Lyell was converted rather slowly to Darwin's camp. Ten years after the appearance of the "Origin", he still refused to accept that Man's intellectual abilities had evolved by natural selection. In the last few chapters in his book "The Antiquity of Man" (1873) he tries to explain how Darwin's arguments may finally convince him. Among the reasons for his change of opinion he lists the discovery of fossils, which narrow the gaps between classes[658].

> The theory of transmutation, as first annunciated by Lamarck, was impugned on the ground that no adequate causes were adduced which could bring about the necessary modifications. Mr. Darwin, by the theory of Variation and Natural Selection, supplied these causes. But still the opponents of transmutation urged that no proofs were to be obtained in the fossil world of these transitions which are assumed to have taken place. These proofs we now see are gradually presenting themselves, few and far between as might be expected[659].

Darwin's greatest supporter, Huxley, disagreed with Darwin's insistence that nature only progresses gradually ("Natura non facit saltum"). Huxley wrote to Lyell:

> I know of no evidence to show that the interval between two species must necessarily be bridged over by a series of forms, each of which occupy... a fraction of the distance between A and B. On the contrary, in the history of the Ancon sheep, and the six-fingered Maltose family given by Reamur, it would appear that the new form appeared at once in full perfection... I have a sort of notion that ...in passing from species to species, "natura facit saltum"[660].

George Campbell, the 8th Duke of Argyll, objected to the title of Darwin's book as inadequate:

---

656 Chapter 7.

657 Lyell, K.M. 1881: Lyell's Life and Letters, II: p. 363.

658 The first specimen of Archaeopteryx was discovered in 1860. See Richard Owen, "on the archaeopteryx of von Meyer..", Phil. Trans. R. Soc. 153: 33-47, 1863.

659 Lyell, 1973. The Antiquity of Man, (1863) p. 103.

660 Huxley L., Huxley's Life and Letters, p. 173; II: 232.

> Strictly speaking, Mr. Darwin's theory is not a theory on the origin of species at all, but a theory on the causes which lead to the relative success or failure of such forms as may be born into the world... It does not even suggest the law under which... such new forms are introduced.

Campbell completely rejected Darwin's theory in his book "The Reign of Law" (1867). The core of his objections was the same as that of Paley[661]: wherever there was in nature an indication of design – a contrivance fulfilling a function – this was proof of the presence of a designer. Without divine planning and control, the world would return to chaos. Like Chambers[662], he claimed that "the natural laws" – which according to biologists regulate the world – were designed and created by the Higher Intelligence.

Darwin as his own Critic

Darwin was well aware of the weak points and the difficulties that his theory encountered. Two chapters in later editions of "The Origin of Species" were dedicated to the many objections raised by some of the critics of the early editions[663], to which he responded as best he could.

> I had, also, during many years, followed a golden rule, that whenever a published fact, a new observation or thought, came across me, which was opposed to my general results, to make a memorandum of it without fail. For I had found by experience that such facts and thoughts were far more apt to escape from the memory than favourable ones[664].

Intermediate Stages between Evolving Species: One of the most difficult issues, repeatedly raised by Darwin's critics, is the absence in the fossil record of intermediate stages in the transition from the ancestor to the descendent species.

> As according to the theory of natural selection, an interminable number of intermediate forms must have existed... why does not every collection of fossil remains afford plain evidence of the gradation and mutation of the forms of life[665]?

Darwin suggested four possible reasons for the rarity or absence of intermediate forms. Firstly, the geological record is incomplete: only a small fraction of the earth's surface has been explored geologically, and vast areas which could potentially hold fossils of intermediate stages have not been searched at all. Secondly, only forms having a solid skeleton or shell, such as mollusks or vertebrates, can leave a permanent record in the rocks as fossils. Thirdly, many species could have become extinct without leaving a trace, because the period in which the changes occurred could have been far shorter than the period in which no change occurred[666]. Fourthly, the rarity of intermediate forms may result from natural selection: a rare intermediate individual, slightly different from the typical parent form, may be less well adapted to the existing conditions, and be selected against[667]. The rarer a form is, the less likely it is to be found by geologists.

Transmutation: How can natural selection gradually change a species' characteristics to become another species – for example, how did flying bats evolve from some unknown flightless animal? No intermediate ancestors of bats are known. But, Darwin argues, there is a similar case in a different

---

661 Chapter 2.
662 Chapter 10.
663 Darwin 1898 (1868), Chapters VI, VII.
664 Barlow, Darwin's Autobiography, p. 123.
665 Darwin 1898 (1868), The origin of species, 6th edition. p. 125.
666 Ibid. p. 357. This argument was revived many years later when Gould and Eldredge (1977) proposed their theory of "punctuated equilibrium".
667 Ibid. p. 125.

group of mammals – the Rodents. In the flying squirrels, there are species with different grades of flight ability, from simple gliders to more skilled forms, with each form adapted to its present habitat. Why not assume that a similar gradation had occurred in the ancestors of bats?

The Vertebrate Eye: A particularly difficult case to uphold was that of the evolution of the vertebrate eye – which Paley used to argue its design by the Creator[668]. Darwin confessed that the evolution of the eye by natural selection was a tough problem, but nevertheless it must be true:

> To suppose that the eye, with all its inimitable contrivances for adjusting the focus to different distances, for admitting different amounts of light, and for the correction of spherical and chromatic aberrations, could have been formed by natural selection seems, I freely confess, absurd in the highest degree. [But] when it was first said that the sun stood still and the world turned around, the common sense of mankind declared the doctrine false. But the old saying, "vox populi, vox dei", cannot be trusted in science[669].

Darwin argued that primitive "eyes" do exist in invertebrate species, and apparently serve the same function of detecting light. These structures must be useful for the animals in their struggle for existence. Consequently, slight changes that improved the structures could have occurred in the evolution of the vertebrate eye (Darwin recognizes that the Cephalopod eye, although of a very sophisticated structure, cannot be considered an intermediate in the evolution of the vertebrate eye).

The Social Insects[670]: In hive bees, ants, and termites, the workers are not sexual forms and do not reproduce ("neuters"). How could they have transmitted their instincts, habits, and even adapted morphologies for different jobs [pollen collection by worker bees for example?]

> The case of neuter social insects… is by far the gravest difficulty which I have encountered. So grave, that to anyone less fully convinced than I am of the strength of the principle of inheritance & of the slowly accumulating action of natural selection, I do not doubt the difficulty will appear insurmountable[671].
>
> The Lamarckian doctrine of all modifications of structure being acquired through habits & being thus propagated, is false. For whatever may have been the habits of our neuters, they never leave offspring to inherit the effects of habit or practice. For my part, although I do not doubt that use & disuse may affect structure & be inherited, yet long before thinking of this case of neuter insects, I had concluded that the effects of habit were of quite subordinary importance[672]. It might be argued that the hive [bee] neuters have retained by inheritance from an early progenitor, certain normal characters which the queen bee lost[673].

Neutral Characters: How did characters evolve which apparently have no effect on the fitness of their carriers (such as variation in color patterns on butterfly wings or mollusk shells)? In the Origin, Darwin suggested that characters which have no effect on fitness may be important for the systematist. Because they are not affected by the environment and free from natural selection, they may be used to demonstrate the affinity between taxa due to common descent[674]. Such characters may be important for the organism, even though we cannot detect their value, and may have evolved due to correlation with other, useful characters.

668 Chapter 2.

669 Darwin 1898 (1868) p. 133.

670 This argument, written by Darwin in preparation of the Big Book" in 1857-8, was omitted from the Origin and published in Natural Selection (Stauffer 1975), p. 365.

671 Stauffer 1975, p. 373.

672 Stauffer 1975, p. 365.

673 Ibid. p. 367.

674 Darwin 1898 [1868 p. 164].

In his later work, Darwin suggested an alternative selective mechanism for the evolution of some such characters: Sexual Selection[675].

Fleeming Jenkin's Rejection (1867)

Fleeming Jenkin[676] was one of the toughest critics of Darwin's theory. In a very long review article of "The Origin of Species" in 1867, he attacked Darwin's arguments one by one. He dismissed some of the most basic assumptions of the theory as unsupported by facts.

One of Jenkin's most important objections was to the concept of unlimited time, which, Darwin suggested, allowed small favorable changes slowly to accumulate by natural selection, until a species could be changed into another. Jenkin wondered:

> The difference between six years and six myriads, blinding by a confused sense of immensity, leads man to say hastily that if six or sixty years can make a pouter out of a common pigeon, six myriads may change a pigeon to something like a thrush; But this seems no more accurate than to conclude that because we observe that a cannon-ball has traversed a mile in a minute, therefore in an hour it will be sixty miles off, and in the course of ages that it will reach the stars... the rate of variation in a given direction is not constant... it is a constantly diminishing rate, tending to a limit[677].

Jenkin argued that variation of a character within a species may be likened to the inside of a sphere. The center of the sphere is the mean of the character. Individuals vary in all directions, but most variants will be near the mean. To become a new species an individual must cross the envelope – as a "sport", which is a very rare occurrence. It is therefore not likely that a new species will be produced by natural selection. Jenkin most severely criticized Darwin's suggestion – in the early editions of the Origin – that rare "sports", if they have a characteristic favorable in the struggle for existence – may be picked up by natural selection and become the beginning of new species.

> This theory of the origin of species... simply amounts to the hypothesis that from time to time, an animal is born differing appreciably from its progenitors, and possessing the power of transmitting the difference to its descendants. What is this but stating that, from time to time, a new species is created[678]?

Jenkin based his objection on the common belief – shared by Darwin – that the characters of the parents "blend" in the offspring. To illustrate his point, Jenkin suggests the following metaphor:

> Suppose a white man to have been wrecked on an island inhabited by negroes, and to have established himself in friendly relations with a powerful tribe... Suppose him to possess the physical strength, energy, and ability of a dominant white race... Grant him every advantage which we can conceive a white to possess over the native; concede that in the struggle for existence his chance of a long life will be much superior to the native chiefs. Yet from all these admissions there does not follow the conclusion that, after a limited or unlimited number of generations, the inhabitants of the island will be white. Our shipwrecked hero would probably become king... He would have a great many wives and children... Yet he would not suffice in any number of generations to turn his subject's

[675] Chapter 16.

[676] Henry Fleeming Jenkin, a Scottish engineer (1833-1885), was employed in planning submarine communication cables, and later became a professor of engineering at the universities of London and Edinburgh.

[677] Jenkin 1867, p. 2.

[678] Ibid. p. 6.

> descendants to be white… the advantage of structure possessed by an isolated specimen is enormously outbalanced by the advantage of numbers possessed by others[679].

[Jenkin noted that if the characters of the parents did not "blend" – but were retained in all descendants, then selection could in fact be effective: but this kind of inheritance was contrary to the common knowledge on heredity by contemporary scientists].

Jenkin also criticized Darwin's conclusion that similarity between species indicates a community of descent. There could be other reasons for similarity: all organisms are made of the same elements, and are likely to be more or less similar to each other[680]. Jenkin concluded that Darwin's theory was a speculation unsupported by facts:

> What can we believe but that Darwin's theory is an ingenious and plausible speculation… containing… some faint half-truths, marking at once the ignorance of the age and the ability of the philosopher. Surely the time is past when a theory unsupported by evidence is received as probable, because in our ignorance we know not why it should be false, though we cannot show it to be right. Yet we have heard grave men gravely urge, that because Darwin's theory was the most plausible known, it should be believed[681].

St. George Jackson Mivart

One of the most important critics of Darwin's theory was St. George Jackson Mivart[682].

> Mivart has recently collected all the objections which have ever been advanced, by myself and others, against the theory of natural selection… and has illustrated them with admirable art and force[683].

Mivart accepted the concept of gradual evolution, but suggested that natural selection alone is insufficient for driving the changes in the living world: it must be supplemented by Creation.

> Creation is not a miraculous interference with the laws of nature, but the very institution of these laws[684].

In his book, entitled "The Genesis of Species" (1871), Mivart listed many biological facts which are difficult to explain by means of Darwin's theory. Mivart declared that many of these difficulties would disappear if Darwin would allow that new species may be formed suddenly, rather than gradually[685].

---

679 Ibid. p. 5. Francis Darwin, who edited his father's letters, noted that Darwin's copy of "The North British Review", the journal with Jenkin's article, was heavily annotated, indicating that Darwin had read it several times. Darwin did not find a suitable answer to this criticism, and in the fifth and subsequent sixth editions of the Origin he changed the emphasis from the "sports" to selection on small, individual variations as the source of new species.

680 Ibid. p. 11.

681 Ibid. p. 14.

682 St. George Jackson Mivart (1827-1900) was trained as a biologist at Oxford University (an Anglican institution), but converted to the Catholic faith in 1844 and obtained his degree in a Catholic college. Appointed a lecturer at St. Mary's School of Medicine, he published research articles on carnivorous and insectivorous mammals for many years (but was excommunicated by the Church in 1855).

683 Darwin 1898 [1868] p. 164.

684 Mivart, G.J. 1871, p. 19. A similar objection was raised by Chambers in 1844. Mivart heralded Robert Chambers (Chapter 10) as the English interpreter of Lamarck's transmutation theory.

685 Ibid. p. 112.

Some of the most difficult cases were that the early, initial stages of the appearance of a new character must be too minor to affect the fitness of the organism, and natural selection must be unable to affect them. Mivart lists many examples to support this claim.

> [Natural selection] utterly fails to account for the conservation and development of the minute and rudimentary beginnings, the slight commencement of structures, however useful those structures may afterwards become[686].

Darwin dedicated two whole chapters in later editions of "The Origin of Species" to Mivart's objections and attempted to reply to them as best he could. A few of the many examples presented by Mivart follow, with Darwin's attempts to explain the difficulties.

The Giraffe's Neck: If the giraffe has evolved from short-necked grazing savannah mammals, such elongation of the neck also meant a great increase in the animal's size and, consequently, in the animal's energy costs. What was the great advantage of a longer neck that would offset the need to procure a greater amount of food for a greater expenditure of maintenance energy? And if the advantage was so great, why was the giraffe the only species of African herbivorous mammals to get a longer neck?

[Darwin replied that the long neck, in addition to helping to obtain getting food, had a further advantage: it enabled a better chance of early detection of approaching enemies, such as the lion[687]. There is no need for other species to "endeavor" to lengthen their necks to reach the higher food source, if a taller species already exists there.]

> In South Africa the competition for browsing on higher branches of the acacias and other trees must have been between giraffe and giraffe, and not with other ungulate animals[688].

Baleen in Whales: Mivart agrees that improvement in the baleen (whale bone) on the jaws of whales must have been advantageous, since it facilitated the acquisition of food in plankton-feeding animals. But, he argued, what selective value could this possibly have when it was just formed as minute growths on the jaws?

[There are no known animals with intermediate stages of development of baleen. Darwin raised the analogous case of the beak structure of mud-sifting birds. Ducks filter the water for plankton according to the same principle as whales, and in some species of Anseriformes (geese) there is a gradation of beak structures to fit different modes of life. The evolution of whale bones may have developed along similar lines – although no trace of this development is apparent among either living or fossil Cetacea].

Wings and Flight: Mivart wonders what advantage could be attributed to the first appearance of changes in the forelimbs of the ancestors of birds. There would be no survival value to rudimentary wings until they could be used for flying. [A similar objection applies to the evolution of bats. No intermediate fossils between an insectivorous mammal and a fully winged bat are known].

> It is difficult to believe that the avian limb developed in any other way than by a comparatively sudden modification of a marked and important kind[689].

---

686 Ibid. p. 26.
687 Darwin 1898 [1868) p. 166.
688 Ibid. p. 167.
689 Mivart 1871, p. 121.

[Darwin did find an analogous case of supposedly gradual evolution of flight in mammals – the "flying squirrels". In this group, different species exist with varying development of wings and flight – from mere gliders jumping between trees to species that cover quite long distances by gliding – with each species adapted to its particular mode of life.]

The Eyes of Bottom-dwelling Fish: In the adult sole, which lie on their side on the sandy bottom of the sea, both eyes are located on one side of the fish's body. This may be advantageous for the fish, perhaps for better detection of predators. The young fry swim freely and have a normal fish appearance, one eye on each side. Mivart argued that intermediate positions of the eyes could have no advantage either for the fry or the adult fish, and could not have developed by natural selection.

[Darwin replied that young fish swim normally for a while, and then fall to the bottom. Their skull is soft and flexible and they can move the eye at an angle of 70°. This situation may be advantageous for the young fish, and may have been further modified by selection.]

Mivart raised many objections to Darwin's interpretation of the structure of orchids and of climbing plants. Mivart's arguments, as well as Darwin's replies, illustrate the characteristic form of the 19th-century debate, when philosophical "armchair biology" prevailed rather than experiments or observations in nature.

Lord Kelvin and the Age of the Earth

The famous physicist, William Thomson (better known as Lord Kelvin) raised, in 1866, a strong and persuasive argument against Lyell's idea of uniformitarianism and unlimited time. Kelvin argued that the earth is too young to allow for the origin of biological variety by "descent with modification" and natural selection[690].

> The uniformitarian theory – as proposed by Charles Lyell – that the earth's internal heat is due to chemical production of thermo-electric processes that in turn decomposes the chemical products, forming a perpetual cycle – violates the principles of natural philosophy in exactly the same manner, and in the same degree, as to believe that a clock constructed with a self-winding movement may fulfil the expectation of its ingenious inventor by going for ever (*perpetuum mobile*).

Kelvin considered that the earth had become separated from the sun when a large comet had collided with it, and that it had no source of heat other than the original heat it possessed as a fragment of the sun. He tried to estimate the age of the earth by calculating the cooling time of a red-hot body, until it reached a temperature which life could tolerate. His calculations were based on data from the temperature changes in deep mines, and estimates of the initial size and composition of the earth (assumed to be uniform). Kelvin calculated that the earth could not be more than 400 million years old – and later reduced this estimate even further:

> But I think we may with much probability say that the consolidation [of the earth] cannot have taken place less than 20,000,000 years ago, or we should have more underground heat than we actually have. Nor more than 400,000,000 years ago, or we should not have so much as the least observable underground increment of temperature.
>
> During the 35 years which have passed, since I gave this wide-ranged estimate, experimental investigation has supplied much of the knowledge… we have good reason for judging that it was… no less than 40,000,000 years ago, and probably much nearer 20 than 40[691].

690 See S.J. Gould, 1985; "False premises, good science"; in The Flamingo Smile, Norton, New York. p. 126.
691 Lord Kelvin Quotations, 1868.

Lyell (and Darwin) disagreed, claiming that the rate of deposition of sediments to form geological strata required vastly greater lapses of time. But Physics was considered a much more powerful and exact science than Biology. Even Huxley and Wallace tended to consider the "younger-earth" argument and allow that evolutionary rates were faster than previously thought. The dispute continued to the end of the century.

The solution was found in 1903, when Pierre Curie discovered that radioactive decay produces heat. The heat generated in the earth's core compensates for heat loss by radiation and slows down the cooling process. The age of the earth is conventionally estimated as 4.5 billion years, and life has existed on its surface for the last 3.5 billion years[692].

692 Gould, S.J., 1985; in "The Flamingo's Smile", pp. 126-138.

15

Darwin: The Descent of Man (1871)

"And God created man in his own image, in the image of God created He him. Male and female created He them"[693] Genesis 1: 27)

One of Darwin's critics put the alternatives in strong terms by asking, whether we are to believe that man is modified mud or modified monkey[694].

The famous verbal exchange between Bishop Wilberforce and Huxley had brought the issue of the position of Man into a matter of public discussion, but diverse opinions were already being expressed early in the 19th century. In the "notes" to his poem, "The Temple of Nature", Erasmus Darwin wrote:

> These philosophers ...seem to imagine that mankind arose from one family of monkeys on the banks of the Mediterranean, who accidentally had learned the use of that strong muscle which constitutes the ball of the thumb, and draws the point of it to meet the points of the fingers, which common monkeys do not. And that this muscle increased in size, strength and activity in successive generations. And by the imported use of the sense of touch, these monkeys acquired clear ideas, and gradually became men[695].

At about the same time, Jean Baptiste Lamarck suggested an imaginary scenario for the evolution of Man from "a family of monkeys" who left their arboreal habitat and started walking on their hind legs[696].

"The Descent of Man" (1871)

> Just as the geocentric conception of the universe – namely, the false opinion that the earth was the center of the universe... was overthrown by Copernicus and his followers, so the anthropocentric conception of the universe – the main delusion that man is the center of terrestrial nature, and that its whole aim is to serve him – is overthrown by the application – attempted long since by Lamarck – of the theory of Descent of Man[697].

Darwin's book "The Descent of Man", appeared about 12 years after the "Origin of Species". The book is bound together with Darwin's book on Sexual Selection (Chapter 16) and Darwin considered both as a single, continuous contribution, for reasons stated at the end of the "Descent".

693 Genesis 1: 27.
694 Lyell, K.M. 1881, Charles Lyell's Life and Letters II: 376.
695 Darwin, E.,"The Temple of Nature" (1803), notes, p. 541.
696 See above, p. 90.
697 Haeckel 1876, II: 264.

That Man developed from a primate ancestor was very clear to Darwin all along, and his opinion on this issue was apparently no secret, judging from the strong opposition expressed by his adversaries[698].

> Thus we understand how it has come to pass that Man and all vertebrate animals have been constructed on the same general model, why they pass the same early stages of development, and why they have certain rudiments in common. Consequently we ought frankly to admit their community of descent. It is only our natural prejudice, and that arrogance which made our forefathers declare that they were descended from demi-gods, which leads us to demur from this conclusion[699].

In his "Autobiography", Darwin explained why he had refrained from expressing these ideas in the "Origin":

> It would have been useless and injurious to the success of the book to have paraded, without giving evidence, my convictions with respect to his [Man's] origin. But when I found that many naturalists fully accepted the doctrine of the evolution of species, it seemed to me advisable to move up such notes as I possessed and to publish a special treatise on the origin of Man[700].

During the interval between the publication of the "Origin" and that of the "Descent", Darwin's theory was debated in the press and in scientific and other circles. Huxley's book, "Man's Place in Nature"[701] was published during that interval, providing strong support for Darwin's thesis. Darwin himself did not participate in the public discussions, but he corresponded widely with colleagues in England and abroad, especially in Germany, as documented in Darwin's "Life and Letters"[702].

Darwin thought it likely that Man had descended from primitive primates, although intermediates connecting between man and other primates have yet to be discovered. Darwin considered that classifying Man as a separate vertebrate order on the basis of his highly developed brain, as suggested by Cuvier, was not justified. The majority of Man's organs are very similar to the organs of other primates, and many of the differences are consequences of the upright posture – being adaptive characters which should not be considered when determining descent relationships. Judging from the anatomical similarities, Man belongs to a branch of Old-World apes, together with the African chimpanzee and gorilla and the Asian orangutan and gibbon. Darwin describes his ideas as to the common ancestor of man:

> We thus learn that Man is descended from a hairy, tailed quadruped, probably arboreal in its habits, and an inhabitant of the old world[703]. But we must not fall into the error of supposing that the early progenitor of the whole simian stock, including Man, was identical with or even closely resembling any existing ape or monkey[704]. In a series of forms, graduating insensibly from some ape-like creature to Man as he now exists, it would be impossible to fix on any definite point when the term man ought to be used[705].

---

[698] Chapter 14.
[699] Darwin, The Descent of Man. 1952, p. 410.
[700] Barlow, Darwin's Autobiography, p. 131.
[701] Chapter 12.
[702] F. Darwin 1887; interestingly, Francis Darwin only included letters from Darwin to colleagues, but not the letters that Darwin received and to which he replied, and it is often difficult to determine what they were arguing about.
[703] Darwin, C. 1952: The Descent of Man, p. 911.
[704] Ibid. p. 541.
[705] Ibid.

Darwin insisted that Natural Selection had driven the evolution of Man, just as it had done in the evolution of primates. Natural selection does not seem nowadays to be active in changing human characters, but – aided, in some cases, by the inherited effects of "use and disuse" – it must have had a strong effect in the past:

> Some of the most distinctive characters of Man have, in all probability, been acquired – either directly or, more commonly, indirectly, through natural selection[706]. We may infer that when, at some remote epoch, the progenitors of Man were in a transitional state… natural selection would probably have been greatly aided by the inherited effects of increased or diminished use of different parts of the body[707]. Modifications acquired and continually used, during past ages, for some useful purpose, would probably become firmly fixed, and might be long inherited[708].

Darwin states that the growth rate of extant primitive human societies is far lower than in modern, developed nations, due to the heavy mortality of infants and young in the former. These mortality factors must have been even more severe in the early stages of human societies. Infanticide, which is known in monkeys, may also have existed in ancient human societies.

The transition to bipedal walking, which freed the hands for other purposes, greatly improved the performance and survival of humans. The rudiments of the use of tools for procuring food can be found today in some apes, Man's closest relatives[709]. The early human ability to invent new weapons, to build rafts and boats, to use fire and learn how to start it when needed – all of these must have given them a strong selective advantage.

Darwin attributed the loss of the tail and of the hairy body cover to "disuse". Disuse also accounts for the fact that the eyesight of Europeans is less keen than that of the primitive "savages" (whom Darwin had met in his travels around the world in the "Beagle"). Eyesight and color discrimination may have been important for the "savage", for example in collecting suitable food, but had become less so in the modern way of life.

Wallace's arguments, that Man was a being "above nature", did not convince Darwin. He insisted that Man is subject to the same processes which affect the animal world. Moreover, early humans must have lived in groups, and appropriate social behavior was first expressed within groups and selected for:

> As the reasoning power and foresight of the members became improved, each man would soon learn that if he aided his fellow men, he would commonly receive aid in return[710]. There is not the least inherent improbability, as it seems to me, in virtuous tendencies being more or less strongly inherited[711].

Appropriate social behavior could evolve by means of two social responses *within* a group: encouragement and praise on the one hand, and blame on the other. Even dogs learn to behave by means of these two responses – the savages must have responded similarly. Darwin suggested that the rules of proper conduct had first developed and been applied within family or tribal groups. With the advance of civilization, when small groups coalesced to become nations, men had to broaden the circles of application of the rules of conduct to include people outside their family or tribe. The extension of the application of the rules to include all humanity merely required the crossing of an artificial barrier.

---

[706] Ibid. p. 411.
[707] Ibid. p. 421.
[708] Ibid. p. 418.
[709] Ibid. p. 430.
[710] Ibid. p. 499.
[711] Ibid. p. 492.

As supporting evidence for the heritability of moral (or immoral) conduct, Darwin reports that he had heard of a tendency to steal, found in members of certain upper-class families. Since stealing is a rare crime among the already wealthy, it is unlikely that the occurrence of this immoral behavior in two or three members of the same lineage would be due to chance alone.

Darwin suggested that social behavior could also have survival value in the competition among groups.

> Although a high standard of morality gives but a slight or no advantage to each individual man and his children over other men of the same tribe, yet an increase in the number of well-endowed men and an advancement in the standard of morality will certainly give an advantage to one tribe over another[712].

A group of more clever and intellectual men, with better relationships among its members, would have out-competed the less-endowed groups, in the same way that modern "advanced" nations replace the less-advanced indigenous peoples "wherever the climate enables the advanced nations to take hold".

Relaxed Selection in Civilized Nations: In primitive human societies, the weak and sickly were naturally selected against, and the survivors tended to be the stronger and healthier. In modern societies, medical aid and the "poor laws" keep the weak and sickly alive and allow them to reproduce. Anyone involved in the breeding of domestic animals knows that relaxation of the strict rules of breeding will cause a rapid deterioration of the breed[713]. Moreover, in humans, the strong and healthy are recruited to the army and may die in battle, while the weak and sickly, who are exempt from military service, may survive and produce progeny[714]. The poor and the heedless people marry early and produce more generations per unit of time as a burden on society, while the benefits to society are contributed by the educated – who leave fewer offspring:

Darwin, Origin of Man

> Great lawgivers, the founders of beneficial religions, great philosophers and discoverers in science aid the progress of mankind in a far higher degree by their works than by leaving a numerous family[715].

True, the denial of help to the sick and the poor is contrary to proper social behavior, but it is nonetheless desirable that the poor should reproduce less. In the struggle for existence, the lower race will dominate – due not to its virtues, but to its faults[716]!

---

712 Ibid. p. 500. This "group selection" argument has been controversial in 20th century evolutionary biology.

713 Darwin may have shared these ideas with his cousin, Francis Galton, who bred dogs. Galton was one of the founders of the Eugenics movement in England (see Chapter 21). Darwin cites Galton's statement that the Catholic Church caused great damage to the Spanish population when the Inquisition killed and burned the most clever and courageous of the population – a thousand men per year for 300 years.

714 This sentiment was also expressed by Ernst Haeckel (Chapter 19).

715 Darwin 1952, Descent, p. 504.

716 Ibid. p. 505.

How Many Species and Races of Man Exist? The geographical races of man differ from each other in many characteristics. These differences remain stable for many generations and must therefore be heritable[717]. Darwin cites references to different authors who claimed that there are 2, 3, 4, 5, 6, 7, 8, 11, 15, 16, 22, 60, or 63 races[718]. Nevertheless there is no reason to regard the races as separate species – as some researchers do – especially because the distinctive characters vary widely within each race, and there is no agreement on how different a population must be to be regarded as a separate species.

> It is a hopeless endeavor to decide this point until some definition of the term species is generally accepted. All the races agree in so many unimportant details of structure that these could be accounted for only by inheritance from a common ancestor[719]. Those naturalists… who admit the principle of evolution will feel no doubt that all races of man are descended from a single primitive stock, whether or not they may think fit to designate the races as distinct species[720].

The Evolution of Emotions

Darwin mentioned in his "Autobiography" that he had intended to discuss the development of emotions as a chapter in the "Descent of Man", but the amount of material had been too large. In a separate book "On the Expression of Emotions in Animals and Man", published in 1873, Darwin analyzed the similarities of facial and other expressions of fear, anger, affection, and aggression performed involuntarily by animals like dogs, cats, and monkeys, to expression of the same emotions in humans. He included observations and photographs of his own children when very young, and of hired professional actors expressing anger and despair (the photographs are glued to the pages of the book).

> My object is to show that certain movements were originally performed for a definite end, and that under nearly the same circumstances they are still… performed through habit when not of the least use. I have already given, in the case of Man, several instances of movement, associated with various states of mind or body, which are now purposeless, but which were originally of use, like lifting one's arms to protect the face even when falling on a soft surface[721].

Darwin lists three principles which seem to account for the application of most of the expressions in Man as well as in animals. The sharing of these principles demonstrates that Man was derived from mammalian progenitors – not only physically but also emotionally. These principles are: a) "The principle of serviceable associated habits" (when the same state of mind is induced, there is a tendency for the same movement to be performed, although they may not be of the least use); b) "The principle of antithesis" (when a directly opposite state of mind is induced, there is a strong tendency to perform a movement of directly-opposite nature, although it is of no use); and c) "The principle of action due to the constitution of the nervous system", independently of the will or habit[722].

Darwin provides many examples to substantiate his claim that emotions are expressed similarly by man and animals. Very similar expressions of emotions are observed when young children and young primates are compared:

[717] Ibid. p. 529 ff.
[718] Ibid. pp. 536, 537.
[719] Ibid. p. 910.
[720] Ibid. p. 537.
[721] Darwin 1873, p. 42.
[722] Ibid. p. 28.

> Young children, when in violent rage, roll on the ground on their backs or bellies screaming, scratching or biting everything within their reach… so it is as we have seen with the young of anthropoid apes[723]. The appearance of dejection in young Orangs and Chimpanzees, when out of health, is as plain …as in the case of our children[724].

A very common expression of horror or sudden fear in animals is the bristling of the hair, which makes the animal look larger than it really is. It is surprising that Man, who is relatively hairless, still retains this involuntary response under similar situations:

> With respect to the involuntary bristling of the hair, we are led to believe that Man has retained, through inheritance, a relic of them, now become useless. It is certainly a remarkable fact, that the minute unstriped muscles by which the hairs, thinly scattered over Man's almost naked body, should have been preserved to the present day. And that they should still contract under the same emotions, namely terror and rage, which cause the hairs to stand on end in the lower members of the order to which man belongs[725].

However, some expressions are unique to Man and have not been seen in primates. These must have been acquired after the human and simian branches separated.

> Frowning is due to the contraction of the corrugators by which the eyebrows are lowered and brought together. Both the Orang and the chimpanzee are said to possess this muscle, but it seems rarely brought into action[726]. In no case did a monkey keep its mouth open when it was surprised. This fact is surprising, as with mankind hardly any expression is more general than a widely-open mouth under a sense of astonishment[727]. Blushing is the most peculiar and the most human of all expressions. Monkeys redden from passion, but it would require an overwhelming amount of evidence to make us believe that any animal could blush[728]. The keepers in the Zoological Gardens assured me that they never heard a sob from any kind of monkey[729] [this habit] must have been acquired since the period when Man branched off from the common progenitor of the genus Homo and the non-weeping Anthropoid apes[730].

The message Darwin delivers throughout the book is that the evidence strongly supports the claim that Man has descended from a common mammalian [primate] progenitor. These emotions and their expressions are the same in all races of man everywhere in the world.

This fact is interesting, and affords a new argument in favour of the several races being descended from a single parent stock, which must have been almost completely human in structure, and to a large extent of mind, before the period at which the races diverged from each other[731].

---

[723] Ibid. p. 241.
[724] Ibid. p. 137.
[725] Ibid. p. 309.
[726] Ibid. p. 143.
[727] Ibid. p. 145.
[728] Ibid. p. 310.
[729] Ibid. p. 157.
[730] Ibid. p. 154.
[731] Ibid. p. 361.

16

Charles Darwin: Sexual Selection

[Sexual selection] does not depend on any superiority in the general struggle for life; But on certain individuals of one sex, generally the male, being successful in conquering other males, and bearing a larger number of offspring to inherit their superiority than do the less successful males[732].

Darwin's book on "Selection in Relation to Sex" is bound together with his volume "The Descent of Man". At the end of the latter volume, Darwin explains why he decided to do so:

> It can further be shewn, that the differences between the races of Man, as in colour, hairiness, form of features, etc. – are of a kind which might have been expected to come under the influence of sexual selection. But in order to treat this subject properly, I have found it necessary to pass the whole animal kingdom in review[733].

In this comprehensive review of the animal world, Darwin offered an explanation for the evolution of characters that seem not to be related to the fitness of individuals in the struggle for life and are therefore unlikely to have evolved by natural selection. Such characters are, for example, the color patterns on the wings of butterflies and birds, and the sexual dimorphism in colors and shape in many other organisms. Darwin suggested that such characters evolved as a result of within-species competition among the *males* for possession of the females. The winners of that competition thus obtained an advantage in producing most of the progeny, and thereby passing on their characters. Darwin called this mechanism *Sexual Selection.*

Rationale

In sexually-reproducing species, the sexes often differ in characters not directly involved in reproduction ("secondary sexual characters"). The males vary in the expression of these characters, some of which may help them to attract the attention of the females. Characters that make a male more attractive in competition with other males (e.g. the intensity of color or the tail-length in peacocks) increase their chances of producing progeny. Darwin suggests that males were selected for attractiveness, while the females developed acute senses to detect the differences among competing males – and this has led to the brilliant coloration and other characteristics of the males. This form of selection differs from natural selection:

> The males have acquired their present structure, not from being better fitted to survive in the struggle for existence, but from gaining an advantage over other males, and from having transmitted this advantage to their male offspring alone… It was the importance of this distinction which led me to designate this form of selection as "sexual selection"[734].

---

[732] Darwin 1874, p. 863; 1952, p. 575.
[733] Darwin 1952, p. 556.
[734] Ibid. p. 569.

Characters which have evolved by sexual selection are, for example, organs used for fighting or defense; boldness; various decorations; "musical" chords (for singing in birds and some mammals); scent glands; and other characteristics. The main driving force is *selection* of the most attractive males *by the females*. The females decide which of the males will father their brood. A sexually-selected male should leave more offspring than other males, and his male offspring should carry the characteristics that made their father more successful[735]. These offspring should be more successful than the male progeny of other fathers[736]. Sexual selection differs from natural selection in that it does not normally end in the death of the loser:

> Female birds, in a state of nature, have by long selection of the most attractive males, added to their beauty and other attractive characters[737]. Sexual selection acts in a less-rigorous manner than natural selection. The latter produces its effects by the life and death at all ages of the more or less successful individuals. Death, indeed, not rarely ensues from conflict of rival males. But generally, the less successful male merely fails to obtain a female, so they leave fewer progeny[738].

Darwin reasoned that sexual selection could have been much more effective had the number of males greatly exceeded that of the females, so that the less successful males would be unlikely to find mates. He collected data on the sex ratio in many organisms including Man, and found it generally to be approximately 1:1. In this case even the less successful males may eventually find mates and reproduce. On the other hand, data on domesticated animals reveal that the males are the more variable sex, so that the females may be able to choose and select the best or most attractive male.

In "Selection in Relation to Sex", Darwin systematically surveys the animal world, presenting and interpreting examples of sexual dimorphism. He consistently held to his pet theory of sexual selection as the main force in the evolution of sexual dimorphism in size, color, and pattern. However, his explanations are sometimes based on faith alone. In some cases he included Wallace's alternative suggestions in his book: Wallace was consistent in assigning a central role to natural selection, and offered reasonable explanations for the same phenomena, but Darwin did not accept them.

The Examples

Invertebrates: In many species of invertebrates – such as crustaceans – the male is larger than the female. Darwin explained the evolution of this difference:

> With most animals when the male is larger than the female, he seems to owe his greater size to his ancestors having fought with other males during many generations[739].

In many species of beetles, for example dung beetles (Scarabaeidae), the males have "horns" or other prominent structures which are smaller or absent in females. Darwin admitted that his explanation is weak:

---

[735] NOTE: The main difficulty in the theory of sexual selection is that the selected traits must be transmitted exclusively to the *male* offspring of the winner. Darwin recognized this problem, but his knowledge of heredity provided no clue to how this might happen, and he expresses his ideas in a general, ambiguous, statement: (the characters) will become permanent if the exciting cause acts permanently, and in accordance with a frequent form of inheritance they may be transmitted to that sex alone in which they first appeared.

[736] Ibid. p. 572; Darwin 1874, p. 230.

[737] Ibid. p. 570.

[738] Ibid. p. 583; Darwin 1874, p. 242.

[739] Darwin 1952, p. 619; 1874, p. 286.

> The most obvious conjecture is that they [the horns] are used for fighting together, but the males have never been observed to fight[740].

Sexual dimorphism is particularly striking in species of butterflies: the males are the more colorful. Males and females develop as caterpillars on the same plants. How then did the difference evolve? Darwin suggested that the males were selected for their bright colors, because the brightest males were more attractive to the females.

> On the whole, it seems probable that most of the brilliantly coloured species of Lepidoptera owe their colours to sexual selection[741]. If the females habitually, or even occasionally, prefer the most beautiful males, the colours of the latter will have been rendered brighter by degrees… the process of sexual selection will have been facilitated[742].

The dimorphism in butterflies seemed to Wallace an excellent example of protective coloration evolved by natural selection[743]. Darwin cited Wallace's alternative explanation in detail, but insisted that the dimorphism is the result of sexual selection.

Fishes, Amphibians, and Reptiles: Darwin and Wallace disagreed on the question of the sexual dimorphism of certain species of *fish* in which the male is more colorful and conspicuous than the female. These differences become more pronounced during the breeding season. There exist detailed descriptions of inter-male aggression, and of the male guarding and protecting the female, the nest site, and the eggs. All these facts convinced Darwin that the colors in fish had evolved by sexual selection[744].

Wallace suggested that the bright colors protected the fish from predators in the colorful environment of the coral reefs where they live, and therefore evolved by natural selection. Darwin objected, contending that colorful fishes can be found in tropical rivers, where no corals exist. Nor do the colors function as warning, since the colorful fish are edible and preyed upon by water birds and by other fishes.

In the class *Amphibia* there is no sexual dimorphism: Darwin agreed that the bright colors of some frogs and toads must be warning colors, advertising their being poisonous or noxious. The green coloration of most amphibians is protective and evolved by natural selection – but, Darwin explained, the males have strong and clear mating calls, which must have evolved by sexual selection!

The only explanation Darwin was able to offer to explain the bright coloration of some species of snakes is that the colors evolved by sexual selection:

> It does not… follow that they [female snakes] should be endowed with a sufficient taste to admire bright colours in their partners, so as to lead to the adornment of the species through sexual selection. Nevertheless, it is difficult to account in any other manner for the extreme beauty of certain species[745].

Birds: Male birds differ from females in a number of secondary sexual characters – particularly color and voice. There are good reasons to study the evolution of these characters in birds:

---

[740] Darwin 1952, p. 644; 1874, p. 315. 130 years later, Moczek and Emlen (1999, 2000) observed the behavior of these beetles in their underground tunnels. In one species – which Darwin refers to specifically – the males do fight for possession of the females, and the horns are in fact used in these conflicts – the larger males usually win. Darwin's conjecture was, after all, correct!

[741] Darwin 1952, p. 664; 1984, p. 339.

[742] Ibid. p. 662; Ibid. p. 336.

[743] Ibid. p. 658; Ibid. p. 339; see Chapter 11.

[744] Ibid. p. 684; Ibid. p. 362.

[745] Darwin 1952, p. 692; 1874, p. 372.

> Birds appear to be the most aesthetic of all animals excepting of course Man, and they have nearly the same taste for the beautiful as we have. This is shown by our enjoyment of the singing of birds, and by our women… decking their heads with borrowed plumage and using gems[746]. It would appear that… slight changes for the sake of change, like change of fashion with us, are intelligible only on the principle of novelty having been admired for its own sake[747].

Male birds appear to be very aggressive towards other males, and quarrels are commonly observable. The well-known courtship displays of birds serve, Darwin suggests, to emphasize the differences in performance among competing males and allow the females to select the most beautiful or attractive performer. The fights are often just for show, with no casualties, but they do indicate fierce competition, as suggested by the theory of sexual selection:

> Almost all male birds are extremely pugnacious, using their beaks, wings and legs for fighting together. The peacock with his long train appears more like a dandy than a warrior, but he sometimes engages in fierce contests[748].

To support this claim, it is necessary to show that the females do in fact observe the display of individual males and select their mates responding to characters that the males display. Darwin had no such proof. He gives some examples from the reported behavior of domestic birds – one example is of a female that "fell in love" with a male of a different species, but it is not clear what were the characteristics involved in the female's choice:

> With respect to female birds feeling a preference for particular males, we must bear in mind that we can judge of choice being exerted only by analogy. [if we were watching a group of young rustics at a fair courting a pretty girl, and quarrelling about her… we would …infer that she had the power of choice. Now with birds, [here Darwin describes the evidence in a half page] are we not justified in believing that the female exerts a choice, and that she receives the addresses of the male who pleases her most? It is not probable that she [the female bird] consciously deliberates, but she is most excited or attracted by the most beautiful, or melodious, or gallant male[749].

Color Patterns in Birds

How did the spectacular ornamental patterns of birds, like those on the peacock's tail or the ball-and-socket patterns in the male Argus pheasant, evolve? The origin of the peacock is unknown, but, Darwin argues, there must have existed intermediate species between Asian Galliformes and the present-day peacock. The color patterns cannot be assumed to have arisen by chance:

> That these ornaments should have been formed through the selection of many successive variations, not one of which was originally intended to produce the ball-and-socket effect, seems as incredible as that one of Raphael's Madonnas should have been formed by the selection of chance daubs of paint, made by a long succession of young artists, not one of whom first intended to draw the human figure[750].

---

746 Darwin 1952, p. 697; 1874, p. 378.

747 Darwin 1952, p. 814; 1874, p. 515.

748 Darwin 1952, p. 699; 1874, p. 379; Ibid. p. 701; 1874, p. 383; Ibid. p. 717; 1874, p. 413.

749 Darwin 1952, p. 749; 1874, p. 441; Ibid. p. 750; 1874, p. 441. Japanese scientists observed the behavior mating of peacocks in a large nature reserve during several years. The males and the females were individually tagged. Several morphological characteristics were measured on each male, including the length and color patterns on the tail. There was no evidence that the females used any of these ornamental characters in their mate choice [Takahashi et al. 2008].

750 Darwin 1952, pp. 759-761.

Darwin dedicated several pages in his book to the controversy with Wallace on the evolution of color patterns. Darwin recognized that some of Wallace's arguments are justified, but insisted that the males evolved by sexual selection, and not by natural selection as Wallace maintained; while the females remained similar to the primitive progenitor[751]. Darwin suggested that the matter should be decided upon by examining the transmission of the color and pattern between generations: if the male characters are transmitted equally to both sexes of the progeny, it is unlikely that natural selection could fix them in one sex only. Darwin stated that this must be determined by "the laws of heredity" [but he clearly did not know what these laws were]. Sexual selection, if added to natural selection, can easily magnify a character in the male, while in the female it will remain under the control of natural selection[752].

Sexual Selection in Mammals

Darwin noted that male mammals are often bigger and stronger than the females. Males often have horns or large fangs, while the females have smaller ones or lack them entirely. Male horns and fangs have no other uses than in contests and fights with other males: these weapons must have evolved by sexual selection. Farmers and breeders of domestic animals have often testified that males are ready to mate with any available female. Darwin considered that sexual selection led not only to the evolution of male physical traits, but also to the differences in character of males and females in mammals:

> There can hardly be a doubt that these [horns and fangs] serve for fighting with other males, and that they were acquired through sexual selection, and were transmitted to the male sex alone. They were of no use to the female, and consequently they would have tended to be eliminated in the female through natural selection[753]. Male quadrupeds are also more courageous and pugnacious than the females. There can be little doubt that these characters have been gained, partly through sexual selection, owing to a long series of victories, by the stronger and more courageous males over the weaker, and partly through the inherited effects of use[754].
>
> In mammals, we do not at present possess any evidence that the males take pains to display their charms before the females. With mammals the male appears to win the female much more through the law of battle than through display of his charms[755].

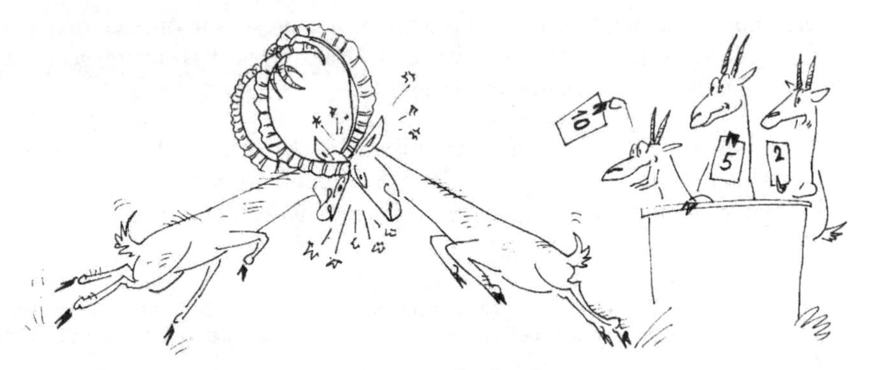

Darwin, Sexual Selection

[751] Darwin 1952, pp. 795-800.
[752] Ibid. pp. 772, 773.
[753] Darwin 1952, p. 820; 1874, p. 524.
[754] Darwin 1952, p. 831; 1874, p. 538.
[755] Darwin 1952, p. 853; 1874, p. 563; Ibid. p. 818; 1874, p. 522.

Darwin had some doubts about the evolution of deer antlers. He considered that the antlers had not evolved as weapons, because they seem to be inefficient for fighting – they tend to become entangled when two males are engaged in combat, and occasionally both contestants die when unable to separate. He thought, rather, that the horns may have evolved by sexual selection as ornaments, but he had no proof for this suggestion[756].

In the "Origin of Species" Darwin suggested that the giraffe's long neck had evolved by natural selection. However, in Sexual Selection, he added that sexual selection – not natural selection – may have been a driver of neck elongation: the males may be using their long necks in male-to-male contests. He reported watching the behavior of giraffes in the London Zoo:

> The giraffe uses his short, hair – covered horns... in a rather curious manner: for, with his long neck, he swings his head to either side, almost upside down, with such force that I have seen a hard plank deeply indented by a single blow[757].

Sexual Selection in Human Evolution

Men differ from women, Darwin wrote, in morphological and sexual characters. As in other mammals, men are more energetic, bolder, and pugnacious than women. Women are more gentle and less selfish than men[758]. Absolute brain size in men is larger than in women, and it is not clear whether this difference is due to the greater body size of men[759]. A comparison of the achievements of men and women, according to Darwin, illustrates the mental superiority of men:

> [Man] has a more inventive genius. It is probable that sexual selection has played a highly important part... The average mental power of man must be above that of woman. The chief distinction in the intellectual power of the two sexes is shewn by man attaining to a higher eminence ...than can woman – whether requiring deep thought, reason or imagination[760].

Darwin was aware of the different social standing of women, preventing them from getting higher education. In order for a woman to get an equal standing with man, she ought to be trained "to energy and perseverance, and to have her reason and imagination exercised to the highest point". Women who will excel in that training, and will marry and have children, may be able to transmit their qualities to their daughters. Only in this way the womenfolk may be improved[761].

Darwin wondered how these mental characters came to be expressed in one sex only – when they are transmitted to both sexes. However, "the law of equal transmission" is advantageous in reducing the difference between the sexes to reasonable levels:

> It is fortunate that the law of equal transmission of characters to both sexes prevails with mammals. Otherwise it is probable that man would have become as superior in mental endowment to woman, as the peacock is in ornamental plumage to the peahen[762].

---

756 Darwin 1952, p. 827; 1874, p. 532. In 2005, Malo et al. reported on research on deer held in parks, where periodical male culling is practiced for population control. The size of male horns was correlated with the quality and quantity of sperm in the male testes.

757 Darwin 1952, p. 825; 1874, p. 530. More than 100 years later, two conflicting reports were published: Simmons and Scheepers (1996) found supporting evidence for Darwin's suggestion, while Mitchell et al. (2009) found no difference between male and female skeletal dimensions and thus no basis for the role of sexual selection in giraffes.

758 Ibid. p. 873; 1874, p. 586.

759 Darwin 1952, p. 867; See also S.J. Gould 1981.

760 Darwin 1952, p. 867; 1874, p. 586; Ibid. p. 873; 1874, p. 587.

761 Darwin 1952, p. 874; 1874, p. 588.

762 Darwin 1952, p. 874.

Darwin suggested that "the law of battle" operated in human evolution in the past as it did in other mammals – although it does not operate now. Stronger men could have more wives and produce more children. There are reports of wars between primitive savages for the purpose of acquiring wives[763]. The social structure of primitive human societies was suitable for the operation of sexual selection: women could influence sexual selection by preferring men with certain characters, such as a beard[764].

Darwin's Conclusions: Darwin was aware of the speculative nature of his theory of sexual selection. But he believed that it should stimulate discussion and thus promote science:

> Many of the views here advanced are highly speculative and some no doubt will prove erroneous... False facts are highly injurious to the progress of science, for they often endure long. But false views, if supported by some evidence, do little harm, for everybody takes a salutary pleasure in proving their falseness. And when this is done, one path towards error is closed, and the road to truth is often at the same time opened[765].

Sexual Selection: Criticism

Darwin's theory of sexual selection was severely criticized. Thirty years after its publication and more than twenty years after Darwin's death, T.H. Morgan fiercely attacked the theory – although he appreciated the compilation of information in Darwin's book:

> Nearly the whole of the animal kingdom is passed in review by Darwin from the point of view of the sexual selection theory. There is brought together a large number of extremely interesting facts, and if the theory did no more than hold them together, it has served in this respect a useful end[766]. This loose way of guessing as to a possible explanation is characteristic of the whole hypothesis of sexual selection. First one, and then another, guess is made as to the causes of the difference between the sexes[767]. As long as the theory of selection, in any of its forms, appears to offer a satisfactory solution, we find the facts used in support of this theory, but as soon a difficulty arises, the Lamarckian theory is brought to the front[768].

The operation of sexual selection necessarily involves a considerable loss of life within a species: for example, if a secondary character like the beard in male humans – which in itself has no effect on fitness – is to evolve by sexual selection, it requires that only, or almost exclusively, bearded men should survive and leave offspring, while many other individuals – who are otherwise fit – would die. This mortality will reduce the competitive ability of the population in the struggle with other populations of the species:

> If, in order to bring about or to maintain the results of this selection, such a tremendous elimination of individuals must take place, it is surprising that natural selection would not counteract this by destroying those species, in which a process so useless for the welfare of the species is going on[769].

---

[763] Note: See Judges: 21. This still happens in Nigeria, Africa in the 21st century.
[764] Darwin 1952, p. 901; 1874, p. 621.
[765] Darwin 1952, pp. 908, 909; 1874, pp. 629, 630.
[766] Morgan 1903, Adaptation and Evolution, p. 175.
[767] Ibid. p. 180.
[768] Ibid. p. 205.
[769] Ibid. p. 217.

Morgan concludes:

> The theory meets with fatal objections at every turn… In the light of the many difficulties that this theory of sexual selection meets with, I think we shall be justified in rejecting it as an explanation of the secondary sexual differences in animals[770].

[770] Ibid. pp. 220-221.

17 Darwin in the Vegetable Kingdom

Although the popular impression of Charles Darwin's work is strongly influenced by his early concentration on the evolution of animals, including Man, he was, in fact, even more successful as a botanist, despite his professed lack of training in that subject[771].

The Botanical Work

Charles Darwin was not the first botanist in the family: his grandfather, Erasmus, had published in 1800 a textbook entitled "Phytologia" on the structure and function of plants[772]. Darwin himself published six volumes (and various short papers) of botanical works. Three of them deal with flowers and pollination – a subject briefly mentioned in the later editions of the Origin. An important volume deals with the effects of self- and cross-fertilization in plants. The book on insectivorous plants, which began as an interesting curiosity, resulted in quite unexpected physiological discoveries. Two other books deal with the power of movement in plants.

These books reveal Darwin in a different light to that created by his "Origin", "Domestication", "Descent of Man", and "Sexual Selection". In the botanical books Darwin emerges as a meticulous experimenter – planning his experiments, taking quantitative measurements, and reporting the results in detail, even when an experiment failed to provide the expected outcome.

Experiments with plants necessarily require time. This gave Darwin respite from the efforts of writing his "Big Book": he could posit a hypothesis, put in motion an experiment to put the hypothesis to test, then do other work while waiting for the results. He had his gardeners to help carry out the experiments, and he had a hothouse constructed for working with tropical plants in the cold English winters. He could continue some of this work even during his frequent spells of illness, when the plants would be placed near the window in his bedroom. His children – in particular his sons George and Francis[773] – gave him a hand with field observations.

Darwin's botanical works paved the way for many later studies. Modern botanical research has confirmed many of Darwin's conclusions (and rejected some others).

On the Evolution of Flowers

In the early 19th century, the role of the anthers and pistils, and the sexual reproduction of plants, were not fully recognized. Erasmus Darwin, in his poem "The love of the plants", suggested that insects acted "like the anthers of the flowers". The poem was criticized and ridiculed by his contemporaries.

---

771 H.G. Baker 1986, in the introduction to Darwin's book on heterostyly.

772 See Chapter 2.

773 Francis Darwin later taught botany at Cambridge. Interestingly, when he edited his father's "Life and Letters" (1887), Francis assembled his botanical correspondence in a special section at the end of Vol. III. Darwin's son George became a professor of mathematics and astronomy in Cambridge.

In the 6th edition of the "Origin of Species", Darwin had this to say about flowers and insects:

> Flowers rank among the most beautiful productions of nature; but they have been rendered conspicuous in contrast with the green leaves, and in consequence at the same time beautiful, so that they may be easily observed by insects. I have come to this conclusion from finding it an invariable rule, that when a flower is fertilized by the wind, it never has a gaily-coloured corolla. Hence we may conclude that, if insects had not been developed on the face of the earth, our plants would not have been decked with beautiful flowers, but would have produced only such poor flowers as we see in our fir, oak, nut, and other trees which are fertilized through the agency of the wind[774].

A.R. Wallace had found in New Zealand relatively few beautifully colored flowering plants, and also a relatively poor insect fauna compared with the Malay Archipelago. He referred to this as proof of Darwin's "beautiful theory" that the colors of flowers had evolved for the purpose of attracting pollinators, and not – as others claimed – for satisfying Man's or God's aspirations for beauty in nature[775].

Darwin suggested a possible scenario for the association of flowers with insects: some plants perhaps exuded a sweet liquid, in order to eradicate certain harmful metabolites. The quantities were small, but insects were fond of the liquid. Next, suppose that this liquid (nectar) was exuded near or inside a flower in certain individual plants. Insects searching for this nectar will become covered with pollen and may transfer it to other plants. These cross-pollinated plants may produce larger offspring, with better chances of reproduction in competition with other plants. Individual plants with larger or more conspicuous nectaries may attract more insects and have an advantage (even if 90% of the pollen is consumed by the visiting insects)[776].

> We certainly owe the beauty and odour of our flowers, and the storage of a large supply of honey, to the existence of insects[777].

The fact that many flowers emit sweet odors indicates that they may be fertilized by nocturnal moths, and that insects are guided to the flowers by odor in addition to color. Darwin hypothesized that the means of attracting insects to the flowers had evolved before the means of avoiding self-fertilization:

> The means for favouring cross-fertilisation must have been acquired before those which prevent self-fertilisation; as it would manifestly be injurious to a plant that its stigma should fail to receive its own pollen, unless it had already become well adapted for receiving pollen from another individual[778].

Darwin admitted that one difficulty still remained: some species in nature regularly self-fertilize – although their flowers seem adapted to accommodate insect pollination. The structure of papilionaceous plants attracted his attention. He wrote to the American botanist Asa Gray:

> Fertilisation often occurs in the closed bud – but I must believe that the structure of the flowers was built – at least partly – for visiting insects. I cannot understand how the insects can avoid transferring pollen from other individuals of the plant[779]... I fully believe that

[774] Darwin 1898, p. 147. Was this Darwin's original idea? See Friedman, 2009.
[775] Wallace 2003 I: 463.
[776] Darwin 1898, p. 72.
[777] Darwin (1877) 1891 (Cross-and Self-Fertilisation), pp. 372-374; p. 382.
[778] Ibid. p. 38.
[779] Darwin to Gray, 1857.

> the structure of all irregular flowers is governed by the relation with insects. Insects are the lords of the floral world[780].

Some self-fertilizing species (e.g., *Ophrys apifera*) occur among the orchids, which Darwin considered as the ultimate case of design for cross-pollination! Some other species have small and inconspicuous flowers, although they carry large quantities of nectar.

> I must believe that plants now bearing small and inconspicuous flowers profit by their still remaining open, so as to be occasionally intercrossed by insects. It has been one of the greatest oversights in my work that I did not experimentise on such flowers, owing to the difficulty of fertilizing them, and to not having seen the importance of the subject[781].

"The Abominable Mystery"

Wind-pollinated trees – cycads and pines – evolved when plants began to colonize the land, before the appearance of insects. Paleontological data on fossil plants confirm that wind-pollinated monocotyledons (cycads, pines, and other groups) dominated the fossil record of the Carboniferous (coal) era. Flowering plants seem to have appeared quite suddenly in the Cretaceous (chalk) era, with no traces of predecessors or intermediate forms. This seemed like "saltation", contrary to Darwin's basic premise of a gradual evolution. In a letter to his botanist friend, Joseph Hooker, Darwin referred to this problem as "the abominable mystery"[782]. It had been suggested that the level of CO_2 in the atmosphere during the Carboniferous was 20 times as high as it was in the present, and that the monocotyledonds – but not the dicotyledoneous plants – could tolerate this level. Darwin and Hooker found no data to support this suggestion. Darwin offered his own solution to the mystery of the sudden appearance of flowering plants:

> Plants of this great division must have been largely developed in some isolated area, whence owing to geographical changes they at last succeeded in escaping and spread quickly over the world[783]. I have fancied that perhaps there was during long ages, a small isolated continent in the S. Hemisphere which served as the birthplace of the higher plants – but this is a wretchedly poor conjecture[784].

In 1877, a French botanist, Gaston de Saporta, suggested that the appearance of insects enabled the proliferation of flowering plants:

> Insects and plants have therefore been simultaneous cause-and-effect through their connection with each other, plants not being able to diversify without insects, and the latter not being able to provide much pollen and nectar [to] feeders so long as the plant kingdom remained poor in arrangement and was composed almost exclusively of entomophilous plants[785].

Darwin enthusiastically endorsed the idea:

> I am surprised that the idea never occurred to me, but this is always the case when one first hears of a new and simple explanation of some mysterious phenomenon[786].

---

[780] Darwin to Gray, 1862.
[781] Ibid. p. 389.
[782] Darwin to Hooker, 1879.
[783] Darwin to Oswald Heer, 8 March 1875.
[784] Darwin to Hooker, 22 July 1879.
[785] Cited in Friedman 2009, p. 17.
[786] Darwin to Saporta, 24 December 1877.

"Cross- and Self-Fertilisation in the Vegetable Kingdom"

In this book, published in 1876, Darwin summarized experimental research carried out over 11 years[787].

Darwin's interest in the effects of crossing started from an unexpected observation. He had two plots of flax (*Linaria*) sown with seeds from the same individual plant. In one plot, the flowers were self-pollinated, while the seeds in the other plot formed in flowers that had been cross-pollinated by hand with pollen from a different plant. Darwin was surprised to observe that the seedlings in the second plot were twice as large and vigorous as the seedlings in the first plot. He wrote to his friend, the American botanist Asa Gray:

> If I could establish this fact… in some fifty cases, with plants of different orders, I think it will be very important, for then we shall positively know why the structure of every flower permits, or favours, or necessitates an occasional cross with a distinct individual[788]. I am experimenting on a very large scale on the difference in power of growth between plants raised from self-fertilized and crossed seeds. And it is no exaggeration to say that the difference in growth and vigour is sometimes truly wonderful. Lyell, Huxley and Hooker have seen some of my plants and been astonished[789].
>
> I therefore began a long series of experiments, continued for about ten years, which I think conclusively show the good effects of crossing two distinct plants of the same variety, and the evil effects of long-continued self-fertilisation[790].

Darwin reports the results of experiments with 57 plant species belonging to 30 families. He endeavored to give the seedlings from crossed and selfed flowers equal starting conditions: he germinated the seeds on wet sand in his room, and planted selfed and crossed seedlings which germinated on the same day on either side of the same pots. He measured and recorded the height of each plant and the number of seed capsules it produced, and the seeds inside them. Detailed tables of the results are given in the book[791]. He states his main conclusion even before presenting the data:

> The most important conclusion at which I have arrived is that the mere act of crossing by itself does no good. The good depends on the individuals which are crossed, differing slightly in constitution, owing to their progenitors having been subjected to slightly different conditions or to what we call in our ignorance, spontaneous variation[792].

Some results were particularly interesting to Darwin. The plant *Mimulus* (Papilionaceae) was allowed to self-fertilize for seven generations, after which Darwin fertilized it with pollen from a new individual "bought in London". The crossed plants were twice as tall and produced many more seed capsules than the selfed plants. The next winter was very severe, and all the self-fertilized plants died; but the crossed plants survived and reproduced. The crossed plants were more cold-hardy than the selfed ones:

787 F. Darwin 1887, III: 290.
788 Darwin to Gray, 1866. F. Darwin 1887, III: 291.
789 Ibid. Darwin to Bentham.
790 Darwin, C., Domestication II: 108.
791 Darwin, "Cross- and Self-Fertilisation" 1876, p. 356ff.
792 Ibid. p. 27.

> Better evidence could hardly be expected of the potent influence of a cross with a fresh stock, on plants which had been self-fertilized for eight generations ...in comparison with plants self-fertilized for nine generations continuously[793].

Papilionaceous plants, especially common garden varieties like peas and beans, presented a problem: although the flower structure suggested cross-fertilization by insects, the growers reported that five different varieties could be grown side by side and harvested without fear for the purity of the seed.

> We may conclude that the varieties of sweet pea rarely or never intercross in this country. And this is a remarkable fact, considering firstly the general structure of the flowers, secondly the large quantity of pollen produced – far more than is requisite for self-fertilisation, and thirdly the occasional visits of insects[794].

The structure of the flowers of the cultivated Papilionidae, such as peas, beans, and sweet peas, seemed to accommodate fertilization by insects, but Darwin confirmed that these flowers are self-pollinated, although their structures seem adapted for insect visits, and they produce copious amounts of pollen[795]. He suggested that perhaps self-fertilization had been selected for by humans during domestication of these varieties.

Detailed experiments with the flowers of *Viola tricolor* yielded impressive results. The structure of these ornamental flowers is such that self-fertilization is very unlikely, and only pollen transfer by insects (mainly bees) produces seeds. Two plots were sown side by side with selfed and crossed seeds in 1869, and a year later the crossed plants were twice as tall and had produced ten times as many seed capsules as the selfed plants.

> The crossed plants ...had grown and spread so much more than the self – fertilized, that any comparison between them was superfluous... The extraordinary superiority of the crossed plants over the self-fertilized...was no doubt due to the crossed plants at first having had a decided advantage over the self-fertilized, and then robbing them more and more of their food during the succeeding seasons. But we should remember that the same result would follow in a state of nature even to a greater degree[796].
>
> The ensuing winter was very severe, and in the following spring... all the self-fertilised were now dead... On the other hand, all the crossed plants without exception were growing vigorously[797].

Not all experiments resulted in an advantage for the crossed plants, and Darwin reported those failures as well as the successes. The initial experiments with Dianthus were not convincing due to the small sample sizes[798]; but after three generations the plants were fertilized with pollen from a distinct plant brought from London.

> We thus see how greatly the offspring from the self-fertilized plants of the third generation, crossed by a fresh stock, had their fertility increased, whether tested by the number of capsules produced or by the weight of the contained seeds – this latter being the more trustworthy method[799].

---

793 Ibid. p. 261.
794 Ibid. p. 155.
795 Ibid. p. 155.
796 Ibid. p. 126.
797 Ibid. p. 127.
798 E.g. Ibid. p. 134.
799 Ibid. p. 139.

Darwin provides detailed tables with the data from experiments with plants belonging to 57 species in 30 families. He adds some general conclusions, the principal one being that the advantage of crossed plants depends on "constitutional" differences between the parent plants. The relative disadvantage of self-fertilized plants is a result of the parents lacking such constitutional differences, not from the weakness of the experimental plants[800].

> There is the clearest evidence... that the advantage of a cross depends wholly on the plants differing somewhat in constitution. And that the disadvantage of self-fertilisation depends on the two parents, which are combined in the same hermaphrodite flower, having a closely similar constitution.
>
> There can hardly be a doubt that the differences of all kinds between the individuals and the varieties of the same species depend largely, and as I believe exclusively, on their progenitors having been subjected to different conditions, though the conditions to which individuals of the same species are exposed in a state of nature often falsely appear to us the same[801].

Darwin recognized that cases occur in nature which do not fit this general pattern. He described these cases in his books on heterostyly and on orchids. At the end of the book, Darwin surveys the literature and lists plants which are particularly dependent on insect pollination, and others which do not need insects and produce seeds by self-fertilization[802]. He summarizes these cases here:

> We may therefore conclude from the facts now given, that varieties sometimes arise which when self-fertilised possess an increased power of producing seeds and of growing to a greater height, than the intercrossed or self-fertilised plants of the corresponding generation... The appearance of such varieties is interesting, as it bears on the existence under nature of plants which regularly fertilise themselves, such as *Ophrys apifera* and other orchids... Some observations made on other plants lead me to suspect that self-fertilisation is in some respects beneficial; although the benefit thus derived is as a rule very small compared with that of a cross with a distinct plant[803].

In large trees bearing plenty of flowers, pollinating insects move from one flower to the next, and there is no doubt that some – or the majority – of the flowers receive pollen from their neighbors, which possess the same "constitution". But Darwin suggests the possibility that most of these flowers are not fertilized and drop without forming seed, and the only flowers which fruit and carry seed are those few that have received pollen from a different individual tree.

> When we behold our orchard-tree covered with a white sheet of bloom in the spring, we should not falsely accuse Nature of wasteful expedience, though comparatively little fruit is produced in the autumn[804].

On the Evolution of Cross-fertilization: Darwin suggested that wind-pollination had preceded insect-pollination of plants. The enormous quantities of pollen produced by wind-pollinated trees[805] are necessary because the trees are dioecious and the distances between individuals are wide. Wind-pollination has all the advantages of cross-pollination, except the need for huge quantities of

---

[800] Ibid. p. 448. "Constitution", in Darwin's usage, stands for some unknown heritable property.
[801] Ibid. p. 254.
[802] Ibid. p. 356ff.
[803] Ibid. p. 350.
[804] Ibid. p. 401.
[805] Ibid. pp. 407-410.

pollen. When insects came on the scene, they may have first fed on the pollen, but in doing so also transferred it from tree to tree.

The change from dioecious to hermaphrodite plants may have occurred because it reduced the chance of individuals not receiving pollen and not producing seeds.

> Why the descendants of plants which were originally dioecious, and which therefore profited by always intercrossing with another individual, should have been converted into hermaphrodites, may perhaps be explained by the risk which they ran, especially as long as they were anemophilous, of not being always fertilised, and consequently of not leaving offspring. This latter evil, the greatest of all to any organism, would have been much lessened by their becoming hermaphrodites, though with the contingent disadvantage of frequent self-fertilisation[806].

In some cases crossed plants seemed to have no advantage. Some plants produce small and inconspicuous flowers – even subterranean ones – that do not attract insects. Darwin was certain that even these plants benefit from an occasional cross. But, he wrote, he regrets not having paid more attention to these plants in his research[807].

His tables[808] illustrate that crossing confers an advantage, sometimes an extraordinary one. The great height, weight and fertility of the crossed seedlings may be attributed to their possessing greater "constitutional vigour", inherited from their parents.

> There can hardly be a doubt that the differences of all kinds between the individuals and varieties of the same species depend largely, and I believe exclusively, on their progenitors having been subjected to different conditions, though the conditions to which individuals of the same species are exposed in a state of nature often appear to us the same[809].

"The Various Contrivances by Which Orchids are Fertilised"

Orchid flowers fascinated Darwin. A few wild species grew near his home. In 1860 he first noticed that the pollen in these flowers was not scattered in the anthers, as it was in other kinds of flowers, but is sticky and condensed in a pair of pollinia. He wrote to his friend, the botanist Joseph Hooker:

> The sticky glands of the pollinia are congenitally fused to a saddle-like organ… which attaches to any thorn or proboscis in an admirable way. I have never seen anything so beautiful… I have lately observed the orchid, and I declare that I think the adaptations of every part of the flower are clear and beautiful no less than in the woodpecker, and perhaps even more so[810].

The book on Orchids was the first of Darwin's botanical books (an early edition was published in 1862; the final form was delayed until 1877) but the adaptation of the flowers for cross-fertilization was already clear to him:

> The object of the following work is to show that the contrivances by which orchids are fertilised are as varied and almost as perfect as any of the most beautiful adaptations in the

[806] Ibid. p. 412.
[807] Ibid. p. 389.
[808] Ibid. p. 238ff.
[809] Ibid. p. 254. Note: the dramatic advantage of crossed plants that Darwin described became later recognized as hybrid vigor (heterosis). It is a short-term advantage, which breaks down in the second and later generations.
[810] Darwin to Hooker, 1960; F. Darwin 1887, III: 263.

> animal kingdom; and secondly, to show that these contrivances have for their main object the fertilisation of the flowers with pollen brought by insects from a distinct plant[811].

The adaptations of the flowers attracted Darwin. A visiting insect may transfer one or both pollinia – therefore every flower can fertilize, at most, two other flowers. This restriction seemed to be the reason for the evolution of a strong dependence of orchids on insect pollinators:

> When we notice how abundant is pollen in general, it seems strange that a flower which can at most fertilise two others, can survive! I consider in this fact the explanation for improving the contrivances by which the pollen – being so important from being scarce – is transferred from flower to flower[812].

Darwin examined in detail the anatomical structure of many species of orchids. Some he obtained from Kew Gardens and grew in his hothouse, others were sent to him from different parts of the world. He described the structure of the flowers' reproductive parts, illustrated in many woodcuts with the petals and sepals cut away[813]. Overviewing the entire study, Darwin was excited by the variation in flower structure that he observed.

> The flowers of the orchids, in their strange and endless variety of shape, may be compared with the great vertebrate class of fish, or still more appropriately with tropical homopterous insects, which appear to us as if they had been modeled in the wildest caprice, but this no doubt is due to our ignorance of their requirements and conditions of life[814]. What an amount of modification, cohesion, abortion and change of function we have here!

Despite the wide variation among species in size, form, and structure of the reproductive parts, he noted a unifying principle – the adaptation of the flower structure for cross-pollination by insects. The orchid flowers were for Darwin an extreme example of adaptations for avoiding self-fertilization. Darwin wrote to a German botanist:

> I would be surprised if you do not reach the same conclusion as I have, considering so many beautiful contrivances [in the orchid flowers] that all plants must, for an unknown reason, be from time to time fertilised by pollen from a different plant[815].

Darwin reasoned that the differences in flower shape and structure among species arose because each species is adapted to a different pollinator, although in many cases the pollinators were unknown. He suggested that flowers with a long nectary must be fertilized by moths with a long proboscis, while flowers with an open corolla should be fertilized by bees and flies. Aided by his children, Darwin spent many hours watching pollinators of wild species of orchids near his home. He listed the pollinators and noted the behavior of the insects on the flower. He confirmed that the visiting insect, in trying to reach the nectar, detached one or both pollinia from a flower (and interpreted the absence of pollinia as evidence that the flower had been previously visited). He removed some pollinia by inserting a needle into the flower (imitating a proboscis) and described changes in the pollinia which, when removed from the flower, bent forward "to meet the stigma":

---

811 Darwin 1984, p. 1. The 1984 edition is a facsimile of Darwin's 1877 book.

812 Darwin 1861, in F. Darwin 1887, III: 265.

813 NOTE: Darwin rarely illustrated the entire flower (let alone the beautiful colors). Unless the reader is familiar with the species in question, it is difficult to guess what the flower looked like. Darwin must have assumed that the book would be read by gardeners or orchid fans like himself.

814 Darwin 1984, pp. 224-5.

815 Darwin to Hildebrandt, 1866, in F. Darwin 1887, III: 281.

> A poet might imagine that whilst the pollinia, adhering to an insect body, were borne through the air from flower to flower, they voluntarily and eagerly placed themselves in that exact position in which alone they could hope to gain their wish and perpetuate their race[816].

To complete the detailed description, Darwin made sections in the flowers and examined them under his microscope [following the methods in earlier work by Robert Brown][817]. The anatomical study of the finer structure convinced Darwin of the homology of the flower parts and led to his conclusion that all the species must have had a common origin:

> We know that there are [in all flowers] fifteen bundles of vessels, arranged in alternating groups of three-within-three, which must have existed from very early times... Is it not a simple and conceivable view, that all the common characteristics of orchids stem from some monocotyledonous plant, which like many plants of this class had fifteen flower parts, arranged in five alternating groups of three-within-three, and that this wonderful structure of the flower is the result of gradual adaptations, which were useful to the plant during the unceasing changes in the organic and inorganic world[818]?

Darwin wondered why most of the native English orchid species were widely scattered geographically. He devised a method for estimating the number of seeds produced per plant and the results surprised him – 6,000 seeds per capsule, 24,000 per plant.

> The final end of the whole flower, with all its parts, is the production of seeds. And these are produced by orchids in great profusion... at the same rate of increase, the great-grand-children of a single plant would nearly clothe with one uniform green carpet the entire surface of the land throughout the globe... What checks the unlimited multiplication of the Orchidae throughout the world is not known[819].

Some particular species of orchids required Darwin's special attention. The flowers of some common English orchids did not justify the general rule that they should be cross-fertilized. In particular, *Ophrys apifera* – morphologically similar to a bee (hence its name) – produced seeds even when protected from visiting insects by a net. It was clearly self-fertilized: with any light wind, pollen would fall on the stigma of the same flower. Darwin reasoned that originally the plant had been insect-pollinated, and selfing was adopted in cases in which insects were scarce:

> Are we to believe that these adaptations for cross-fertilisation in the Bee Ophrys are absolutely purposeless?... it is however, just possible that insects, although they have never been seen to visit the flowers, may at rare intervals transport the pollinia from plant to plant... The whole case is perplexing in an unparalleled degree, for we have here in the same flower elaborate contrivances for directly opposed objects[820].

The case of the orchid *Angraecum sesquipetale* excited Darwin's imagination. This flower, sent to him from Madagascar, had a nectary 11.5 inches (~30 cm) long – but only the terminal 2 inches contained nectar. Darwin imagined how the species had evolved:

---

816 Darwin 1984, p. 79.

817 Darwin 1984, pp. 232-236.

818 Ibid. p. 246.

819 Ibid. pp. 277-279. Darwin did not know that in order to develop the tiny orchid seeds require obligatory association with a specific fungus which occurs in the soil. The fungus supplies the germinating seed with nitrogen and carbon, since the seed lacks storage tissues.

820 Ibid. p. 57.

> It is surprising that any insect should be able to reach the nectar... But in Madagascar there must be moths with proboscides capable of extension to a length between ten and eleven inches. This belief of mine had been ridiculed by entomologists[821]... Thus it would appear that there had been a race in gaining length between the nectary of *Angraecum* and the proboscis of certain moths. But the *Angraecum* had triumphed, for it flourishes and abounds in the forests of Madagascar, and still troubles each moth to insert its proboscis as deeply as possible in order to drain the last drop of nectar[822].

Despite the general similarity in structure, every species he studied presented Darwin with special adaptations. Some could not be understood because the specific pollinator was unknown. Darwin summarizes his studies by recapitulating the main conclusions:

> It has, I think, been shown that the Orchidae exhibit an almost endless diversity of beautiful adaptations... the regular course of events seems to be, that a part which originally served for one purpose, becomes adapted by slow changes for widely different purposes[823]... That cross-fertilisation, to the complete exclusion of self-fertilisation, is the rule with the Orchidae cannot be doubted from the facts already given in relation to many species in all the tribes throughout the world[824].
>
> It apparently demonstrates that there must be something injurious in the process [of self-fertilisation] of which fact I have elsewhere given direct proof. **It is hardly an exaggeration to say that Nature tells us, in the most emphatic manner, that she abhors perpetual self-fertilisation[825].**

"On the Different Forms of Flowers on Plants of the Same Species"

Although it was first published in 1877, almost all the conclusions that Darwin drew from the data he collected are still valid more than one hundred years after their promulgation[826].

In some familiar plants – in particular *Primula varis* [cowslip] and *Linum* – the forms of the flowers vary among individual plants of the same species: in some individuals the flowers have anthers longer than the style (referred to as "male" flowers) and in other individuals the anthers are shorter than the style ("female"). This phenomenon is termed *heterostyly.*

> The two or three forms, though all are hermaphrodites, are related to one another like the males and females of ordinary unisexual animals[827]... The two kinds of flowers are never found on the same individual plant[828].

Darwin thought that this represented a transition in plant evolution, between hermaphrodite (the usual situation, in which a flower has both female and male organs) and unisexual plants (male and female flowers on different plants)[829]. To test his hypothesis, Darwin measured the pollen grains of the two forms and counted the number of seeds they produced.

---

[821] Ibid. p. 163.

[822] Ibid. p. 166. The moth was in fact discovered three years after Darwin's death. It was named *Xanthopan morgani predicta* (referring to Darwin's prediction that it should exist).

[823] Ibid. p. 282.

[824] Ibid. p. 290.

[825] Darwin 1984, p. 293

[826] H.G. Baker, 1986, in the "Foreword" to "the different forms of flowers"

[827] Darwin 1986 (1877), p. 2; Ibid. p. 29

[828] Ibid. p. 18

[829] Darwin later abandoned this idea [Darwin 1986, p. 257]

> I cannot help suspecting [that] the cowslip is in fact dioecious… it would be a fine case of gradation between a hermaphrodite and unisexual condition[830]… I have marked a lot of these plants, and expected to find the so-called male plants [short-styled, long-anthered] barren. But judging from the feel of the capsules, this is not the case and I am very surprised at the difference in the size of the pollen. If I should prove that the so-called male plants produce less seed than the so-called females [long-styled], what a beautiful case of gradation from hermaphrodite to unisexual condition it will be! If they produce about equal numbers of seeds, how perplexing it will be[831].

In a field study, Darwin collected 522 flowers of Primula – each from a distinct individual plant – and concluded that the two forms were present in the wild population in equal proportions, and there were no intermediates. The flowers of both forms secreted plenty of nectar and did not set seed unless visited by insects – net-covered plants did not yield any capsules[832].

> The other day I had time to weigh the seeds, and by Jove, the plants of primroses and cowslip with short pistils and large pollen grains [male plants] are rather more fertile than those with long pistils and small-grained pollen [female]. I find that [these flower forms] require the action of insects to set them [seed], and I never shall believe that these differences are without meaning[833].

Darwin transferred pollen between distinct plants of the two forms of the cowslip in all four possible combinations – and the results surprised him:

> The fertilisation of either form with pollen from the other form could conveniently be called a legitimate union… and that of either form with its own-form pollen as illegitimate union… The superiority of a legitimate over an illegitimate union admits of not the least doubt[834]. The benefit which heterostyled dimorphic plants derive from the existence of the two forms is sufficiently obvious, namely, the intercrossing of distinct plants is thus ensured[835].

In 1861, Darwin examined another distylous plant – flax (*Linum grandiflorum*). The flowers of the two forms are very similar apart from the relative lengths of the anthers and pistils. Darwin self- and cross-pollinated the flowers by hand, and was surprised to find that the self-pollinated flowers produced no seed at all:

> The same two long-styled plants produced, in the course of the summer, a vast number of flowers, and the stigmas were covered with their own pollen, but they all proved absolutely sterile[836]. We have the clearest evidence that the stigmas of each form require for full fertility that pollen from the stamens of corresponding height belonging to the opposite form should be brought to them[837].

He then tried putting self and foreign pollen simultaneously onto the stigmas of the same flower of *Linum*. His enthusiasm at the results is expressed in a letter to Asa Gray:

830 Darwin to Hooker, May 1860.
831 Darwin to Gray, June 1860.
832 Darwin 1986, p. 18. Ibid. p. 22.
833 Darwin to Hooker, December 1860 in F. Darwin 1887, III: 299.
834 Darwin 1986, p. 24, Ibid. p. 28.
835 Ibid. p. 30.
836 Ibid. p. 83.
837 Ibid. p. 92.

> I have lately been putting the pollen of the two forms on the division of the stigma of the same flower. And it strikes me as truly wonderful, that the stigma distinguishes the pollen; and is penetrated by the tubes of one and not those of the other[838].

This was the beginning of a detailed investigation, carried out intermittently for some 15 years. Darwin ascertained in many experiments that cross-pollination between forms ("legitimate" crosses) resulted in greater fertility than pollen transfer between flowers of the same form ("illegitimate"), or in the latter being entirely sterile.

A much more complex situation was discovered in *Oxalis* and *Lythrum*: three distinct forms differing in the relative lengths of the anthers and pistils. There were 18 different mating combinations possible. After describing at length the morphology of the three forms, Darwin tried to carry out all 18 mating combinations, pollinating the plants by hand. The results are reported in a series of tables[839].

> I am almost struck mad over *Lythrum*. If I can prove what I fully believe, it is a grand case of trimorphism, with three different pollens and three stigmas. I have castrated and fertilised almost ninety flowers, trying all the eighteen distinct crosses which are possible within the limits of this one species[840].

Darwin was convinced that heterostyly had evolved because cross-fertilization was advantageous to the plants.

> These heterostyled plants are adapted for reciprocal fertilisation. So that the two or three forms, though they are all hermaphrodites, are related to one another almost like the males and females of ordinary unisexual animals[841]; We may feel sure that plants have been rendered heterostyled to ensure cross-fertilisation, for we now know that a cross between distinct individuals of the same species is highly important for the vigour and fertility of the offspring[842].

But heterostyly posed a difficult question: in a distylous plant, one half of the possible pollinations are illegitimate and thus infertile. In a tristylous plant, two thirds (12 of 18) are illegitimate and thus sterile or infertile. How could a situation involving such frequent sterility have evolved by natural selection? Darwin did not have a convincing answer:

> Although it may be beneficial to an individual plant to be sterile with its own pollen – cross-fertilisation being thus ensured – how can it be of any advantage to be sterile with half its brethren, that is, with all individuals belonging to the same form? It is a more probable view that the male and female organs in two sets of individuals have been specially adapted for reciprocal action; and that the sterility between the individuals of the same form is an incidental and purposeless result[843].

Darwin returned to the subject of the evolution of self-sterility in his 1876 book on "Cross and Self-Fertilisation".

---

838 Darwin to Gray, July 1862.
839 Darwin 1986, pp. 151-157.
840 Darwin to Gray, August 9 1862.
841 Darwin 1986, p. 2.
842 Ibid. p. 258.
843 Ibid. p. 265.

> We are not therefore justified in admitting that this peculiar state of the reproductive system has been gradually acquired by natural selection; but we must look at it as an incidental result, dependent on the conditions to which the plants have been subjected... I do not, however, wish to maintain that self – sterility may not sometimes be of service to a plant in preventing self-fertilisation; but there are so many other means by which this result might be prevented or rendered difficult... that self-sterility seems as almost superfluous acquirement for this purpose[844].

This explanation was severely criticized by T.H. Morgan (1903) (see Chapter 25).

Insectivorous Plants

Darwin's interest in insectivorous plants began as a chance observation. Vacationing with the family in a sea resort in 1860, Darwin noticed many insects stuck to the leaves of the sundew, *Drosera rotundifolia*, which grew wild in the area. He observed that the surface of the leaves abounded with hairs with sticky glands. This observation led to a long series of experiments, which were finally published as a book in 1875.

Many experiments indicated that the application of particles of materials not containing nitrogen had no effect on the sticky hairs[845]; but when a suitable nitrogenous particle was placed on a gland with a thin brush, all the hairs bent in that direction. A dead insect, or better still a live one, triggered a faster response[846]. Darwin performed meticulous quantitative experiments, trying to understand what caused the hairs to move towards a captured prey.

In July, 1860, Darwin wrote to Hooker that he enjoyed observing these plants[847]. In September of that year, he wrote to Asa Gray:

> I have been infinitely amusing myself by working at *Drosera*. The movements are really curious. And the manner in which the leaves detect certain nitrogeneous compounds is marvellous. You will laugh, but it is at present my full belief (after endless experiments) that they [the glands] detect, and move in consequence of, the 1/2880th part of a single grain[848] of nitrate of ammonia. But the muriate or sulphate of ammonia bother their chemical skills and they cannot make anything of the nitrogen in these salts[849].

Two months later he wrote to Hooker:

> I have been working like a madman at *Drosera*. Here is a fact for you which is certain as you stand where you are, though you won't believe it, that a bit of hair 1/78000 of one grain in weight placed on a gland, will cause one of the gland-bearing hairs of *Drosera* to curve inwards[850].

Hundreds of tests with litmus paper proved that the glands secrete an acid liquid, including a "ferment" which dissolves the nitrogenous material[851]. Darwin considered that these glands not

---

844 Darwin 1986, p. 346.
845 Darwin 1875, p. 35.
846 Ibid. p. 15.
847 Ibid. p. 1.
848 NOTE. 1 grain-the standard weight of a single grain of wheat, ~ 65 mg. Darwin did not possess such small weights to use on his balance. He had to improvise. He weighed accurately a piece of yarn 1 yard long; then cut the yarn into pieces of known length. He estimated the weight of the pieces from their measured length. These he used as weights. 1/2880 of a grain [65/2880] is approximately 0.022 mg.
849 Darwin to Gray, in F. Darwin 1887, III: 318.
850 Ibid. III: 319. The surprisingly small weight is approximately 0.00083 mg.
851 Darwin 1875, p. 132.

only trap the insects, but also digest and absorb part of them. The similarity of the activity of the glands to the digestive process in animals impressed him:

> Last night I put a row of little flies near one edge of two youngish leaves, and after 14 hours these edges are beautifully folded over, so as to clasp the flies, thus bringing the glands into contact with the upper surfaces of the flies. And they are now secreting copiously above and below the flies… no doubt the glands are absorbing the delicious soup[852].
>
> A plant of *Drosera*, with the edges of its leaves curled inwards, so as to form a temporary stomach, with the glands of the closely inflected tentacles pouring forth their acid secretion, which dissolves animal matter, afterwards to be absorbed, may be said to feed like an animal[853]. That a plant and an animal should pour forth the same, or nearly the same, complex secretion, adapted for the same purpose of digestion, is a new and wonderful fact in physiology[854].

After his 1960 burst of enthusiasm, Darwin's work on *Drosera* was suspended for ten years[855]. Before publishing his notes, he double-checked his previous data:

> When I read over my notes in 1873, I entirely disbelieved them, and determined to make another set of experiments with scrupulous care… Notwithstanding the care taken and the number of trials made [73 leaves], when in the following year I looked merely at the results… I again thought that there must have been some error, and thirty-five fresh trials were made with the weakest solution… Hence, after the most anxious consideration, I can entertain no doubt of the substantial accuracy of my results[856].

In the renewed study, Darwin attempted to discover how the stimulation of even a single glandular hair by a tiny cube of meat or a minute insect, was transmitted to the entire leaf and caused all the glands to bend in the same direction – although the plant did not have a nervous system. Darwin first thought that the stimulus was transmitted along the leaf veins. But this idea was proved wrong: the motor impulses radiated in all directions, and whichever side of the tentacle was first stimulated, that side contracted and the tentacle consequently bent towards the site of excitement. Darwin could find no alternative explanation and the phenomenon remained unsolved[857].

> Perhaps sensing that he could make no further progress with this plant, Darwin felt the need to expand his study of insectivorous plants to other species. He started with an American plant of the same family – the Venus fly trap, *Dionaea*. From the information published by American observers on the trapping mechanism of the leaves of this plant, he learned that when an insect lands on the leaf, and accidently touches three filaments positioned perpendicular to the surface of the leaf, the leaf instantly folds along its middle vein, and the insect is prevented from escaping by the strong bristles that line the edge of the leaf.

Although the trapping mechanism in *Dionaea* differs from that of *Drosera*, Darwin suggested – and proved – that what happens once an insect is trapped is physiologically the same in both species. The surface of the *Dionaea* leaf is covered with glands which secrete an acid fluid that digests nitrogenous compounds. If the leaf snapped close over some inorganic material it reopened after

---

[852] Darwin to W.T. Dyer, 1873 in F. Darwin 1887, III: 324.
[853] Ibid. p. 18
[854] Darwin 1875. p. 135.
[855] F. Darwin 1887, p. 322.
[856] Darwin 1875, p. 154.
[857] Ibid. p. 246.

less than 24 hours; but if it captured a live insect it remained closed for a long time – up to several days. The entire leaf then secreted an acid liquid which digested the prey[858].

> The facts just given plainly show that the glands have the power of absorption. For otherwise it is impossible that the leaves should be so differently affected by non-nitrogeneous and nitrogeneous bodies[859]. How living insects, when naturally caught, excite the glands to secrete so quickly as they do, I do not know, but I suppose that the great pressure to which [the insects] are subjected forces a little excretion from either extremity of their bodies[860]. A large crushed fly (*Tipula*) was placed on a leaf, from which a small portion at the base of one lobe has previously been cut away so that an opening was left, and through this the secretion continued [to drop] for nine days[861].

> When a leaf closes over any object, it may be said to form itself into a temporary stomach... and the glands on its surface pour forth their acid secretion, which acts like the gastric juices of animals[862].

The importance of the bristles that line the edge of the leaf only became clear to Darwin late in his study. He observed that small insects often escaped from the closed trap. At his request, an American correspondent sent him 14 leaves collected in nature, which had remained closed and enclosed the insect that had triggered the closure. Ten of the leaves contained large insects, such as beetles[863].

> ...marginal spikes, which form so conspicuous a feature in the appearance of the plant... at first seemed to me in my ignorance useless appendages[864]; It would manifestly be a major disadvantage to the plant to waste many days in remaining clasped over a small insect... far better to wait for a time until a moderately large insect [is trapped], allowing the little ones to escape. And this advantage is secured by the slowly intercrossing marginal spikes[865].

Darwin studied many other species of insectivorous plants, sent to him from tropical countries. Different trapping mechanisms were used by the various species. Darwin offers the following general consideration for the evolution of insectivorous plants:

> As it cannot be doubted that this process [digestion of animal matter] would be of high importance to plants growing in very poor soil, it would tend to be perfected through natural selection. Therefore any ordinary plant having viscid glands, which occasionally caught insects, might thus be converted, under favorable conditions, into a species capable of true digestion[866].

Climbing Plants

> It has often been vaguely asserted that plants are distinguished from animals by not having the power of movement. It should rather be said that plants acquire and display this power when it is of some advantage to them[867].

[858] Ibid. p. 309.
[859] Ibid. p. 299.
[860] Ibid. p. 300.
[861] Ibid. p. 296.
[862] Ibid. p. 301.
[863] Ibid. p. 312.
[864] Ibid. p. 366.
[865] Ibid. p. 362.
[866] Ibid.
[867] Darwin, "Climbing Plants", 1875, p. 206.

Climbing plants probably attracted Darwin's attention during his time in the tropical forests of South America. Plants climb to reach the light, investing as few resources as they can in doing so. "Having thin stems, which cannot support a heavy load, they use other plants as support for the purpose"[868].

Darwin grouped the climbing plants into three main categories (a fourth category was "others"): 1) twining plants; their entire stem wraps around the support; 2) leaf climbers; their leaves are sensitive to touch, and catch and hold on to the support; 3) tendril climbers; the tendrils are modified leaves or flower stalks, which catch and hold on to any support. Darwin experimented with many plant species of each category – some wild, others ornamental or cultivated varieties, always trying to measure the movement quantitatively and understand the mechanism.

Darwin often had to invent the methods required for his study, and showed great ingenuity. In order to trace on a two-dimensional sheet of paper the movement in three dimensions of a plant part – he placed two mirrors vertically, one on either side of the investigated plant, and a third mirror horizontally above them. He glued a thin fiber with a small blob of black wax to the tip of the moving stem, and placed a white card with a fixed black dot behind the plant. Every hour or so the position of the moving plant was recorded on the paper, when the two black dots – one fixed, the other mobile – lay on the same line with the observer's eye. To measure the strength of the stimulus which was causing a tendril to bend, he accurately measured a length of yarn from a source of known weight, cut it into measured lengths made into loops, and calculated the weight of each loop which he hung on the tendril.

Twining Plants: Darwin conducted many experiments with twining plants, recording the time and tracing the length of the movement[869]. The experiments required continuous observations, day and night. When he was sick – as he frequently was – the plants were placed in his bedroom. As a standard twining plant Darwin used hops (*Humulus lupatus*), which twines around long poles in commercial plantations for the beer industry. He noticed that only the top two internodes of the stem actually move, independently of the lower parts of the stem. This spontaneous movement presents a circular pattern ("circumnutation" is the term coined by Darwin to describe it). Darwin concluded that twining is a combined result of circumnutation and the growth of the stem, and does not require a special sensitivity of the stem to touch.

> I allowed the top [of a twining plant] to grow out almost horizontally to the length of 31 inches… the whole revolved in a course opposed to the sun… at rates between 5 hrs 15 min, and 6 hrs 45 min for each revolution. It was an interesting spectacle to watch the long shoot sweeping this grand circle, day and night, in search of some object to twine[870]. The revolutions carried on day and night, a wider and wider circle being swept as the shoot increases in length. This movement likewise explains how the plants twine; for when a revolving shoot meets with a support, its motion is necessarily arrested at the point of contact, but the free projecting part goes on revolving… and thus the shoot winds around the support… If a man swings a rope around his head and the end hits a stick, it will coil around the stick according to the direction of the swinging movement. So it is with a twining plant[871].
>
> I conclude that twining stems are not irritable. And indeed it is not probable that they should be so, as nature always economizes her means, and irritability would have been superfluous.

[868] Ibid. p. 193.
[869] Ibid. Table pp. 22-26.
[870] Ibid. p. 6.
[871] Ibid. p. 15.

Leaf-climbing Plants: Darwin studied leaf-climbing plants belonging to eight families[872]. His standard experimental plants were *Clematis* and *Tropeolum*.

The shoots of plants in this category "circumnutated", but, unlike twining plants, the leaves of these plants were sensitive to touch. A very slight stimulus – a few milligrams of thread, sometimes even a brief finger touch – were sufficient for the leaf to bend in the direction of the irritation.

In *Clematis*, the climbing tool was the leaf petiole. When a petiole touched a support, it bent towards it and wrapped itself around it in less than 24 hours – then became rigid[873]. In *Tropeolum*, very young leaves, resembling thin filaments, responded to slight irritation or to touch within three minutes. In older plants these leaves disappeared and the petioles of the leaves responded instead.

In *Solanum jasminoides*, kept in the hothouse, a stick was clasped by the petioles within 7 hours. Darwin examined the anatomical structure of the petiole under the microscope:

> The flexible petiole... which has clasped an object for three or four days increases much in thickness, and after several weeks becomes so wonderfully rigid that it can hardly be removed from its support. It is a singular morphological fact that the petiole should thus acquire a structure almost identical with that of the axis; and it is a still more singular physiological fact that so great a change should have been induced by the mere act of clasping a support[874].

Tendril-Climbing Plants: Plants in this category were the most interesting to Darwin. He examined tendril-climbing plants belonging to ten families – in particular Papilionaceae, Cucurbitaceae, and Vitaceae[875].

> A clever gardener, my neighbor, who saw the plant on my table last night, said: I believe, Sir, that the tendrils can see; for wherever I put a plant it finds out any stick near enough[876].

The American botanist Asa Gray, with whom Darwin corresponded frequently, described the movement of tendrils in some plants but did not notice that the entire stem circumnutated. Darwin obtained the American plant, *Echinocystis lobata,* and repeated Gray's observations.

> Having the plant in my study I have been surprised to find that the uppermost part of each branch (excluding the growing tip) is constantly and slowly twisting around, making a complete circle in from one-and-a-half to two hours... The movement goes on all day and all early night; It has no relation to light, for the plant stands in my window and twists from the light just as quickly as towards it[877].

A species of Bignonia was observed in detail. The tendril had three branches, like a bird's foot. When it touched a support, the tips of the tendril became adhesive discs, which hardened, and the entire tendril contracted to add support to the shoot. The tips of the tendril were observed to turn away from the light. Darwin studied this phenomenon in a controlled experiment:

> The whole terminal portion exhibits a singular habit, which in an animal would be called an instinct: for it continually searches for any little crevice or hole into which it inserts itself[878]. I placed the pot in a box open only on one side, and obliquely facing the light. In two days all six tendrils pointed with unerring truth to the darkest corner of the box,

---

[872] Ibid. pp. 45-81.
[873] Ibid. pp. 50-51.
[874] Ibid. pp. 72, 78.
[875] Ibid. p.84ff.
[876] Darwin to Hooker, 1863 in F. Darwin 1887, III: 312.
[877] Ibid.
[878] Darwin 1875. p. 95.

> though to do so each had to bend in a different manner. Six wind-vanes could not have more truly shown the direction of the wind, than did the branched tendrils the course of the stream of light which entered the box[879].

After a further set of experiments, Darwin was even more excited about the results:

> How singular a fact it is, that a leaf is metamorphosed into a branched organ which turns from the light, and that can by its extremities either crawl like roots into crevices, or seize hold of minute projecting items[880]. The perfect manner in which the branches [of the tendril] arrange themselves, creeping like rootlets over every inequality of the surface and into any deep crevice, is a pretty sight[881].

A species of tendril-climbing plant, *Bryonia* (Cucurbitaceae), grew wild near his house. Darwin described the movements of the tendrils in detail. When the plant reached a support and attached to it,

> It thus drags itself onwards by an insensibly slow, alternate movement, which may be compared to that of a strong man, supported by the ends of his fingers to a horizontal pole, who works his fingers forward until he can grasp the pole with the palm of his hand[882].

> I have more than once gone on purpose during a gale, to watch a Bryony growing in an exposed hedge with its tendrils attached to the surrounding bushes… As it was, the Bryony safely rode the gale, like a ship with two anchors down, and with a long range of cable ahead to act as springs as she surges to the storm[883].

Darwin measured the strength of the attached tendrils on an old grape vine adhering by its tendrils to the wall of his house. It had been exposed to the weather for 14-15 years at the time. By hanging weights on the tendrils he was able to determine that each branch of the tendril could support a weight of two pounds. The entire five-branched tendril thus supported ten pounds in weight.

Darwin concluded that all types of climbers had evolved from the primitive form of twining plants, and that the use of leaves and tendrils for support was secondary[884].

The Power of Movement in Plants

In 1880, Darwin's last year of life, he published a comprehensive review of experiments on movements in plants, performed over the years on many plant species. Although Darwin is listed as the sole author, the book is written in the first person plural. In a letter to a German scientist, a year previously, Darwin acknowledges the major part that his son Francis had taken in this work.

> Together with my son Francis, I am now preparing a rather large volume on the general movements of plants, and I think that we made out a good many new points and views. I fear that our views will meet a good deal of opposition in Germany, but we have been working hard for some years on the subject[885].

[879] Ibid. p. 98.
[880] Ibid. p. 103.
[881] Ibid. p. 110.
[882] Ibid. p. 133.
[883] Ibid. p. 164.
[884] Ibid. p. 191.
[885] Darwin to Victor Carus; in F. Darwin 1887, III: 332 Francis Darwin later continued to lecture and publish in Botany in Cambridge.

The book describes observations on different moving parts of the plant, albeit not in systematic order. Considerable space in the book is devoted to the experimental methods. Darwin was aware that it contained many details in which not every reader might be interested. These are printed in small print and, Darwin noted, need not be always read. He explains in detail the use of the many graphic illustrations:

> From the distortion of our figures, they are of no use to anyone who wishes to know the exact amount of movement, or the exact course pursued. But they serve excellently for ascertaining whether or not the part moved at all, as well as the general character of the movement[886].

Heliotropism of Plants [= Movement towards the Light]: Darwin was aware, of course, of the tendency of plants to grow towards the light, and was convinced that this was both necessary and advantageous. He guessed at the possible advantage, although his guess, he felt, was incomplete:

> The young stems place themselves so that the leaves may be well illuminated. They are thus enabled to decompose carbonic acid... The sheath of the cotyledon of some Graminae are not green and contain very little starch... we may infer that they decompose little or no carbonoic acid. Nevertheless they are extremely heliotropic[887].

The heliotropic movement of plants – especially of germinating seeds – was for Darwin analogous to the directional movements of animals, in that they enable the plant to find the shortest way to the light:

> It can hardly fail to be of service to seedlings, by aiding them to find the shortest path from the buried seed to the light, on nearly the same principle that the eyes of most of the lower animals are seated at the anterior ends of their bodies[888].

Radicles of Germinating Seeds: Germinating seeds were attached to the stopper in jars with some water, and could be observed without soil. A table on page 68 illustrates the wide taxonomic range of plants investigated[889]. All radicles moved in the twisting "circumnutation" form.

> This movement can hardly fail to be of high importance, by guiding the radicle along a line of least resistance[890].

The experiment proved that the tip of the growing radicle does not elongate. A layer of cells 1-2 mm wide, above the tip, is sensitive to touch and triggers the growth of the part of the radicle just above it, and directs the tip away from an obstacle. This phenomenon, Darwin claims, has never before been reported. More than 40 such experiments were carried out.

> The fact of the apex of a radicle being sensitive to contact has never been observed. When one side of a radicle is pressed to any object, the growing part bends away from the object, and this seems a beautiful adaptation for avoiding obstacles in the soil... We were therefore led to suspect that the apex was sensitive to contact, and that an effect was transmitted

[886] Darwin 1880, p. 8.

[887] Darwin 1880, The Power of Movement in Plants. p. 449. The physiological mechanism of bending towards the light, in which the phyto-hormone auxin plays a central role, was discovered years after Darwin's death. (Kutschera 2009) One experimental system often used in his research is the sheath of the cotyledons of corn seedlings!

[888] Ibid. p. 484.

[889] Ibid. pp. 89-102.

[890] Ibid. p. 72.

> from it to the upper part of the radicle, which was thus excited to bend away from the touching object[891].

Darwin suggested that two forces directed the movement of the radicle: geotropism and the sensitivity to contact[892]. He measured the force which the tip of the radicle applies to the soil, and described the process of penetration:

> Geotropism does not give a radicle force enough to penetrate the ground, but merely tells it (if such an expression may be used) which course to follow[893]. The growing part does not act like a nail when hammered into a board, but more like a wedge of wood, which whilst slowly driven into a crevice continually expands at the same time by the absorption of water; and a wedge thus acting will split even a mass of rock[894].

Darwin enthused over this power in the radicle, which resembled a burrowing animal:

> A radicle may be compared with a burrowing animal such as a mole, which wishes to penetrate perpendicularly down into the ground. By continually moving its head from side to side… he will feel any stone or other obstacle, as well as differences in the hardness of the soil, and he will turn away from that side… nevertheless after each interruption, guided by the sense of gravity, he will be able to recover his downward course and to burrow to greater depth[895]. It is hardly an exaggeration to say that the tip of the radicle thus endowed, and having the power of directing the movement of the adjoining parts, acts like the brain of one of the lower animals[896].

Hypocotyl and Cotyledons: In many species of dicots (153 plant genera were examined) the young seedling bursts through the soil in the form resembling a bent knee, which has the advantage of combining the force exerted by the two sides of the arch[897]. Darwin suggested that this method of emergence is advantageous because it assists the plant in its efforts to reach the light:

> We are led to believe in adaptation when we see the hypocotyl of a seedling, which contains chlorophyll, bending to the light; for although it receives less light, being now shaded by the cotyledons, it places them – the most important organs – in the best position to be illuminated. The hypocotyl may therefore be said to sacrifice itself for the good of the cotyledons, or rather of the whole plant[898].

Darwin noticed that the cotyledons moved in an elliptical manner and followed a diurnal course: they moved downwards in the forenoon, and upwards in the early afternoon and evening – in some cases reaching a vertical position. The movement is related to the light and is caused by the growth of the stem. Many measurements and observations on stems led to a general conclusion: all growing plants "circumnutate":

> Anyone who will inspect the diagrams… will probably admit that the growing stems of all plants, if carefully observed, would be found to circumnutate to a greater or lesser degree[899]. Circumnutation is so general, or rather so universal a phenomenon, that we

---

[891] Ibid. pp. 130-132.
[892] Ibid. pp. 149-153.
[893] Ibid. p. 73.
[894] Ibid. p. 77.
[895] Ibid. pp. 199-200.
[896] Ibid. Conclusions. p. 572.
[897] Ibid. table p. 111.
[898] Ibid. p. 489.
[899] Ibid. p. 201ff. p. 213.

> cannot suppose it to have been gained for any specific purpose. We must believe that it follows from the manner in which vegetable tissues grow[900].

Movement of Leaves: Darwin's detailed observations on plants belonging to 25 families indicated that the leaves moved diurnally in a vertical plane; they moved upwards in the evening or early night, and downwards in the morning[901]. This diurnal movement was described as the "sleep" of plants[902]. Darwin defined "sleep" as forming an angle of up to 90° above the horizontal. He was convinced that this diurnal movement was acquired because it had an important advantage for the plant.

> That the periodicity is determined by the daily alternation of light and darkness there can hardly be a doubt. The above periodicity should be kept in mind by anyone considering the problem of the horizontal position of leaves and cotyledons during the day, whilst illuminated from above[903]... It is hardly possible to doubt that plants derive some great advantage from such great powers of movement... as it seems to me, that the object gained is the protection of the upper surface from being chilled at night by radiation[904].

To test this explanation Darwin carried out many experiments on nights when the temperature was below freezing. Leaves which were experimentally prevented from moving, suffered more from freezing than leaves that were free to move.

> From the several cases above given, there can be no doubt that the position of the leaves at night affects their temperature through radiation, to such a degree, that when exposed to a clear sky during a frost, it is a question of life and death[905].

Darwin believed that the diurnal pattern of leaf movement was heritable, but he noted that plants imported from different countries responded to the light conditions of the new country, not those of the country of origin.

> We may conclude that nyctitropism, or the sleep of leaves and cotyledons, is merely a modification of their ordinary circumnutating movement, regulated in its period and amplitude by the alternation of light and darkness. The object gained is the protection of the upper surfaces of the leaves from radiation at night[906]... We know that the sleep movements of leaves are to a certain extent inherited independently of light and darkness... It seems probable that they have been acquired for the special purpose of avoiding too intense an illumination[907].

[900] Ibid. p. 263.
[901] Ibid. pp. 228-259.
[902] Darwin, E., Phytologia 1800, Chapter 2.
[903] Darwin 1880, p. 262.
[904] Ibid. p. 280; Ibid. p. 280; Ibid. p. 284.
[905] Ibid. p. 294.
[906] Ibid. p. 413.
[907] Ibid. p. 447.

18

Heredity in the 19th Century: What did Darwin Know?

> What can be more wonderful than some trifling peculiarity, not primordially attached to the species, should be transmitted through the male or the female sexual cells, which are so minute as not to be visible to the naked eye, and afterwards... ultimately appear in the offspring when mature[908].

The mechanism of inheritance of characters was generally unknown in the 19th century, although many people had been busy for many years breeding and improving domesticated animal and plant species for human needs. Controlled crosses of sheep, in particular of the Merino sheep – of Spanish origin – with local breeds were carried out by sheep growers in England, Germany, and Bohemia (now the Czech Republic) in the 18th century in order to improve the quality of the wool – an economically important commodity at the time. Different breeders had their own methodologies – sometimes carefully kept secret – of selecting the best sheep for breeding, but all these methods were empirical and based on the experience and intuition of the breeder[909].

The fact that crossed plants may be strong and vigorous was known to plant breeders, and crossing was practiced to produce better crops. The hybrids, however, were worthless for further breeding, because in the second generation they became variable and their good qualities could not be maintained:

> As a general rule, crossed offspring in the first generation are nearly intermediate between their parents, but their grandchildren and succeeding generations revert to a greater or lesser degree, to one or both of their progenitors[910].

The absence of scientific knowledge regarding heredity was an obstacle to the understanding of natural selection. As late as 1861, after the publication of Darwin's The Origin of Species, Huxley urged the botanists to do something about it:

> Why does not somebody go to work experimentally, and get at the law of variation for some one species of plant[911]?

Darwin's knowledge of the processes of heredity is summarized in two of his books: "Variation of Plants and Animals under Domestication" (hereafter: Domestication), first published in 1868, and "The Effects of Cross- and Self-Fertilisation in the Vegetable Kingdom", published in 1876. "Domestication" is a review of Darwin's wide correspondence with many breeders and scientists

---

908 Darwin (1868) 1898a, Variation under Domestication, I: 446.
909 Wood & Orel 2001.
910 Darwin 1898a (1868) II: 23.
911 Huxley to Hooker, 1861, in Huxley's Life and Letters, p. 227.

throughout Europe, and offers a good representation of the knowledge of heredity in Europe in the 19th century. "Cross- and Self-Fertilisation" is an account of Darwin's own crossing experiments carried out over the course of many years.

Darwin recognized that variations which are not passed on from parents to offspring are of no use for the formation of new species. The entire phenomenon of heredity, however, seemed to him wonderful and mysterious.

> Everyone must have heard of cases of albinism, prickly skin, hairy bodies &c, appearing in several members of the same family. If strange and rare deviations of structure are really inherited, less strange and commoner deviations may be freely admitted to be inheritable... The laws governing inheritance are for the most part unknown. No one can say why the same peculiarity in different individuals of the same species, or in different species, is sometimes inherited and sometimes not so. Why the child often reverts in certain characters to its grandfather or grandmother or more remote ancestor. Why a peculiarity is often transmitted from one sex to both sexes, or to one sex alone, more commonly to the like sex[912].

How do we know that breeds carrying some desirable characters, like milk yields in cows and various characteristics of race horses etc., are heritable? Although some people doubted this, the evidence indicates that pedigrees of the good breeds are kept and sought after: "Hard cash paid down, over and over again, is an excellent test of inherited superiority"[913].

On Variation

In Darwin's time, almost no information was available on variation in natural populations, apart from that pertaining to occasional aberrant individuals caught by collectors.

> Most organic beings in a state of nature vary exceedingly little. The amount of hereditary variation is very difficult to ascertain, because naturalists do not all agree whether certain forms are species or races[914].

Like his contemporaries, Darwin knew of two kinds of variation in domesticated plants and animals: there was variation among individuals – small differences in metric characters which seemed to form a continuum ("fluctuating variation"). And there were rare, occasional individuals that were very different from all others – appearing suddenly and referred to as monstrosities or "sports".

The characters of domestic animals, which were acquired or lost by "use or disuse", were nevertheless heritable – according to Darwin. Examples abound:

> It is notorious ...that increased use or action strengthens muscles... and that disuse, on the other hand, weakens them... There can be no doubt that with our anciently domesticated animals, certain bones have increased or decreased in size and weight owing to increased or decreased use [in Rabbits], the ears have been increased enormously [compared with wild ancestors] through continued selection... and their weight, co-joined probably with the

[912] Darwin 1898 (1872), the Origin of Species, 6th ed. p. 19. Darwin sought to explain all the hereditary phenomena of which he was aware. He did not know that the key to the understanding of heredity was available "on the shelf". In 1865, the Czech monk Gregor Mendel (Chapter 20) had published his paper on crossing experiments with peas, which included a theoretical model for inheritance of discrete (non-blending) characters. Mendel's paper was not appreciated at the time, and was rediscovered in 1900, to become the basis of the new science of genetics.

[913] Darwin 1898a (1868), Domestication, I: 447.

[914] Darwin 1844.

> disuse of their muscles, has caused them to lop downwards... [in pigeons] the shortening of the humerus and radius in the seventeen birds may probably be attributed to disuse[915].

Character Blending, Inheritance, and the Evolution of New Species

The popular concept of heredity in the 19th century was that hereditary characters are carried in the blood[916]. The characteristics ("peculiarities") of the father and the mother reach the sex cells with the bloodstream and are mixed, so that the children will present an average of parental characteristics ("character blending"). Darwin, like everybody else, accepted the notion of blending inheritance, but he was aware of cases that could not be explained in this way.

Darwin lists many examples of breeds originating from single "monstrous individuals" of dogs, horses, pigs, and cats, which were heritable and showed no "blending". Many breeds of dogs probably arose suddenly in individuals, and because they were strictly inherited, were then exploited to form a breed. Darwin noted that such breeds could not have been established without selection by man.

A famous case, cited by many writers of the time, was that of the *ancon sheep*. This was a breed of sheep derived from a single male lamb "with short crooked legs and a long neck, looking like a turnspit dog", which was born in Massachusetts in 1781. The owner of the herd crossed the new male with a normal ewe and found, to his pleasant surprise, that the character did not blend – the offspring either resembled one parent or the other – "even one of twins has resembled one parent and the second, the other"[917]. The owner selected for the new form because the short-legged sheep could not jump over the fences – and the breed was very quickly adopted by other breeders in Massachusetts.

Darwin initially tended to think that if the "monsterous" form gave the individual some advantage – "ever so slightly" – in the struggle for existence, it would increase in frequency and form a new species. In 1867, however, the Scottish engineer Fleeming Jenkin forced Darwin to change his mind[918]. Indeed, Darwin altered a paragraph in later editions of the "Origin", changing the emphasis from "sports" to "fluctuating variation" as the origin of new species. In a letter to Wallace, Darwin admits that Jenkin had made him change his mind:

> I have always thought individual differences more important than single variations. But now I have come to the conclusion that they are of paramount importance... Fleeming Jenkin's arguments have convinced me[919].

Darwin's son, Francis, noted:

> The point on which Fleeming Jenkin convinced my father is the extreme difficulty that single individuals, which differ from their fellows in the possession of some useful character, can be the starting point of a new variety. Thus the origin of a new variety is more likely to be found in a species which presents the incipient character in a large number of individuals[920].

---

915 Darwin 1898a (1868) I: 286; 289; 135; 186, respectively.

916 Some expressions in common usage reflect this belief, such as "blue blood" for families of royalty, or "bad blood" for someone born into a family of criminals.

917 Darwin 1898a (1868) I: 104. Mendelian genetics provides the solution: if the mutant male was a recessive homozygote and the ewe dominant, the offspring should look normal but be heterozygous. In a cross of the mutant male and a heterozygous ewe half the offspring should be of either phenotype.

918 See Chapter 14.

919 Darwin to Wallace, 22 January 1869.

920 Reported by Francis Darwin. Darwin, Autobiography, pp. 289-290.

Hereditary Phenomena that Darwin could not Explain

Reversion: It was often claimed in the 19th century that a child resembled a grandparent, an uncle, or a more distant relative in various peculiarities that were not present in the parents. The phenomenon was known as "reversion" [in later literature, the term was "atavism"].

> That a being is born resembling in certain characters an ancestor removed by two or three, in some cases by hundreds or even thousands of generations, is assuredly a wonderful fact. In these cases the child is commonly said to inherit such characters directly from... more remote ancestors. But this view is hardly conceivable[921].

Cases of "reversion" were known to breeders of domesticated animals – for example horses[922]. Breeders disagreed on the number of generations of crosses with a purebred that needed to be performed in order to obtain a reversion-free breed.

Darwin collected information from breeders in order to explain the phenomenon of "reversion". He then offered an explanation for which he had no proof: that many characters can be transmitted in a dormant stage, without being expressed in the individual carrying them.

> If this view is correct, we must believe that a vast number of characters capable of evolution, lie hidden in every organic being, and that [the germ] is crowded with invisible characters, proper to both sexes... and to a long line of male and female ancestors separated by hundreds or even thousands of generations from the present time, and these characters, like those written with invisible ink, lie ready to be evolved when the organization is disturbed by certain unknown conditions[923].

> We can thus understand how it is possible for a good milking cow to transmit her good qualities through the male offspring to future generations, for we may confidently believe that these qualities are present, though latent, in the males of each generation[924].

The Sudden Appearance of "Monstrosities" during Domestication

Darwin described in detail aberrant individuals ["sports"] which appeared in domestic animals. Those that appeared to the breeder to suit his needs or his fancy were selected and used to start new breeds. Darwin concluded that lines must be selected continuously for many generations to produce breeds with the desired characteristics, especially if they were originally derived from an inter-breed cross.

> Some of the peculiar characteristics of several breeds of dogs have probably arisen suddenly, and though strongly inherited, may be called monstrosities. A peculiarity suddenly arising and therefore in one sense deserving to be called a monstrosity, may however be increased and fixed by man's selection[925].

Darwin was aware of certain sudden, heritable changes in humans. The case of the "porcupine man" was publicized as far back as 1733: a boy born with crust-like skin (except on his face and the palms of his hands) which was shed periodically, like the skin of reptiles. None of his brothers

---

921 Darwin 1898a (1868), II p. 25.

922 Ibid. p. 14-19. See also Galton's law of inheritance, Chapter 21.

923 Ibid. II: 59. Less than fifty years later, Mendelian genetics – and after fifty more years, the coding of characters in DNA – showed that Darwin had guessed correctly.

924 Ibid. II: 27.

925 Ibid. The famous "ancon sheep", I: 104.

had this kind of skin. When he matured the boy married and had six children, all of whom bore the "porcupine" skin and five of whom died young. Darwin reported that the character passed through four subsequent generations and that only males expressed it[926].

Darwin attempted to generalize the observed facts known to him and classify them into groups, which were "explained" by principles or "laws of inheritance. For example, *albinism* was observed in several species, and inherited as a fixed form (with no intermediates known between the albino and the normal form). Therefore

> This apparently depends on a law, which generally holds good, namely that the characters common to many species of a genus – and this, in fact, implies long inheritance from the ancient progenitor of the genus – are bound to resist variation[927].

Another "principle", which Darwin refers to over and over again, is the "principle" of correlation of growth: for example, most domestic breeds of pigeons that have short legs – also have short beaks:

> By this term I mean that the whole organization is so connected, that when one part varies, other parts vary. But which of two correlated variations ought to be looked at as the cause and which the effect, we can seldom or never tell[928].

Many generalizations appear in Darwin's book on sexual selection[929]. For example, the principle of "inheritance at corresponding periods of life": a new character which appears in a young animal, will in general reappear in the offspring at the same age and last for the same time[930]. The principle of "inheritance at corresponding periods of the year" refers to the periodic changes in coat color of birds and mammals. Darwin was not sure of this, but suspected that these tendencies are indeed inherited:

> Innumerable instances occur of characters appearing periodically at different seasons [e.g. antlers in deer, breeding feathers in birds etc.]. Although I do not know that this tendency to change the color of the coat is transmitted, yet it probably is so[931].

Inheritance Limited by Sex

This phenomenon was an enigma for Darwin. Characters which first appeared (under domestication) in one sex – e.g. horns in sheep – tended to reappear only in the same sex, although both sexes participate in the production of the offspring.

> In sheep there is a strong tendency for characters, which have apparently been acquired under domestication, to become attached either exclusively to the male sex, or to be more highly developed in this than in the other sex[932].
>
> Characters are somewhat commonly transferred exclusively to that sex in which they first appeared... Why some characters should be inherited by both sexes, and other characters by one sex alone... is in most cases quite unknown... The following two rules seem often to hold good: that variation which first appeared in either sex in a late period of life, tend

926 Moore 1986, p. 598.
927 Darwin 1898a (1868) I: 114.
928 Ibid. p. 177; p. 229.
929 Darwin 1874 (1872).
930 Ibid. p. 585.
931 Much of the evidence on this point may have been obtained from a paper by Blyth [1837].
932 Darwin 1898a (1868) I: 99.

to be developed in the same sex alone; While variations which first appeared in either sex early in life tend to be developed in both sexes[933].

Characteristics of Selected Breeds

Breeds of domestic animals are characterized by clearly-defined differences in form and structure. Darwin examined and measured 150 "true-breeding" breeds of domestic pigeons – all descended from the primitive ancestor, the wild rock pigeon *Columba livia* – and found clear and consistent differences among them:

> There can be no doubt that if well-characterized forms of the several races had been found wild, all would have been ranked as distinguished species[934]... When differently colored birds [= pigeons] are crossed, the opposed forces of inheritance apparently counteract each other, and the tendency which is inherent in both parents to produce slaty-blue offspring becomes predominant[935].

Some of the differences among breeds may be due to the disuse of wing and leg muscles in captivity, and the corresponding effects on beak dimensions may be due to the "law of correlation". The cause of many changes, however, is unknown. What is clear is that the preservation of such characteristics requires continuous selection by the breeder:

> We are profoundly ignorant of the cause of each sudden and apparently spontaneous variation... all such variations appear to be the direct result of changes of some kind in the conditions of life... But without selection, all this would produce only a trifling or no result... This may be called methodical selection, for the breeder has a distinct object in view, namely to preserve some character[936].

Darwin lists many examples from the breeding of domestic fowl – among them a case of his own research: the cross of one black rooster with several pure white hens. To exclude the possibility of uncontrolled mating he destroyed all the other fowl in his possession. The cross yielded mostly black chicks – but some were white. For example, one white hen produced four black and seven white chicks. Assuming "blending" inheritance, these results were totally unexpected. Darwin did not try to explain the "anomaly"

The Inheritance of Aberrant Characters in Humans

Darwin knew about rare and "monstrous" heritable characters appearing and recurring in the same human families for several generations. It was unknown how these characters arise:

> We are driven to conclude that such peculiarities are not directly due to the action of the surrounding conditions, but to unknown laws acting on the organization or constitution of the individual[937].

Extra digits on the hands and feet were reported to recur for five generations in some cases, although the original parents had normal hands and feet. In one case the character appeared again after being absent for three generations. The case is all the more remarkable ("says Professor Huxley")

---

933 Darwin, Sexual Selection in Darwin 1952, pp. 586; 587; 588.
934 Darwin 1898a (1868) I: 139.
935 Ibid. I: 22.
936 Ibid. I: 223; 224.
937 Ibid. I: 448.

since it is ascertained that the affected person(s) were married to spouses who did not show this character[938].

Albinism: Darwin knew of a case of two brothers, who married two sisters. There were seven children in the two families – all of them albinos. Darwin believed that this was a case of reversion, although no case of albinism was previously known in the history of these families.

Sex-linkage: Darwin was aware of characters in humans that appeared to be transmitted differently in males and females (sex-linked characters) – such as hemophilia and color-blindness. He described accurately the pattern of transmission of such characters between generations, but offered no explanation:

> Generally... the sons never inherit the peculiarity directly from their father, but the daughters alone transmit the latent tendency, so the sons of the daughters alone exhibit it. Thus the father, grandson, and great-great-grandson will exhibit a peculiarity, the grandmother, daughter, and great-granddaughter having transmitted it in a latent state[939].

Heritable Characters in Plants

Edible plant varieties – wheat, corn, cabbage, and peas – are the result of long periods of selective breeding by man. This process is not different from selection in animals[940], but some phenomena are peculiar to plants. Variations can suddenly appear in different buds of fully-grown trees. Such "bud variations" can be propagated by cutting and grafting.

> With respect to the more curious cases of full-grown peach trees suddenly producing nectarines ("smooth peaches") by bud variation (or sports as they are called by gardeners), the evidence is superabundant. There is also good evidence of the same tree producing both peaches and nectarines[941].

Bud variation was a mystery. It could not be well explained as "reversion", since the character was unknown in ancestors. It could not have been caused by changes in environmental conditions, because the tree has thousands of buds, and only one of them produces the variation.

> Many cases of bud variation... cannot be attributed to reversion, but to so-called spontaneous variability, as is so common with cultivated plants raised from seed... The laws of inheritance seem to be nearly the same with seminal and bud variations. Notwithstanding the sudden production of bud varieties, the characters thus acquired are sometimes capable of transmission by seminal production[942].

Crosses of Varieties: Darwin experimented for years with crosses of plant varieties in his garden and greenhouse, paying special attention to the effects of self-versus cross-fertilization (inbreeding versus outbreeding)[943]. A condensed account of these results appears in "Domestication".

---

938 Ibid. I: 457. Robert Chambers, and his brother William, were both hexadactylous – having six digits in their hands and feet. Eiseley 1961, p. 138.

939 Ibid. II: 49.

940 Ibid. I: 326.

941 Ibid. I: 361. "Bud variation" is the result of somatic mutations. The cells in the meristematic tissues within the bud retain their juvenile characters and divide, and mutations may occur during cell division.

942 Ibid. I: 442.

943 See Chapter 17.

Mating between related individuals or varieties, carrying similar traits, are useful to the breeder, because in this way the desirable traits may be preserved, but prolonged inbreeding has undesirable effects. Viability and productivity may be reduced, and there is a tendency to the appearance of malformations. Nevertheless, the advantage to the breeder from inbreeding may outweigh the negative effects, because the negative effects accumulate slowly and at unequal rates in different characters and breeds. On the other hand, the positive effect of outbreeding is indisputable. An occasional cross with unrelated stock benefits all individuals:

> The evidence... convinces me that it is a great law of nature that all organic beings profit from an occasional cross with individuals not closely related them in blood. And, on the other hand, long-continued interbreeding is injurious[944].

Darwin seems to have been so focused on the advantages of cross-fertilization that he ignored the inheritance of other characters, affected by his crosses. He lists many examples of crossing experiments, from his correspondence as well as his own work. The following experiment puzzled him – he could offer no explanation for the strange outcome:

> A stunted variety of *Tagetes signata* growing in the midst of the common variety by which it was probably crossed... Most of the seedlings raised from this plant were intermediate in character, only two perfectly resembling their parent. But seed from these two plants reproduced the new variety so truly, that hardly any selection has since been necessary[945]. Many analogous facts may be given showing how apparently capricious is the principle of inheritance.

Pea varieties preserve their characteristics for long periods because fertilization occurs in the closed flower. Cross-pollination of pea varieties by hand yielded "strange and surprising" results. For example,

> Mr. Masters raised from a plant... four different sub-varieties, which bore blue and round, white and round, blue and wrinkled, and white and wrinkled peas. And although he sowed these four varieties separately during several successive years, each kind always reproduced all four kinds mixed together[946].

In a cross of two plants (*iberis*?) differing in the color of their flowers, he noticed that the first-generation plants were not homogenously intermediate between the parents, as expected from "blending" inheritance – the common belief at the time: while the flowers of 24 of 30 plants were the same as the pollen-donor parent, six had flowers like the female plant. Darwin did not discuss the numerical results, but only commented:

> This case offers a good instance of a result, which not merely follows from crossing varieties having differently coloured flowers, namely, the colours do not blend, but resemble perfectly those of either the father or the mother plant[947].

Darwin crossed two varieties of the snapdragon (Antirrhinum) – one with normal and the other with peloric flowers, and then intercrossed the hybrids. The results puzzled him and he reported them in detail: From the selfed flowers of the peloric plants he raised 16 plants, all bearing peloric flowers. The F1 hybrid plants – of which he had at least 90 – invariably produced normal flowers. Selfing of these hybrids (F1 x F1) yielded 127 plants: 88 with normal flowers, 37 peloric, and two

[944] Darwin 1898a (1868), II: 94.
[945] Ibid. I: 463.
[946] Ibid. I: 349.
[947] Ibid. I: 105.

"intermediates". Darwin did not have any clue to the explanation of the numerical results. Instead he refers the reader to his theory of "pangenesis" (see below)[948].

Darwin noticed that the initially-variable colors of *Mimulus* flowers and of three other species of plants, which were self-fertilized for some generations, became gradually more homogeneous.

> The colour of the flowers was a point to which I did not at first in the least attend, and no selection whatever was practiced. Nevertheless, the flowers produced by the self-fertilized plants of the above four species became absolutely uniform in tint, or very nearly so... When self-fertilized plants of the later generations were crossed with a fresh stock, and seedlings were raised, these presented a wonderful contrast in the diversified tints of their flowers compared with those of the self-fertilized seedlings... I kept no record, as the point did not interest me until I was struck with the uniform tint of the flowers on the self – fertilized plants of the fifth generation[949].

Pangenesis

Inheritance, and the "capricious" behavior of the "principles" governing it, seemed mysterious to Darwin. He apparently felt that he needed a unifying theory that would explain all the phenomena with which he was familiar. He called his theory "Pangenesis".

The theory suggests that every cell and organ in the body sends a representative particle – a gemmule – into the blood. These gemmules multiply by fission, like bacteria, and are distributed through the bloodstream into every part of the body, including the sex cells. The gemmules of both parents are mixed at fertilization and are transmitted to the offspring. The gemmules issued from a given organ have a tendency to attract each other, which ensures that the organization of the body of the offspring will resemble that of the parents. However, the arrangement of the gemmules is affected by external conditions, thereby enabling the transmission of acquired (adaptive) characters. The gemmules are transmitted in a dormant state, and may remain dormant for more than one generation. This accounted for reversion and for the transmission of male characters to the grandson via the mother, in whom these characters are not expressed. Darwin had to invoke "tendencies" of the gemmules which were not very convincing, but the theory of pangenesis provided an explanation for many hereditary phenomena – such as the blending of male and female characters, artificial selection, and regeneration of missing or lost organs (as the foreskin in circumcised Jewish boys).

Darwin showed his theory to his friend Huxley, who was not enthusiastic about it:

> Somebody rummaging among your papers half a century hence may find "Pangenesis" and say, "see this wonderful anticipation of our modern theories, and that stupid ass Huxley prevented his publishing them..." But all I say is, publish your views, not as much in the shape of formed conclusions, as of hypothetical developments of the only clue at present accessible[950].

Darwin accepted Huxley's advice and published his theory as a "provisional hypothesis" as the last chapter in "Domestication"[951].

---

[948] Ibid. II: 46. All these difficulties could have been explained, had Darwin been aware of Gregor Mendel's paper, published in 1865. But that paper was only recovered in 1900.

[949] Ibid. I: 306-7; I: 309.

[950] Huxley to Darwin, 1865 in Huxley's Life and Letters, p. 268.

[951] Darwin 1898a (1868), II Chapter 7.

Shortly after publication of the theory, Darwin's cousin Francis Galton[952] demonstrated experimentally that the theory was wrong. Galton transfused the blood between two female rabbits differing in color. After the blood had circulated between the rabbits for some time, he disconnected them and then bred the females. He expected that the color of the offspring would be intermediate, but instead the brood of each female had the same color as that of the mother. The results indicated that no gemmules of the color were therefore carried in the blood. (Galton did not entirely reject the theory, but suggested a modified version).

Towards the end of the 19th century, August Weismann[953] – a devoted Darwinian – wrote that Pangenesis was a valuable contribution of Darwin's, although it was not supported by facts: Darwin had offered it as a model, to illuminate the problems that needed explanation – without however testing whether his assumptions were valid[954]. Moreover, Pangenesis is a revival of the oldest theory of heredity, that of Democritos in Ancient Greece. Democritos had suggested that particles – called Pangens – were recruited from all parts of the body and assembled in the semen – with each pangen being responsible for the formation in the child of the part from which it came[955].

The Pangenesis theory was resurrected and presented in a new light by Hugo De Vries in his "Intracellular pangenesis" theory[956].

[952] Chapter 21.
[953] Chapter 20.
[954] Weismann, On Heredity, 1892, p. 80; p. 168.
[955] Ibid. p. 589.
[956] De Vries 1889; Chapter 24.

19

Ernst Haeckel: Embryology and Phylogeny in Evolution

If Huxley was Darwin's bulldog, then Ernst Haeckel was surely his German Shepherd[957]

Haeckel and Darwin

Ernst Haeckel (1834-1919) had studied medicine, but was attracted to zoology. After practicing medicine for some time, he traveled to Italy and tried his hand at art (his considerable artistic talent is evident in his scientific publications on protozoa and marine invertebrates). He read Darwin's "Origin of Species" in 1859 and soon became the most devoted herald and proponent of the theory of natural selection in Germany. In addition to his research and teaching of biology at the University of Jena, he wrote and lectured about evolution to the public.

> The highest triumph of the human mind, the true knowledge of the most general laws of nature, aught not to remain the private possession of a privileged class of savants, but aught to become the common property of all mankind[958].

Haeckel portrays himself as a great admirer of Darwin and of his theory. The full title of his book (1876) explains the purpose of his efforts: "The History of Creation, or the development of the earth and its inhabitants, by the action of natural selection: a popular exposition of the doctrine of evolution in general, and of Darwin, Goethe and Lamarck in particular". Haeckel held Darwin in the highest esteem:

> [Darwin's theory] is one of the greatest achievements of the human mind, and it may be placed on a level with Newton's theory of gravity – indeed, it even rises higher than Newton's theory[959]. Proud as England may be to be called the Fatherland of Newton… yet may she with even greater pride reckon Charles Darwin among her sons, he who solved the yet harder problem of bringing the complicated phenomena of organic nature under the sway of the same natural laws[960].

As noted above (Chapter 6), Haeckel suggested that the theory of "Development" [= evolution] is in fact a combination of Lamarck's Theory of Descent and Darwin's Theory of Natural Selection. At the end of the second volume of his book, Haeckel lists the biological phenomena best explained by the theory of descent: the paleontological history of the animal world, the anatomical similarity and embryonic development of different species, the geographical distribution and ecology of animals [the term ecology was coined by Haeckel] – and the possibility of arranging the entire animal world in the form of a single branching tree (see below). Natural Selection is, in turn, a mathematical outcome from the facts of the struggle for existence and variation.

---

957 Sapp 2003, p. 36.
958 Haeckel, 1876, "The History of Creation", p. 4.
959 Ibid. p. 125.
960 Ibid. Preface.

> It is not for zoologists and botanists to accept or reject this [Darwin's] as an explanatory theory as they please. They are rather compelled and obliged to accept it, according to the general principle observed in all natural sciences, that we must accept and retain, in the explanation of phenomena, any theory which is compatible with the actual facts – until it is replaced by a better one[961].

Haeckel corresponded with both Darwin and Huxley. Following the publication of his book in German," Generelle Morphologie", which was reviewed by Huxley, Haeckel was invited as a guest to Darwin's home and he stayed there more than once.

Faith and Science

Like Darwin and Huxley, Haeckel too strongly rejected the hegemony of the Church and its doctrines: "Where faith commences, science ends"[962].

> In the whole series of these suppositions the Creator is nothing but an almighty man, who ...amuses himself with planning and constructing most varied toys in the shape of organic species. After having diverted himself with these for thousands of years, they become tiresome for him, he destroys them by a general catastrophe of the earth... and ...calls a new and more perfect animal and vegetable world into existence[963].

No supernatural belief in Creation, argued Haeckel, can explain the mystery of embryonic development: Why take the trouble to create organisms as embryos that must undergo a complex development – when it could have been just as easy to create them as adults? How can Creation explain the presence of rudimentary organs? If one accepts the Theory of Evolution, the explanation is clear. Darwin's theory is the best theory to explain the appearance of new species:

> The struggle for existence is a biological fact, which with mathematical necessity follows from the general disproportion between the average number of organic individuals and the numerical excess of their germs. But as adaptation and inheritance in the struggle for life are in continual interaction, it inevitably follows that natural selection... must produce new species[964].

Haeckel attacked furiously the opponents of the Theory of Evolution. Most of them, he wrote, have no scientific knowledge at all, but still dare to express their opinion on scientific matters. Some of these opponents are zoologists or botanists,

> but these are mostly old stagers, who have grown grey in quite opposite views and whom we cannot expect, in the evening of their lives, to submit to reform in their conception of the universe[965].

The Social Implications of Natural Selection

Haeckel held strong views on human society; views which agreed with his belief in Darwin's theory. He saw natural selection as the main force shaping human society, preserving the characteristics most favorable for survival – just as in the animal world. His social views on eugenics and education

[961] Ibid. I: 29.
[962] Ibid. p. 9.
[963] Ibid. I: 29.
[964] Ibid. II: 355.
[965] Ibid. II: 346. Darwin, in a letter to Haeckel, expressed concern that using such strong words against his opponents may alienate scientists from his theory.

became highly influential. His book "The Riddle of the Universe" was translated into 25 languages and sold hundreds of thousands of copies (reprinted in 10 editions) in Germany.

> The theory of selection teaches us that in human life, exactly as in animal and plant life, at each place and time only a small privileged minority can continue to flourish. The great mass must starve and more or less prematurely perish in misery[966].

Haeckel strongly objected to militarism – quite an outstanding position in 19th-century Germany, when frequent wars between nations raged in Europe. Wars created a strong selective force against the young and healthy:

> While the healthy flower of youth dies on the battlefield, the feeble remainder enjoys the satisfaction of reproduction and of transmitting all their weaknesses and infirmities to their descendants. The more useless, weaker, or infirmer a youth is, the greater his prospect of escaping the recruiting officer and of founding a family[967].

Haeckel observed that the approach used in Sparta in ancient Greece, which selected against the weak and sickly, would be rejected by any modern humane society, but sarcastically commented that such society, which does not hesitate to send thousands of young men to die on the battlefield, prevents the execution of dangerous criminals in the name of liberalism[968]. Haeckel was an idealist, however, and foresaw a good future for mankind:

> The organ which, above all others, in man becomes most perfect by the ennobling influence of natural selection is the brain. The man with the most perfect understanding, not the man with the best revolver, will in the long run be victorious. Thus we may justly hope… that the progress of mankind towards freedom… will, by the happy influence of natural selection, become more and more certain[969].

Embryology

Important observations on the morphological similarity of different vertebrates during their embryological development were reported by several scientists in Germany in the early 19th century. Impressed by the anatomical similarity of certain stages in the embryogenesis of mammals (Man in particular) to the adult forms of certain "lower" organisms, some were led to the notion that human embryos pass first through the form of a worm, then that of an insect, then a fish, ending up in the shape of a mollusk. In particular, Meckel, in a series of papers between 1808 and 1821, advocated the general interpretation:

> …that the embryo of higher forms, before reaching its complete development, passes through many stages that correspond to those at which the lower animals appear to be checked through their whole life… In fact, the embryos of higher animals, the mammals, and especially man, correspond, in the form of their organs.. to those of animals standing below them[970].

Haeckel accepted Meckel's interpretation of embryological development. He believed that the human embryo, initially identical to the embryos of other mammals, later undergoes a series of changes

966 Cited by Pearson. 1892.
967 Haeckel 1876, History of Creation I: 172.
968 Ibid. I: 173.
969 Ibid. I: 174.
970 Morgan 1903, p. 59; p. 58ff; see also Dobzhansky 1955, p. 236.

– including possession the gills of a fish[971] – and he even went so far as to draw, for comparison, the human embryo side by side with the embryo of a dog in order to illustrate the similarity[972]. Haeckel thought that new, "advanced" species develop when new features and organs are added to previously existing ones at the end of embryogenesis of their simpler, more primitive ancestors ("terminal addition") – like legs added to the form of a fish embryo in the evolution of amphibia. To keep embryogenesis within a reasonable time limit, the early steps of development were considered to be condensed and shortened during the earliest stages of the embryo of the "advanced" species[973].

In 1866, Haeckel formulated the "Biogenetic Law" – referred to also as the "biogenetic fundamental principle", or more commonly, "the principle of recapitulation":

> Ontogeny is nothing but a short and quick repetition, or recapitulation, of phylogeny, dependent on the laws of transmission and adaptation[974].

Many have interpreted this "law" as supporting Darwin's Theory of Evolution – although in fact Haeckel's interpretation of embryology was quite contrary to Darwin's. Early in the 19th century, K.E. von Baer discovered that early-stage embryos of all vertebrate classes are rather similar morphologically – although as adults, the different species greatly differ from each other. Von Baer thought that the reason is that each species develops from a simple to a more complex form. At first all species have simple forms. Characters which distinguish the class (and are common to all members of the class) appear first, then characters which distinguish the order within the class, then the family, and only towards the end of embryogenesis does the animal assume the characteristics of the particular species. Darwin thought that the similarity of species stems from descent. He quoted von Baer's interpretation at length in the Origin: the embryos of different species will be more similar, the more closely they are by descent from a common ancestor[975].

Phylogeny

Almost the entire second volume of Haeckel's book "The History of Creation" is dedicated to a bold attempt to reconstruct the phylogeny of the biological world. The basis for this attempt was Haeckel's conviction that Lamarck's (and Darwin's) Theory of Descent, which is supported by comparative anatomy, paleontology, and embryology, just had to be true:

> The true cause of the intimate agreement in structure can only be the actual blood relationship. Hence we may, without further discussion, lay down the important proposition that all animals and plants belonging to one and the same circle or type must be descended from one and the same primary form[976].

In 1868, Haeckel described an entire hypothetical assemblage of primitive, unicellular organisms, which he called Monera, and which he placed at the root of the Tree of Life. He described how these organisms multiplied by cell division and illustrated them in beautiful drawings in a large monograph. The problem was that these simple organisms – simpler than the known amoeba – had left no paleontological record (his critics accused him of inventing these animals purely from imagination, aided by his artistic talent). He hypothesized that these Monera had evolved from

[971] Creationists and anti-abortionists later accused Haeckel of advocating the destruction of human life: if early embryos are in a "fish" stage, early abortion was not considered homicide and, therefore, was morally acceptable.

[972] Haeckel 1876, II: 306-8.

[973] See Gould (1977) for a rejection and comprehensive review of "terminal addition".

[974] Ibid. I: 33; it was Haeckel who coined the terms Ontogeny, Phylogeny, and Ecology – so widely used in the biological literature since.

[975] See Chapter 8.

[976] Haeckel 1876, II: 122.

more primitive protoplasmic creature without a nucleus, to which he gave the scientific name *Protoamoeba primitiva*.

For a while this scheme seemed acceptable. T.H. Huxley, while examining some dredged mud material from the sea bottom, found something that answered to the description of "protoamoeba" and named it *Bathybius haeckeli*.

> This paper is about a new "Moner" which lies at the bottom of the Atlantic... I have christened it *Bathybius Haeckelii*, and I hope that you will not be ashamed of your god-child[977].

Later this material proved to be a non-biological artifact, and Huxley publicly apologized for his mistake at the meeting of the British Association in 1879[978].

Haeckel described and illustrated in greater detail parts of the great phylogenetic tree. He described the evolutionary descent of different forms of sponges in an illustrated monograph with color pictures drawn by his own hand. Later he did the same for the crustaceans, emphasizing that their embryonic and larval stages (nauplius, zoea) are very similar, although the adult forms are extremely different[979].

Despite its faults, Haeckel's attempt at phylogeny must be given due credit, for connecting together all the groups of plants and animals within one comprehensive scheme[980].

Human Evolution

> Some chronological data suggest that the primitive origin of man was in an ancient land, which is now submerged under the Indian Ocean. No fossil remains of this hypothetical ancient man exist, but he probably was similar to the black African race. His body was more hairy than any living human, his arms were longer and his legs crooked, shorter and thinner, and his posture bent[981].

Haeckel suggested that the main difference between man and ape was the ability to speak. He postulated an intermediate animal, looking more human than the apes but who could not speak, and gave to the non-existent animal a scientific name – *Pithecanthropus alatus* (an artist drew the hypothetical animal for him). Stimulated by this prediction, a Dutch investigator – Dubois – searched for and discovered the remains of pre-human creatures in Java a few years later. In the middle of the 20th century the first paleontological records of the ancestors of man were discovered in Africa and assigned to a new genus Australopithecus. Haeckel's "Pithecanthropus" remains an historical curiosity.

977 Huxley to Haeckel; in Huxley's life and letters I: 295.
978 Ibid. II: 6.
979 Haeckel 1876, II: 172.
980 Haeckel's phylogenetic trees raised many objections. The branching sequences that he postulated and illustrated, and regarded as "a sketch of a natural system of organisms based on their descent", was based on speculation and imagination no less than on morphological similarity. "Haeckel... created in his phylogenies a lot of theoretically intermediate lines between the now-existing organisms. The intermediates have lived only on the pages of books in which they were described [Dobzhansky 1955, p. 241].
981 Haeckel 1876, II: 326-7.

20

August Weismann: The Theory of the Germ Plasm and the [Non]-Inheritance of Acquired Characters

August Weismann (1834-1914) was a naturalist from early childhood. He studied medicine, joined the faculty of the university in Freiburg in 1863, and taught zoology and comparative anatomy until he retired. He became the first director of the newly-established Museum of Zoology in Freiburg, and engaged in diverse zoological research, mainly on invertebrates – particularly insect metamorphosis and the sexual cells of Hydrozoa. Some of his early publications dealt with variation in color patterns of butterfly wings and the markings of butterfly larvae, in an attempt to construct a "natural" classification of the Lepidoptera; as well as a series of publications related to parthenogenetic and sexual reproduction in species of Daphnia.

Weismann correctly interpreted the function of *meiosis* in cell division. His observations of the sexual cells of Hydrozoa led to the theory of the *germ-plasm*, which was published first in German (1886) and then in English (1891-1893). His eyesight, which deteriorated intermittently, prevented him from using the microscope and later limited his ability to read, with his students thereby carry out the observations and reading aloud to him. Although Weismann had no evidence in support of some of his theories, he constructed a detailed theory and a model of *hereditary material.*

Weismann was an enthusiastic supporter of Darwin's Theory of Evolution, and correctly interpreted the process of sexual reproduction as a mechanism that creates the necessary variation for natural selection. However, he rejected the then common belief, shared by Lamarck and Darwin, that acquired characters can be heritable, and he demonstrated experimentally that mutilations are not passed on to the offspring.

Weismann's ideas about inheritance – as expressed in his books "Essays On Heredity" and "The Germ Plasm" (1891-1896), constitute perhaps the most advanced ideas that 19th-century science reached on the subject, just prior to the discovery of Mendel's paper at the dawn of the 20th century and the emphasis on particulate, rather than blending, inheritance[982].

Meiosis, Fertilization, Heredity, and Variation

Weismann reviewed at length the opinions of contemporary scientists on inheritance – Galton, de Vries, and Nägeli in particular. It was already known at the time that the nucleus of the egg cell contains the hereditary material, and that the egg and the sperm cells contribute equally to heredity, despite the differences in their size. It was also known that the hereditary characters within the nucleus seem to be concentrated in the chromosomes – and that each species has a characteristic, constant number of chromosomes. The shapes and behaviors of the chromosomes had been described in detail in some species.

It was assumed for a long time that the egg alone is responsible for embryonic development: the entry of the sperm into the egg cell was thought only to bring the dormant egg to life and stimulate it to grow ("rejuvenation"). Weismann declared this idea to be utterly false. At fertilization, the

[982] Chapter 23.

nucleus of the egg cell combines with a sperm from another individual, carrying different hereditary tendencies. However, the total hereditary material of the resulting offspring cannot be double that of the parents. Therefore, the material in each sperm and egg must be halved in the reduction division (meiosis). Fertilization enables the mixing together of hereditary material from two different individuals (*amphimixis*).

> The doubling of the idants {=chromosomes] before the "reducing division" possesses this very significance: it renders possible an almost infinite number of different kinds of germ plasm, so that every individual must be different from the rest. And the meaning of this endless variety is to afford the material for the operation of natural selection[983].

Weismann's ideas on heredity, including his understanding of the role of meiosis and fertilization, were theoretical, based on logical and philosophical arguments, and not on experimental data. Contrary to the opinions of Darwin and Wallace – and to his own earlier ideas – Weismann suggested that hereditary variation in natural populations was not a direct product of exposure to different external environmental conditions, but due to the internal process of sexual reproduction. The contents of the chromosomes of individuals must change with each new fertilization event. As a result, the variation in the population increases, producing the necessary material for natural selection to work on[984]. The role of sexual reproduction is, therefore, to constantly renew the variety of combinations by mixing the chromosomes in the population with new individuals.

> The germ cells of an individual contain many different combinations, and frequent repetitions of amphimixis never indeed result in germ plasm of the same parents containing the same combinations[985]. Offspring and parents are never identical. Variation thus forms an integral part of heredity. After the precedence of Darwin and Wallace, we regard [individual variation] as the foundation of all the processes of natural selection[986].

Nevertheless, the process of amphimixis cannot produce new variations: it is only a process of mixing together existing characters. How new variations come about is not clear to Weismann any more than it was to Darwin:

> The process of amphimixis… is not the primary cause of hereditary variation. By its means those specific variations which already exist in a species may be continually blended in a fresh manner, but it is incapable of giving rise to new variations[987]. The origin of a variation is equally independent of selection and amphimixis, and is due to the constant recurrence of slight inequalities of nutrition in the germ plasm[988].

Many unicellular organisms do not reproduce sexually, but Weismann was convinced that natural selection must nonetheless be at work in unicellular as well as in multicellular organisms. The process of conjugation, observed in some unicellular organisms, must serve the same purpose as sexual reproduction:

> I am convinced that the two forms of amphimixis – namely the conjugation of unicellular, and the sexual reproduction of multi-cellular organisms – are means of producing variation. The process furnishes an inexhaustible supply of fresh combinations of individual variations which are indispensable to the process of selection[989].

---

983 Weismann 1892, p. 135.
984 Weismann, The Germ Plasm, pp. 247-249.
985 Ibid. p. 250.
986 Ibid. p. 410.
987 Ibid. p. 414.
988 Ibid. p. 431.
989 Weismann 1893: The Germ Plasm, p. 413.

The Inheritance of Acquired Characters – Lamarck to Weismann (1809 to 1893)

That the external environment influences the inherited tendencies in the newborn was a common belief in the 19th century. The earliest documented experimental support of this belief actually references the biblical Jacob:

> [Who] took him rods of green poplar, and of the hazel and chestnut tree, and pilled white strakes in them. And made the white appear, which was in the rods. And he set the rods which he has pilled before the flocks in the gutters in the watering troughs when the flocks came to drink. And the flocks conceived before the rods, and brought forth cattle ringstraked, speckled and spotted[990].

The belief that traits which were acquired or modified in an individual's lifetime are heritable (Lamarck's Second Law[991]) – was unchallenged until the late 19th century. Darwin did not hesitate to apply "use and disuse" in cases in which natural selection did not seem sufficient to explain a phenomenon.

> ...Species have been modified during a long line of descent. This has been effected chiefly through the natural selection of numerous, successive, slight, favorable variations, aided in an important manner by the inherited effect of the use and disuse of parts[992].

In "The Descent of Man", Darwin dedicated three pages to the inheritance of characters acquired through use or disuse of organs – adding that it is unknown whether these changes are heritable, but they probably are[993]. Ernst Haeckel, in his "History of Creation", wrote as late as 1876:

> This is an important point upon which very much depends. An organism can transmit to its descendants not only the qualities of form, color, and size which it has inherited from its parents, but it can also transmit changes of these qualities, which it has acquired during its own lifetime through the influence of outward circumstances, such as climate, nourishment, training etc.[994].

Contrary to this general belief, Weismann insisted that acquired characters are not heritable. In a series of lectures in 1883 (entitled "Essays on Heredity", translated into English shortly after their publication in German), Weismann stated that repeated work and exercise in fact improve the performance of the individual's muscles and nerves: for example, repeated exercise certainly improves the performance of pianists – but this improvement is not heritable:

> The increase of an organ in the course of generations does not depend upon the summation of exercise taken during single lives, but upon the summation of more favorable predispositions in the germs... We cannot by excessive feeding make a giant out of a germ destined to make a dwarf... or the brain of a pre-destined fool into that of a Leibnitz or a Kant, by much thinking[995].

---

[990] Genesis 30: 37. That this belief persisted well into the 20th century is shown by T.H. Morgan's following statement: "Jacob's slippery trick with the rods will be long remembered: the world today is filled with old wives' tales of pre-natal influences" (Morgan 1925, p. 157).

[991] Chapter 6.

[992] Darwin 1898b (1868, Origin), 6th ed., Conclusions.

[993] Darwin 1874, p. 418.

[994] Haeckel 1876, p. 158.

[995] Weismann 1891-2, On Heredity, p. 85.

Weismann surveyed the published reports on the transmission of acquired characters[996]. Many of these were stories of female cats with mutilated tails producing tailless offspring, or of a bull which accidentally lost his tail and sired tailless calves, and the like. Weismann recalls a case of inheritance of a mutilated cat's tail that was presented at a scientific conference; but upon further inquiry it turned out that there was no record of the father of the tailless progeny. The supporters of the Lamarckian principle [the transmission of acquired characters] have nonetheless always pointed to the transmission of mutilated tails as one of their principal lines of evidence[997].

Weismann points out that the spontaneous appearance of tailless puppies – regardless of any accident – was described in dogs by Bonnet in the 18th century, and was also known in different strains of domestic cats. He suggested that the reason for the frequent occurrence of taillessness was that the absence of a tail did not impair the fitness of cats or dogs.

An amusing case was brought to his attention. A ewe broke her leg by accident and then recovered. It was claimed that she gave birth to a lamb and, on one leg of this lamb, exactly on the same spot as of its mother's fracture, there appeared a black band of wool. Weismann recalls that he suggested that the black band should have been accompanied by a sign "In memory of the broken leg of my dear mother"[998].

To determine whether mutilations are heritable, Weismann carried out a controlled experiment. He cut the tails off seven female and five male mice, and reared their offspring to maturity. All 333 offspring in the first generation had normal tails. Fifteen of these were treated in the same way (tail amputation). The second generation ($n = 237$), third ($n = 158$), fourth ($n = 138$), and fifth generation ($n = 41$) mice all had normal tails. The experiment continued for 19 generations with the same result: mutilation of the parents was not transmitted to the offspring. Weismann noted that although people may suggest that if he had continued this for 1,000 generations, an effect may have been found, there exists evidence from human populations that this is unlikely:

> Furthermore, the mutilations of certain parts of the human body, as practiced by different nations for time immemorial, have not in a single instance, led to the malformations or reduction of the parts in question. Such hereditary effects have been produced neither by circumcision, nor the removal of the front teeth, nor the boring of holes in the lips or nose, nor the extraordinary artificial crushing and crippling of the feet of Chinese women[999].

Weismann concludes emphatically:

> It has never been proved that acquired characters are transmitted, and it has never been demonstrated that, without the aid of such transmission, the evolution of the organic world becomes unintelligible... The hereditary transmission of acquired characters remains an unintelligible hypothesis, which is only deduced from the facts which it attempts to explain[1000].
>
> The mere fact that the assertion has been hitherto accepted as a matter of course by almost every one, and has been doubted by a very few... cannot be taken as any proof of its validity[1001]... **The truth cannot be decided upon by a consensus of opinion[1002].**

---

996 Weismann 1891, Essays on Heredity, I: 433ff.

997 Ibid. I: 448.

998 Ibid. I: 460.

999 Ibid. I: 446.

1000 Ibid. I: 81; I: 83. A.R. Wallace was impressed by Weismann's arguments that acquired characters are not heritable, and cites a few trivial stories to show how unfounded these stories are (Wallace 1893 a, b).

1001 Weismann 1891, On Heredity, p. 402.

1002 Ibid. p. 396.

The Theory of the Germ-Plasm

> The substance with which [the phenomena of heredity] must be connected... transfers its hereditary tendencies from generation to generation, at first unchanged, and always uninfluenced in any corresponding manner by that which happens during the life of the individual that bears it[1003].

Weismann did not only reject the belief in the inheritance of acquired characters: he constructed a theory that explains why such characters cannot possibly be transmitted. His theory was based on observations on the reproduction of primitive crustaceans and hydrozoa.

Weismann suggested that the hereditary content in the fertilized egg nucleus – called the idioplasm or *germ-plasm* – separates from the rest of the cells of the embryo (the *soma*) very early on in embryogenesis and remains separate. Only the germ-plasm is transmitted to the offspring. This precludes the possibility that any changes which affect the soma during later life – like mutilations – be passed on to the next generation:

> I have attempted to explain heredity by supposing that in each ontogeny, a part of the specific germ plasm contained in the parent egg cell, is not used up in the construction of the body of the offspring, but is reserved unchanged for the formation of the germ cells of the next generation[1004].
>
> By acquired characters I mean those not preformed in the germ, but which arise through special influences affecting the body or individual parts of it. It is an inevitable consequence of the theory of the germ plasm... that somatogenic variations are not transmissible[1005].
>
> I can therefore state my conviction... that all permanent (i.e., hereditary) variations of the body proceed from modifications of the primary constituents of the germ. And that neither injuries, functional hypertrophy and atrophy... or any other influence of the environment on the body can be communicated to the germ cells and so become transmissible[1006]... All these, with use and disuse, may perhaps produce great effects upon the body (soma) of the individual, but cannot produce any effect in the transformation of species, simply because they never can reach the germ cells[1007].
>
> If these views... be correct, all our ideas upon the transformation of species require modification, for the whole principle of evolution by means of exercise (use and disuse) as proposed by Lamarck and accepted in some cases by Darwin, entirely collapses[1008].

Weismann's Bold Hypothesis

Weismann contended that the germ-plasm is continuous between generations: at every stage of ontogeny, the information in the germ-plasm of the fertilized egg cell is retained, unchanged, for building the sex cells of the next generation. The continuity of the germ-plasm enables the transmission of all the species-specific characteristics between generations throughout evolution, unchanged since life first began:

[1003] Ibid. p. 105.
[1004] Weismann 1891-2, Essays on Heredity. p. 170.
[1005] Weismann, 1893. The Germ Plasm, p. 392.
[1006] Ibid. p. 395.
[1007] Ibid. p. 400.
[1008] Ibid. Preface.

> If, as I believe, the... germ plasm has remained in perpetual continuity from the first origin of life, and if the germ plasm and the substance of the body, the somatoplasm, always have occupied different spheres, and if changes in the latter only arise when they have been preceded by corresponding changes in the former, then we can, up to a certain point, understand the principle of heredity[1009].

The Structure of the Hereditary Substance

Heredity means the transfer of the characteristics of an organism to its progeny. Physically, both the characteristics shared by all members of a species, and the particular characteristics of each individual, must be contained in the sex cells:

> The physical cause of all apparently unimportant hereditary habits or structures... must all be contained in the minute quantity of germ-plasm which is possessed by the nucleus of the germ cell – not indeed as the pre-formed germs of structure – the gemmules of Pangenesis – but as variations in its molecular constitution[1010].

Weismann's theoretical model of the structure of the hereditary substance was quite similar in principle to the model adopted by geneticists fifty years later. The basic assumption is that characters have a physical basis and are inherited without "blending".

Weismann, Non-heritable Mutilations

> We must at any rate conclude from the facts of the case, that the characters of one parent may be strictly inherited without any apparent intermingling with the corresponding characters of the other parent[1011].

It is not easy to follow Weismann's theory because the terminology he used differs from the lexicon of current genetics. His model is based on the processes of development as known at the time – and on logical assumptions, not on experimental work. Weismann's model provides explanations for two processes: first, the transmission of characters unchanged from one generation to the next; and second, the regulation, by the germ-plasm, of the differentiation of the somatic cells in every generation to produce the complete embryo (ontogeny).

According to Weismann, the hereditary matter in the nucleus of the fertilized egg (the idioplasm) is composed of units called biophores. "The biophores are not, by any means, hypothetical units: They

---

1009 Ibid. I: 105.
1010 Ibid. p. 194.
1011 Weismann 1893, The Germ Plasm, p. 281.

must exist"[1012]! Functional groups of biophores are arranged in a definite order near each other, and constitute a determinant.

> Biophores determining a cell not only lie close together in the germ plasm... but they combine to form a higher unit. The determinant is not a disconnected mass of different biophores, but a vital unit of a higher order[1013].

Groups of determinants, which control a specific function, are called ids (equivalent to the modern term genes). The chromosomes are strings of ids (for the sake of uniform nomenclature, Weismann suggested that chromosomes should be renamed idants). The order of the ids on the idant is important:

> It is at least possible, and perhaps the rule, for the order and arrangement of the ids in the idant to remain unchanged from the germ cells of the parent to those of the offspring. The chromosomes (idants) only apparently break up during the nuclear resting stage, but in reality persist... I imagine that after the period of the resting stage, they are composed of the same ids, for the most part arranged in series similar to those which existed before the preceding nuclear division[1014].

However, Weismann needed to incorporate in his theory the possibility that the sequence of ids may undergo change, to accommodate new characters:

> We must not regard this constitution [of the idants] as for ever unchangeable. The universally observed change of individuality, which takes place in the course of generations... suggests to my mind an occasional change in the arrangement of the ids within the idants[1015].

Weismann suggested that most of the ids are inactive (dormant), and are transmitted while dormant to the next generation to form the idioplasm of the sex cells of the embryo. This assumption, which Darwin had already suggested in his "Domestication", was necessary to account for cases in which the child resembles only *one* of the parents, while the characters of the other parent are not expressed, or cases in which characters not expressed in the parents appear in later generations ["reversion"], although Weismann doubted the validity of such cases:

> Instances of a descendant of an ancient family reverting to a great-grandfather whose characters were not present in the intermediate generations are certainly occasionally mentioned in novels, but it is only natural to doubt the accuracy of such cases, even when they claim to be true[1016].

Ontogeny

The idioplasm in the fertilized egg cell contains all the information needed for building the body of the embryo, with all its cell types and different tissues. In each division of the fertilized egg cell, the daughter cells have fewer and fewer potentialities to develop into different types of cells: the hereditary message in a given daughter cell becomes more and more specialized, so that ultimately each differentiated daughter cell can give rise to only one type of cell. Only the sex cells retain the ability to give rise to the complete variety of cells. Therefore it is unlikely that any somatic cell may at some stage of ontogeny give rise to sex cells[1017].

[1012] Weismann 1891-2, On Heredity, pp. 43-44.
[1013] Weismann 1893, The Germ Plasm, p. 60.
[1014] Weismann 1891-2, On Heredity, pp. 130-132.
[1015] Ibid.
[1016] Weismann 1893, The Germ Plasm, p. 313.
[1017] Weismann 1891-2, On Heredity, pp. 197-199.

> We suppose that the process in the idioplasm which brings about the ontogeny of multicellular organisms is due to thousands of determinants, which constitute the germ plasm of the fertilized ovum, becoming systematically separated into groups, and distributed among the successors of the egg cell... until each cell contains determinants of one sort only[1018].

> From the moment when the phenomena that precede segmentation commence in the egg, the exact kind of organism which will be developed is already determined... In spite of this, there still remains a certain scope for the influence of external conditions... but this scope is limited[1019].

The ontogenetic changes in the developing embryo are controlled by the hereditary material in the nucleus of the fertilized egg – by means of materials transported from the nucleus to the cytoplasm. Weismann suggested that this transport is due to the disintegration of the determinants into their smaller units, which alone can pass through the nuclear membrane.

> It seems to me that the determinants must ultimately break up into the smallest vital elements of which they are composed, the biophores, and that these migrate through the nuclear membrane into the cell substance.

> It does not seem to me theoretically necessary to assume that the particles which migrate into the cell bodies should themselves be chlorophyll or muscle particles. They may quite well be only the architects of these, that is to say, particles which by cooperation with the elements already present in the cell body give rise to chlorophyll or muscle substance[1020].

Modifications of cell structure occur during ontogeny.

> A struggle for food and space must take place between the protoplasmic elements already present and the newcomers, and this gives rise to a more-or-less marked modification of the cell structure[1021]. The struggle does not occur between the elements of the "reserve germ plasm" which brings about the formation of the germ plasm of the progeny[1022].

Life and Death

In a long article, entitled "The Duration of Life", Weismann argued that unicellular organisms are in one sense immortal: they reproduce by fission, and the concept of death does not apply to lineages in which daughter cells keep re-dividing and giving rise to daughter lineages (death is defined as an irreversible cessation of all processes of life – unlike a temporary stage like a cyst, which resumes life after a resting stage[1023]). The concept of death, suggested Weismann, becomes applicable only when multicellular organisms are formed. Multicellular organisms have a selective advantage due to the principle of division of labor: cell colonies in which different cells perform different functions, enable the increased efficiency of each cell type to perform its special function[1024]. In multicellular organisms the cells preserving the immortality – the sex cells – are separated early, while the rest of the cells (the soma) take over all the other functions of the organism. From this point of view, the

[1018] Weismann 1893, The Germ Plasm, p. 225.
[1019] Ibid, I: 104.
[1020] Weismann 1891, p. 379-780. In modern terminology, regulation is effected by the molecules of mRNA passing through the nuclear membrane into the cytoplasm, to be translated into proteins.
[1021] Weismann 1893, The Germ Plasm, pp. 45-47.
[1022] Weismann 1893, The Germ Plasm, p. 295.
[1023] Weismann 1891-2, On Heredity, p. 114.
[1024] Ibid. p. 76.

soma appears to be a secondary appendage of the real bearers of life – the reproductive cells[1025]. The soma is the part that dies. The death of surplus cells (and of old organisms past the reproductive age) has a selective advantage, because these competed for resources with younger organisms that are able to reproduce and contribute to the growth of the population:

> The somatic cells (the soma of the organism) died, but the characters of the species lived on – transmitted by the sexual cells. This is the only way that the same organisms could appear, generation by generation, with only slight modifications[1026].

Evolution

Weismann suggested that the germ-plasm is continuous and has remained unchanged (within a lineage) since the beginning of life. All variation within populations is a result of amphimixis or conjugation, and natural selection is the force behind species formation. Therefore the origin of variation and differences *among species* must be looked for in the variation among the early, primitive, unicellular lineages, caused by selection and the differential mortality of lineages. The primitive origin of the biological world must have been *polyphyletic!*

> The dissimilarity of individuals must be traced to the primordial organisms... from these organisms the dissimilarity was transmitted to the unicellular forms, which cannot all have originated from one primordial organism, but each species must have arisen polyphyletically from a large number of modified organisms[1027].

According to Weismann, the driving mechanism of evolution in the biological world is, and always has been, natural selection. This removed the least-fit individuals, those that "from the germ were feebly disposed". Useful characters were first found in a few individuals (less-useful characters were to be found in many individuals, but would have been removed by selection). Thus natural selection, by reducing the variation *within* populations, in this way augmented the differences *among* populations and species.

Criticism of Darwin's Views

Pangenesis: Weismann admired Darwin and was a dedicated supporter of his theory, but he totally rejected the idea of *Pangenesis*. The early separation of the germ – plasm from the soma precludes the possibility that the "gemmules", if they existed in the blood stream, will reach the germ cells. Respectful of Darwin, Weismann offered an honorable dismissal of the idea:

> I consider the gemmules to be a deliberate invention... which has no claim to reality: the gemmules merely serve to show the sort of suppositions we must make in order to understand the phenomenon of heredity[1028].

Instincts: Darwin believed that small, heritable behavioral changes in individuals gradually accumulate by natural selection and are transmitted from one generation to the next as instincts. But, Weismann objected, this is unlikely because acquired habits cannot be transmitted to the germ-plasm:

[1025] Ibid. p. 142.
[1026] Ibid. p. 136.
[1027] Weismann 1893, The Germ Plasm, p. 419.
[1028] Weismann 1891-2, On Heredity, Remarks on some problems of the day, p. 80.

> The origin of instincts may... be referred to the process of natural selection, although many observers had followed Darwin in explaining them as inherited habits – a view which becomes untenable if habits adopted and practiced in a single life cannot be inherited[1029].

Disuse: Darwin interpreted the evolution of retrogressive development (like the reduction or loss of eyes in cave-dwelling and subterranean organisms) to the inherited effects of disuse. Weismann suggested that such traits should be investigated in great detail, because the evolution of such "degenerate forms" could teach us more about the changes in organic nature than the study of progressive changes[1030]. He explained that for a trait or a limb to be properly functional, natural selection must continuously work for its improvement. The evolution of "degenerative" characters is due to the *suspension* of natural selection in a new environment – when the continuous selection for further improvement of the character is no longer required. When selection is suspended, negative changes may accumulate and the character atrophies[1031].

As an example, Weismann noted the degeneration of normal food-collecting and brood-caring instincts in the evolution of "slave-maker" ants, since they rely on the slave ants for these jobs and no longer do them:

> So it is not only among men that there is a curse upon slavery: even animals become degraded by it[1032].

Plants: Weismann dedicates a long article in his book to inheritance in plants. Many of his opponents claimed that even if not in animals, acquired characters are certainly heritable in plants. Weismann admitted that he was not very familiar with plants since he had not been trained as a botanist, but insisted that the separation of germ and soma must equally be present in plants as in animals.

Germinal Selection: Near the end of the 19th century, the supremacy of natural selection as the major factor in evolution was challenged by important writers, including Herbert Spencer, Hugo de Vries,[1033] and others. Weismann, a devoted Darwinian, published a long paper in defense of natural selection, attempting to answer the critics. In the paper, entitled "Germinal Selection", he sought to explain the apparent long-term directional trends in some characters. He found it difficult to account for the regular improvement of useful structures with time ("the phyletic processes of transformation"), if new random variations have to be selected every generation. Rather, he suggested that there should be some connection between the *utility* of the change in the character (for the individual) and the *probability* of its occurrence:

> My inference is a very simple one: if we are forced by the facts on all hands to the assumption that the useful variations which render selection possible are always present, then some profound connection must exist between the *utility* of the variation and its actual appearance, or in other words, the *direction* of the variation of a part must be determined by its utility. By a selection of the kind referred to, the germ is progressively modified in a manner corresponding with the production of a definitely directed progressive variation of the part[1034].

1029 Weismann 1891-2, Essays on Heredity, p. 401.
1030 Weismann 1891-2, On Heredity, II: 3.
1031 Ibid. II: 15-17. This idea was adopted many years later as "The red queen hypothesis" by Van Valen, 1973, referring to Lewis Caroll's Red Queen in "Through the Looking Glass": "It takes all the running you can do to stay in the same place".
1032 Ibid. II: 27.
1033 Chapter 28 below.
1034 Weismann 1896, [The Monist, 6: 250-293]; cited by Morgan 1903, p. 162.

As an example, Weismann cited the breeding in Japan of a domesticated rooster strain, with an extraordinary six-foot-long tail, by means of persistent selection of small variations in tail length. Weismann suggested that the camouflage color patterns on butterfly wings, or the marvelously close resemblance of some insects to the plant part on which they alight, could have been improved in a similar manner. In Weismann's interpretation, selection of individuals carrying a trait "in a plus or minus direction from the mean" ["personal selection"] triggers a change in the germ-plasm, which subsequently continues in the same direction:

> The initial impulse conditioning the independent maintenance of useful direction of variation in the germ plasm must rather be sought in the utility of the modification itself... For as soon as personal selection favors the more powerful variations of a determinant... at once the tendency must arise for them to vary still more strongly in the plus direction...
>
> Thus, I think, may be explained how personal selection imparts the initial impulse to processes in the germ plasm, which they are once set going, persist in themselves in the same direction, and are, therefore, in no need of the continued supplementary help of personal selection as directed exclusively to a definite part[1035].

In Support of Weismann's Theory

A.R. Wallace reported being convinced by Weismann's arguments concerning the non-inheritance of mutilations: it has long been known to breeders that if a dog or a horse, of a good breed, is hurt or injured – even to the extent that it can no longer perform the kind of work it was bred for – the breeders still keep the injured animal for breeding and propagation of the qualities of the breed, since they know [from experience?] that the injury is not heritable. But the problem discussed by Weismann has a more profound aspect:

> I propose to waste no time on the question whether mutilations are ever inherited, because both parties are now agreed that this is not the point at issue. What we want to know is, whether the effects produced during the lives of individuals, by such natural causes as the use and disuse of muscles or organs, change of food, or change of climate, are transmitted to offspring, so as to accumulate such effects and thus serve as an important factor in evolution[1036].

If such characters are inherited, Wallace argued, we should have been able to detect this effect in the human population: young people should excel their parents in their abilities, and children born later – when their parents have gained more experience in life – should be better in their professions than children born earlier. This is certainly not the case: high skill or proficiency is not heritable.

> Men of exceptional genius or mental power appear suddenly, rising above their immediate ancestors, and they are usually followed by successors who... rarely equal their parents[1037].

Wallace discusses in detail a case elaborated upon by Herbert Spencer, who had argued strongly in support of the inheritance of acquired characters. Pupae of a species of moth (*Saturnia*) were shipped from Texas to Switzerland and bred there. The larvae hatching from their eggs were fed a different food to that of their parents in Texas. When these larvae matured to adults they were so different from their Texan ancestors that they were considered a different species. Spencer

---

[1035] Weismann 1896, p. 276. This seems to be a reversal of Weismann's statement, that changes in the germ plasm must have preceded any changes in the soma. Weismann 1891, p. 105.

[1036] Wallace 1895, in Fortnightly Review, Vol. 53.

[1037] Ibid.

considered that the novel adult characters had been acquired due to the change in diet and rearing conditions. Wallace, in contrast, concluded that

> Being a change produced in the body by the environment, it is not hereditary. In Weismann's phraseology, it is somatic variation, not a germ variation. Our conclusion: no case has yet been made out of the inheritance of individually acquired characters[1038].

Objectors to Weismann's Theory

George J. Romanes: Weismann's theories evoked a great deal of antagonism. The biologist George J. Romanes dedicated a book – entitled "Weismannism" – to criticism of Weismann's theories, in particular the idea of the stability and continuity of the germ plasm since the origin of life[1039]. The book, published just before the re-discovery of Mendel's paper, reflects the "armchair biology" that prevailed in the 19th century – including most of Weismann's published works: philosophical and logical arguments with little, if any, use of facts from observation or experiment. Referring to Weismann's model of hereditary material, Romanes wrote:

> In reading the account one is reminded of that which is given by Dante of the topography of the inferno. For not only is the "sphere" of germ plasm now composed of nine circles (molecules, biophores, determinants, ids, idants, idioplasm, somatic idioplasm, morpho-plasm, apical-plasm)... We return to the fields of science with a sense of having been indeed in some other world[1040].

Romanes noted that Weismann, in his later publications, had modified his germ-plasm theory and conceded that external conditions may have a slight effect on heritable characters. This modified theory was similar to the theory of the "stirp" expressed ten years earlier by Darwin's cousin Francis Galton[1041]. Romanes, being English, rose to the defense of his countryman's priority rights:

> Galton... had what I regard as the sound judgement to abstain from carrying his theory of stirp to any such transcendental "sphere" as that which is occupied by Weismann's theory of the germ plasm[1042]. Stirp is continuous, but unlike the germ plasm, is not necessarily or absolutely so. Again, stirp is stable, but unlike the germ plasm, is not perpetually or unalterably so[1043]. I must repeat that it makes a literally insurmountable difference whether we suppose, with Galton, that the Lamarckian factors may sometimes and in some degree assert themselves, or whether we suppose with the great bulk of Weismann's writings and in accordance with the logical requirements of his theory, that they can never possibly occur[1044].

Thomas Hunt Morgan: Morgan (1903), in a critical review of the theory of natural selection, objected strongly to Weismann's theory of utility-directed variation:

> This is indeed the old methods of the Philosophers of Nature: an imaginary system has been invented which attempts to explain all difficulties – and if it fails, then new inventions are to be thought of. Weismann is mistaken when he assumes that many zoologists object

---

[1038] Ibid.
[1039] Romanes 1899, p. 86.
[1040] Ibid. p. 118.
[1041] Chapter 21. The hereditary material that was not used in the construction of the body but transferred to the offspring, Galton named *stirp*.
[1042] Romanes 1899, p. 63.
[1043] Ibid. p. 66.
[1044] Ibid. p. 69.

> to his methods because they are largely speculative. The real reason is that the speculation is so often of a kind that cannot be tested by observation or experiment[1045].

20th-century Comments

Research in the field of science education in the 20th century suggested that the belief that acquired effects of "use and disuse" are heritable, so commonly accepted in the 19th century, is understandable in the absence of knowledge of the hereditary transmission of characters before the re-discovery in 1900 of Mendel's (1865) paper[1046]. Schoolchildren, high-school students, and even university undergraduates in biology and medicine – when asked before they had been taught or instructed in the Theory of Evolution – tended to explain biological phenomena in Lamarckian terms. For example:

> Only 18% of first-year university students with an advanced – level Biology background were consistently able to apply this concept [=natural selection] to common environmental problems[1047] [In a survey of 150 first-year medical students in the USA]. These results clearly demonstrate that the majority of these otherwise very able science students leave school believing that evolutionary change occurs as a result of need, i.e., Lamarckian view. Students seem to be extrapolating from changes (which they call adaptations) seen within the lifetime of an individual to account for changes seen in populations selected over many generations[1048].

1045 T.H. Morgan, 1903, p. 166.
1046 Chapter 25 below.
1047 Brumby 1979.
1048 Brumby 1984.

21

Francis Galton: Quantitative Measurements of Heredity

If I wanted to know how to put a saddle on a camel's back without chafing it, I should go to Francis Galton. If I wanted to know how to manage the women of a treacherous African tribe, I should go to Francis Galton. If I wanted an instrument for measuring a snail, or an arc of latitude, I should appeal to Francis Galton. If I wanted advice on any mechanical, of any geographical, or any sociological problem, I should consult Francis Galton. In all these matters, and many others, I feel confident he would throw light on my difficulties. And I am firmly convinced that, with his eternal youth, his elasticity of mind, and his keen insight, he can aid us in seeking an answer to one of the most vital of our national problems: how is the next generation of Englishmen to be mentally and physically equal to the past generations, which have provided us with the great Victorian statesmen, writers, and men of Science. [Carl Pearson][1049].

His Life and Science

As a young man, Darwin's cousin Francis Galton (1822-1911) studied medicine and even acquired some medical experience by working in hospitals. He interrupted his university studies for a year and traveled to southern Europe and the Middle East, including the Holy Land and Egypt. When his father died, leaving him a large fortune, he decided that he did not have to earn his living as a doctor and quit his medical studies at Cambridge altogether.

In 1845, Galton organized two research expeditions to then unknown areas in western Africa. The expeditions did not reach their respective goals, but he published the experience gained from them in an illustrated book on the necessary preparations for expeditions in undeveloped countries[1050]. The book covers methods of setting up camp, building a comfortable tent, hunting and fishing for food, lighting a fire, crossing difficult terrain such as cliffs and rivers, finding one's way back to the caravan after being lost, and instructions for preparing tea properly at the right temperature. He recommended that a herd of cattle and sheep driven by local shepherds should be arranged to accompany the travelers as a source of meat and milk. The book earned Galton membership in the Geographical Society. His later contributions to science earned him membership in the Royal Society.

Galton was one of the founders of the science of statistics, and initiated the application of statistics to the analysis of heredity. He was the first to use population surveys to collect data on heredity in humans, which he analyzed for his book, "Natural Inheritance". He was interested in the inheritance of outstanding talents of individuals ("geniuses") and collected data from pedigrees of such individuals. His analysis convinced him that these talents were heritable ("Hereditary Genius"). Further statistical analysis led to the "law of regression to the mean", showing that the characters of the offspring of outstanding individuals tend to be, on average, less extreme than those of their parents and closer to the population mean.

[1049] Karl Pearson, 1904, at a meeting of the Society for Sociology, London.

[1050] Galton 1971 (1872). The Art of Travel. Stockpole Books, Harrisburg, Pennsylvania, USA.

The study of twins – again using a population survey – enabled him to estimate the relative contribution of heredity versus the environment to the inheritance of human characteristics ("Nature versus Nurture").

Among other studies of heredity, Galton tested Darwin's Pangenesis hypothesis and disproved it: he performed a blood transfusion between two rabbits and demonstrated that "gemmules" of rabbit color did not circulate in the blood (he offered a modified theory instead; see below). He also examined statistically whether prayers [for better health and long life] increase people's life span. If prayers were effective, then royalty and the clergy should live longer than people of other professions. His findings showed that there was no such trend:

> The public prayer for the sovereign for every state, Protestant and Catholic, is and has been… "Grant her in health long to live"... Now the sovereigns are literally the shortest-lived of all who have the advantage of affluence. The prayer has therefore no efficacy, unless the very questionable hypothesis be raised, that the conditions of royal life may naturally be more fatal, and that the influence is partly neutralised by the effects of public prayer[1051].

Galton established statistically that the patterns of ridges in the fingerprints of humans are unique to each individual. He published a book on the use of fingerprints for identification – a method still in use today in forensic work worldwide.

Galton was one of the founders of the "Eugenics" movement, which sought to protect the human race from the accumulation of deleterious changes. He left a large sum of money in his will for a Chair in Eugenics at University College, London. His friend and collaborator, the scientist Karl Pearson, was the first incumbent of the Chair.

Statistics

Statistics, in the sense of collecting and listing data on human populations, had been used in the 18th century and was treated with some suspicion (Benjamin d'Israeli, former British Prime Minister, was reported to have said that there existed "three kinds of lies: lies, damned lies, and statistics"[1052]. Thomas Malthus[1053] had used the data of population size in the American colonies to support his theory that populations are limited by food supply. Galton, however, was one of the founders of statistics as a science, based on probabilities. He used the tables of the "law of deviations from the average" worked out by the Belgian astronomer Adolph Quetelet (1796-1874). Quetelet based his tables on measurements of the chest circumference of 5,738 Scottish soldiers, and discovered that if they are arrayed from the smallest to the largest, the distribution appeared symmetrical, with most observations around the mean value and fewer cases at either end. Galton checked the distribution using published data on the stature of 100,000 French soldiers and found that the same law could be applied.

Galton devised a simple instrument for demonstrating the law of deviation from the average. This was basically an inclined board, into which regular rows of pins were inserted. Marbles released at the top struck the pins randomly on their way down, until they accumulated in bins placed at regular intervals at the bottom. The resulting distribution was a smooth curve (the normal distribution)[1054]. Galton enthusiastically described the outcome of his studies:

[1051] Galton 1872, "Statistical enquiries into the efficacy of prayer".
[1052] Quoted in Ruff, 1954, "How to lie with statistics", Norton, New York.
[1053] Chapter 1.
[1054] Galton 1889: "Natural Inheritance".

> Whenever a large sample of chaotic elements are taken in and marshaled in order of their magnitude, an unexpected and most beautiful form of regularity proves to have been latent all along[1055].

"Hereditary Genius"

Galton considered that mental characters and special talents, not only morphological traits, might be hereditary. In his book "Hereditary Genius", first published in 1862, he analyzed the pedigrees of outstanding and famous people in Europe – judges, political and military leaders, and church dignitaries, as well as musicians, painters, and scientists, and calculated the frequencies of talented people among their offspring, compared with their frequency among the general public. His analysis revealed that outstanding individuals were more frequent in the pedigrees of "geniuses" than in those of the general public, confirming that genius is indeed heritable.

He described in detail in his book his unconventional approach to the choice of talented people for analysis. To assess the amount of talent with which a person was endowed, he relied on the qualitative assessment of that person's qualities by his contemporaries. This was reflected in the person's reputation:

> By reputation I mean the opinion of contemporaries, revised by posterity. I speak of the reputation of the leader of an opinion, an originator, of a man to whom the world deliberately acknowledges itself largely indebted.
>
> By natural ability, I mean those qualities of intellect and disposition which urge and qualify a man to perform acts that lead to reputation. I mean a nature which, when left to itself, will, urged by an inherent stimulus, climb the path that leads to eminence, and has the strength to reach the summit[1056].

To characterize the persons he perceived as outstanding, he compared their frequency of occurrence with Quetelet's law. The most outstanding persons he defined as "geniuses":

> I am speaking of the very first-class men, prodigies, [their frequency in the general public being] one in a million, or one in ten million.

Galton searched the biographies and history books of the 17th and 18th centuries for pedigrees of more than 300 specially-endowed people [geniuses]. Among them were 109 British high-court justices who were in office between 1660 and 1865, 31 peerages, 32 European army leaders, 52 literary men, 65 scientists, 24 poets, 26 musicians and composers, 26 painters, 33 clergymen, and a few others. He counted the numbers of geniuses among the descendants of his selected persons. His analysis of the pedigrees led to his conclusion that the frequencies of famous [therefore, specially-endowed] people among the close relatives of geniuses was much higher than their frequency among similar age groups in the general population. Therefore the talents must have been inherited[1057].

His analysis of the 31 peerages led Galton to some unexpected conclusions. About 12 of the peerages had become "extinct". A peerage became extinct when the family produced no sons: daughters could inherit the family fortune but not the title. The oldest son in a peerage inherited the title, but often needed money to support his sisters (for a dowry) and his younger brothers. Marrying an

---

[1055] Ibid. p. 66. A similar instrument is used in modern science museums to illustrate the formation of a "normal distribution" when many independent, random events affect a character.

[1056] Galton 1869, "Hereditary Genius", p. 77.

[1057] Galton ignored the possibility that children reared in families of talented people may have been encouraged by their family atmosphere to make greater efforts to improve their performance in whatever subject for which the family was famous. The fact that the children of J. S. Bach became composers may not have been only due to genetics.

heiress was a good solution for both sides: a woman who was the sole issue of her family would have money and property, but no title to transmit to her children – marrying an impoverished peer provided that. However, noted Galton, heiresses had the disadvantage of coming from families that produced no sons and/or very few children:

> We might, in fact, have expected that an heiress, who is the sole issue of a marriage, would not be so fertile as a woman who has many brothers and sisters. Comparative infertility must be hereditary in the same way as other physical attributes[1058].

Further examination supported Galton's conclusion. He selected at random 50 families in which the woman was an heiress. In 11 of them no sons were born, and eight more produced only one son. In a sample of 50 families in which the mother was not an heiress, only one had no sons. The total number of children produced in the latter group was 50% larger than in the families of heiresses. Therefore, in the long run, marriages for economic needs were the cause of the loss of the family title:

> The most highly gifted men are ennobled. Their elder sons are tempted to marry heiresses, and their younger ones not to marry at all, for these do not have enough fortune to support both a family and an aristocratic position[1059].

> Every advancement in dignity is a fresh inducement to the introduction of another heiress into the family. Consequently, Dukes have a greater impregnation of heiress blood than Earls, and Dukedoms might be expected to be more frequently extinguished than Earldoms, and Earldoms to be more apt to go than Baronies. Experience shows this to be most decidedly the case[1060].

As if to balance the negative role of women as regards peerage titles, Galton found a positive contribution of the mothers of scientists to the success of their sons: that the fathers of famous scientists were often simple, uneducated small craftsmen. When the mother was the more talented parent, the child was raised in a more open-minded atmosphere:

> It therefore appears to be very important for success in science, that a man should have an able mother. I believe the reason to be, that a child so circumstanced has the good fortune to be delivered from the ordinary, narrowing, partisan influence of home education; happy are those whose mothers… showed them, by practice and teaching, that inquiry may be absolutely free without being irreverent[1061].

"Natural Inheritance"

Galton was interested in the heredity of human characters. In the absence of relevant data, he set out to build a data-base of the frequencies of individual variants in human populations, which he could analyze with statistical tools. He stated his purpose thus,

> To show that a large part is always played by chance in the course of hereditary transmission, and to establish the importance of an intelligent use of the laws of chance and the statistical methods that are based on them in expressing the conditions under which heredity acts[1062].

[1058] Ibid. p. 179.
[1059] Ibid.
[1060] Ibid, p. 186.
[1061] Ibid. pp. 246-247.
[1062] Galton, 1889, Natural Inheritance, p. 17.

Galton collected data in several ways. First, he sent a questionnaire to many households in England, offering to pay anyone who provided the requested data for all members of his family for at least three generations (responding families, and the money paid to them, are listed in his book "Natural Inheritance"). The 18 variables he asked for were biometrical measurements such as height and weight, eye color, certain descriptive characters such as artistic talents and temper, and the occurrence of certain diseases (e.g., consumption). He was highly critical in interpreting the family records he received. [For example, he disregarded the evaluation of the artistic talents of young children, if the person reporting them was a parent or grandparent of the child, because such evaluations tended to be exaggerated]. Second, Galton opened a laboratory in London, where passersby could, for a few pence, have their physical ability and strength measured. He calculated the means and standard deviations of the variables, and compared the frequency distribution of each variable with Quetelet's tables. He looked not only at the means of the variables but also the variances, and criticized those scientists who consider means only:

> It is difficult to understand why statisticians commonly limit their enquiry to averages, and do not revel in more comprehensive views. Their souls seem as dull to the charm of variety as that of our flat English counties, whose retrospect of Switzerland was that, if the mountains could be thrown into the lakes, two nuisances could be got rid of at once[1063].

Galton recognized that "discrete" (categorical) characters like eye color may be transmitted in a different manner than continuous (biometrical) variables like stature or weight. But he had no doubt that characters of both kinds are heritable:

> A man must be very crotchy or very ignorant, who nowadays doubts the inheritance of either this or any other faculty. The question is whether or no its inheritance follows a similar law to that which has been shown to govern stature and eye colour[1064].

Some characters puzzled him: for example, the heritability of diseases. One of the characters he asked for in his questionnaire was whether any member of the family had died of consumption (tuberculosis). He reports that one of every 6 or 7 persons in England had died of this disease. He found that in families where the mother had died of the disease, there was a stronger tendency for the children to contract it – than when the deceased parent was the father. This result suggested to him that the disease was acquired, not inherited: if it were a hereditary disease, it should not have mattered which of the parents died of it[1065]. [The mother often stayed at home and spent much more time with the children than the father, and therefore was more likely to pass it on to her children if she became sick].

Galton was not free of the prevalent attitude towards women in English society of the 19th century. For example, when he found a negative trend for marriage between people with artistic tendencies, he wrote:

> Every quiet, unmusical man must shrink a little from the idea of wedding himself to a grand piano, in constant action, with its vocal and peculiar social accompaniment. A sensitive

[1063] Galton, Natural Inheritance, p. 62. Galton noticed that populations could be very different even though the average remained the same. In a note appended to the book in 1889, J. Venn explained: "As regards the average, it comes to the same thing whether a quality disappears by an imperceptible faint presence in all the descendants, or by being present in a marked degree in an extremely small percentage of all descendants. But the actual concrete results are extremely different".

[1064] Galton, Natural Inheritance, p. 155.

[1065] The bacillus causing the disease was discovered in 1882 by Robert Koch, who was awarded the Nobel Prize for Medicine for his discovery.

and imaginative wife should be conscious of needing the aid of a husband who has enough plain common sense to restrain her too-enthusiastic and frequently foolish projects[1066].

The Law of Ancestral Heredity

In search of basic laws of heredity, Galton performed some experiments with moths, which failed however to provide large-enough samples for analysis. He then turned to the pedigree data from a breeder of basset hounds, who had imported 93 of them into England and bred them for 20 years. Galton analyzed the data on the inheritance of the coat color in the pedigree. On the basis of these data, as well as theoretical considerations, he formulated the following law:

> It is that the parents contribute between them, on average, one half (0.5) of the total heritage of the offspring, the four grandparents, one quarter (or 0.5^2), the eight great-grandparents, one eighth (or 0.5^3), and so on. The sum of the ancestral contributions expressed by the series, ($0.5 + 0.5^2 + 0.5^3$ + etc) being equal to 1, accounts for the total heritage[1067].

In his published paper, Galton provided a thorough quantitative analysis of the data, together with all the difficulties he had encountered, but concluded that the law of ancestral heredity can be a general law, not limited to a particular data-set. This law – based on the blending of parental characters at mating – had an immediate practical value for breeders, because it enabled them to estimate how long it should take to breed one strain into another in order to obtain a pure-breeding line. They calculated that after four or six generations of inbreeding, the selected strain would contain only a negligible amount of the parental characters and should be free of "reversion", enabling selection to be relaxed.

The "Law of Ancestral Heredity" was enthusiastically endorsed by the physicist, philosopher, and mathematician, Karl Pearson (see next chapter), as an outstanding contribution to science and the theory of heredity:

> In short, if Mr. Galton's law can be firmly established, it is a complete solution – at any rate to a first approximation – of the whole problem of heredity. If Darwinian evolution be natural selection combined with heredity, then the single statement [Galton's Law] must prove almost as epoch-making to the biologist as the law of gravitation to the astronomer[1068].

Pangenesis: An Experimental Examination of Darwin's Hypothesis

Galton put to test Darwin's "temporary" theory of Pangenesis – in which each cell in the parental body is believed to send a representative "gemmule" into the blood. The gemmules accumulate in the sex cells and are transmitted to the forming zygote for the construction of the body of the child.

Galton ran a series of studies before forming a conclusion[1069]. First, a measured quantity of filtered blood was exchanged between a standard, pure-bred strain of rabbits ("silver-grey") and rabbits of other breeds, which proved that the recipient was not damaged by the exchange. Then two rabbits – one standard and the other foreign –were connected by their jugular veins and a cross-transfusion of blood was effected (animals of either sex were used). The operations were performed by an expert surgeon ("my part in this series was limited to inserting and tying the cannules, to making the cross-connections, to recording the quality of the pulse through the exposed arteries,

[1066] Galton 1889, Natural Inheritance, p. 158.
[1067] Galton 1897, Proc. Roy. Soc. Lond. 61: 401-413.
[1068] Pearson, cited in Moore, 1986b.
[1069] Galton 1871: "Experiments in pangenesis, by breeding from rabbits of pure variety, into whose circulation blood taken from other varieties had previously been largely transfused." Proc. R. Soc. Lond. 61: 393-410.

and taking the other necessary notes!"). In preliminary experiments, Galton determined that ten minutes were sufficient for the entire quantity of blood in a rabbit to pass through the entire system. Finally, the partners of transfusion were separated, bred, and the offspring were classified by color. A total of 88 pups were obtained from 13 blood-transfused litters. All of them had retained the silver-grey phenotype – proving that the blood contained no "gemmules" of the other breed.

> The conclusion from this series of experiments is not to be avoided: that the doctrine of pangenesis …as I interpreted it, is incorrect[1070].

Galton's Theory of Heredity

Galton summarized his ideas about heredity in a paper published in 1872, and later – in a more detailed form – in 1875. These ideas had little impact at the time. Strong support for his ideas was first expressed in 1899, by Romanes, as a reaction to Weismann's publications "On Heredity", as if to defend the rights of his countryman.

Galton did not abandon the possibility that some external influences do affect the hereditary material. He was reluctant to give up the Pangenesis theory completely:

> Individual variation depends upon two factors. The one is the variability of the germ… the other is that of all external circumstances… The law of heredity goes no further than to say, that like tends to produce like. The tendency may be very strong, but it cannot be absolute[1071].

According to Galton, the particles which constitute the germ cells carry the characters of the father and the mother and multiply, like Darwin's gemmules. The question as to which of the characters will be expressed in the child is decided by majority rules among competing particles, like in Parliament[1072]. The surplus particles which were not expressed in a given generation are passed on as "stirp" to the next generation. Acquired characters may be heritable but only rarely,

> We may be confident …that acquired modifications are barely, if at all, inherited in the correct sense of the word. I propose, as already stated, to accept the proposition of their being faintly heritable, and to account for them by a modification of pangenesis[1073].

Romanes appears to be angry that Weismann did not acknowledge this fact. He found close similarities between the two theories, and concluded:

> If he [Weismann] were to express his willingness to abandon his theory of evolution for the sake of strengthening his theory of heredity, by identifying its main features with those of Galton's, personally I would have no criticism to pass. Indeed, I was myself one of the first evolutionists who called in question the Lamarckian factors, and ever since the publication of Galton's theory of heredity at about the same time, I have felt that in regards to its main principles – or in those which it agrees with Weismann's – it is probably the right one[1074].

[1070] Ibid. p. 404.
[1071] Galton 1875, cited by Romanes, 1899.
[1072] Ibid. p. 136.
[1073] Ibid. footnote on p. 60.
[1074] Romanes 1899, p. 108.

Galton's Study of Twins[1075]

Galton recognized that the study of twins could shed light on an important hereditary question in which many were interested – in particular, the heritability of mental abilities:

> Their history affords means of distinguishing between the effects of tendencies received at birth, and of those that were imposed by the circumstances of their after lives; in other words, between the effects of nature and of nurture[1076].

Galton recognized the existence of two kinds of twins: identical twins – who are always of the same sex and issue from two "spots" on the same ovum (monozygotic twins); and non-identical twins, who issue from two ova and may be of different sex. The twins of both kinds share the common environment of the maternal uterus, but are different in "constitution".

Galton sent a "circular" to persons who were either twins themselves or near relatives of twins – all of them adults – in England, asking them to describe how similar (or otherwise) they were in their childhood as well as in later life (he added a request to report to him the names of other twins, who might respond if written to). He thereby obtained 80 reports. Thirty-five of these were of identical twins, who reported that they were extremely similar in physical as well as mental characters since their childhood, although the environments they had lived in since they became adults were quite different (Galton lists detailed anecdotes on the difficulties of distinguishing between these twins, by mothers and teachers, and over many pages he gives instances demonstrating the close similarity in behavior and mentality of twins even when living in different localities). The only factor that differed between them was disease, although they often reacted similarly to the common ailments. Galton concluded:

> Here are thirty-five cases of twins who were "closely alike" in body and mind, and who have been reared exactly alike up to their early manhood and womanhood, since then the conditions of their lives changed[1077]... Nature is far stronger than nurture, within the limited range that I have been careful to assign to the latter[1078].

In the data assembled from the circulars, 20 other pairs of twins were, from childhood, reported to be quite different from each other in both physical and mental characters. Galton reasoned that the mental differences were due to home education in early childhood – chiefly by the mother – and that the biological mother should be more influential than a stepmother. As a metaphor, Galton used the cuckoo: although her offspring hatched and were reared always by a "stepmother", they never learned the song, behavior, or other characters of the foster parents[1079].

> There is no escape from the conclusion that Nature prevails enormously over Nurture, when the difference of nurture do not exceed what is commonly to be found among persons of the same rank of society and in the same country[1080].

[1075] Galton 1875: The history of twins, as indicators of the relative powers of "nature" and "nurture" . Fraser's Magazine, Nov. 1875, pp. 391-406.

[1076] Ibid. p. 391. The problem of "nature versus nurture" became the focus of the debate on education and social reform of the Social Darwinians in the 20th century, and Galton's conclusion became the center point in the Eugenics movement.

[1077] Ibid. p. 401.

[1078] Ibid. p. 392.

[1079] Ibid. p. 405.

[1080] Ibid. p. 404.

The Law of "Regression to the Mean"

Following the results of his enquiry in "Heritable Genius", Galton endeavored to learn more about the transfer of parental qualities to their offspring. Because no reliable quantitative data were available, Galton chose to approach the issue experimentally – in a controlled experiment with peas, "with the help of several country friends"[1081]. He carefully weighed individual seeds of sweetpeas and packed them in seven groups according to weight, from large (K) to small (Q). Each packet contained 10 seeds of the same weight. The packages were sent to friends with "precise" instructions to sow them in marked plots, harvest the pods when they matured, and send the offspring seeds to him accordingly marked. Galton weighed the individual seeds and plotted the offspring against their parental weight. Surprisingly, the relationship was not as expected:

> It appeared from these experiments that the offspring did not tend to resemble their parent seeds in size, but to always be more mediocre than they – to be smaller than the parents, if the parents were large; to be larger than the parents, if the parents were very small... This curious result was based on so many plantings conducted for me by my friends living in various parts of the country ...that I could entertain no doubt of the truth of my conclusion[1082].

Galton then decided to search for parent-offspring relationships in human characters. He chose to analyze human stature [height], for which he had measurements of 205 parents and their 930 adult children. His detailed analysis of the data fully confirmed the conclusions from the seed experiments, and is fully reported in his paper. First, he confirmed that the measurements, if statistically arrayed in their order of magnitude, show the beautifully symmetrical regular distribution described in "Natural Inheritance", because stature is a composite variable, a sum of a number of variable elements in the human body. He then found that the stature of the children depended closely on the average stature of the two parents, which he labeled the "midparent value". Next, he defined the variable of interest, which is the deviation of individual children from their midparental value. This variable he called the "deviate". Using these terms, the law of regression is stated thus:

> The height-deviate of the offspring is, on the average, two-thirds of the height-deviate of his mid-parentage... The explanation is as follows. The child inherits partly from its parents, partly from his ancestry. Speaking generally, the further his ancestry goes back, the more numerous and varied will his ancestry become... Their mean stature then will be the same as that of the race, in other words, it will be mediocre.
>
> This law tells heavily against the full hereditary transmission of any gift, as only a few of many children would resemble their mid-parentage[1083].

This law of regression to the mean has important consequences for the transmission of "genius": the more exceptional the parent, the less likely it will be that the child will be as exceptional, or more so. But the law is even-handed:

> If it discourages the extravagant expectation of gifted parents that their children will inherit all their powers, it no less discountenances extravagant fears that they will inherit all their weaknesses and diseases[1084].

[1081] Galton 1870: Regression towards mediocracy in hereditary stature; Appendix. In Anthropological Miscellanea, pp. 246-263.

[1082] Ibid. p. 246.

[1083] Ibid. pp. 252-253.

[1084] Ibid. p. 253.

The number in a population which differ little from mediocracy is so preponderant *that it is more frequently the case that an exceptional man is the somewhat exceptional son of rather mediocre parents, than the average son of very exceptional parents*[1085].

Eugenics

Arguments about the need to save the human race from deterioration through the accumulation of hereditary deleterious variations – which are protected from elimination by natural selection by social and medical procedures – appeared early on in the 19th century. The writer was a physician to the Queen, Sir William Lawrence:

> A superior breed of human beings could only be produced by selections and exclusions similar to those so successfully employed in rearing our most valuable animals. Yet in the human species, where the object is of such consequences, the principle is almost entirely ignored. Hence, all the native deformities of mind and body, which spring up so plentifully in our artificial form of life, are handed down to posterity and tend by their multiplication and extension to degrade the race…
>
> This inattention to breed is not, however, of so much consequences in the people as in the rulers… here, unfortunately, the evil is at its height; laws, customs, prejudices, pride, bigotry, confine them to intermarriages with each other; and thus degradation of race is added to all the pernicious influences inseparable from such exalted stations… the strongest illustration of these principles will be found in the present state of many royal houses in Europe[1086].

Galton was one of the founders of the Eugenics movement, which had a strong impact on sociological thinking and even on legislation in the early 20th century in the USA. The purpose of the eugenicists was to apply the scientific knowledge of heredity to the protection of the qualities of the human race from deterioration, and to find ways to ensure their improvement.

The results of the analysis of genius led Galton to suggest that by proper selection and an organized arrangement of marriages, it may be possible to produce – within a few generations – a more talented strain of humans [Galton himself married rather late, and had no children].

> [the idea] is smiled at as most desirable in itself, and possibly worth of academic discussion, but absolutely out of the question as a practical problem[1087].

Using statistical arguments about the distribution of talent in the human population, and his "law of regression to mediocrity" as supporting evidence, Galton argued that it is highly desirable to encourage talented parents to breed more offspring; and that it should be considered a national economic interest that talented children should be selected, as these are worth any price:

> If such people… could be distinguished as children and procurable by money to be reared as Englishmen, it would be a cheap bargain for the nation to buy them at the rate of many hundred, or some thousands of pounds per head[1088].

Galton listed several ways by which to encourage talented people to marry young and produce many children. *Both* partners should be talented, and not only the male, in order to qualify for support, because his data suggested that this increases the probability of rearing talented children.

[1085] Ibid. p. 254.
[1086] Sir William Lawrence, physician (1783-1867), cited by K.D. Wells. J. Hist. Biol. 4: 397.
[1087] Galton 1901, p. 659 [Nature, 64, 659-665].
[1088] Ibid. p. 661.

Conditions to be provided to such couples were to include cheap rental housing – or endowments from the wealthy.

As a practical starting point, Galton recommended establishing a data bank of the qualities of individuals of all classes in the community, by sending out suitable questionnaires, so that when choosing a partner in marriage all his/her qualities (and those of their parents) should be known. Certificates of quality could be issued to prospective marriage partners.

> An enthusiasm to improve the race would probably express itself by granting diplomas to a select class of young men and women, by encouraging their intermarriage, by hastening the time of marriage of women of that high class, and by the provision of rearing children healthily[1089].

Galton was the keynote speaker at a conference of the London Sociological Society in 1904, which attracted such diverse people as the statisticians Karl Pearson (who chaired the meeting) and Frank Weldon, and the literary men H.G. Wells and George Bernard Shaw. In his lecture, Galton defined Eugenics and established its aims:

> Eugenics is the science which deals with all influences that improve the inborn qualities of a race; also with those that develop them to the utmost advantage. The aim of eugenics is to bring as many influences as can be reasonably employed, to cause the useful classes of the community to contribute more than their proportion to the next generation[1090].

Galton called for dissemination of the knowledge of the laws of heredity, so that people would understand the aims of eugenics; while also calling for moderation in the application of these ideas:

> I see no impossibility of eugenics becoming a religious dogma among mankind, but its details must be worked out sedulously in the study. Over zeal leading to hasty decisions would do harm, by holding out expectation of a near Golden Age, which will certainly be falsified[1091].

Responding to criticism against his proposed planned marriages, as restricting the freedom to choose one's partner and contrary to human nature, Galton argued in a paper that the rules of marriage had varied since biblical times and depend on customs, religion, and other social factors.

The paper was followed by a lively discussion illustrating a divergence of opinion. Some speakers regretted that the laws of heredity were insufficiently known, and were not as simple as those suggested by "the Austrian abbot, Gregory Mendel" [published by Bateson just three years earlier] and considered that the selection of people with the right qualities might not be successful. In particular, H.G. Wells argued that not enough attention was being given to the inheritance of disease, and the mating of two otherwise talented persons could result in sick offspring. Moreover, criminals too are often very bright, and the average criminal may be more talented than the average law-abiding poor person. Wells also suggested a different approach to the problem:

> It is the sterilization of failures, and not in the selection of successes for breeding, that the possibility of improvement of the human stock lies[1092].

[1089] Ibid. p. 663.
[1090] Galton 1904, p. 1; p. 2.
[1091] Ibid. p. 3 [possibly published in 1905, pp. 18-33?].
[1092] Ibid. pp. 7-8. This line of thought was followed even more extremely by Karl Pearson (Chapter 22 below).

The most serious objections came from physicians. Dr. Archibald Reid argued that the key to success in solving the sociological issues should be the application of heredity in the curriculum and practice of the medical profession. The distinction should be made between inborn and acquired traits (Nature vs. Nurture). He noted that many students of heredity denied that acquired traits are heritable – among them many diseases associated with poor living conditions:

> It is generally assumed that changes in the parents do tend to influence the inborn traits of offspring… that much of the degeneracy which is alleged is befalling our race owing to the bad hygienic conditions under which he dwells in our great growing cities… I believe this assumption to be a totally unwarrantable one. It is founded on confusion between inborn and acquired traits… I believe in fact, that while a life in the slums deteriorates the individual. It does not affect directly the hereditary tendencies of the race in the least[1093].

He goes on, over several pages, to list many cases of contagious diseases that have been transmitted by Europeans to isolated native peoples since the time of Columbus, with disastrous consequences, because the natives' inborn traits did not protect them from the imported, acquired maladies.

> Slum life, and the other evil influences of civilization, including bad and insufficient food, vitiated air, and zymotic diseases, injure the individual. They make him acquire a bad set of traits. But they do not injure the hereditary tendencies of the race[1094].

Fingerprints

One important practical contribution by Galton to science was the discovery that each person has a unique fingerprint pattern, which does not change with age. Galton described his discovery in a book (published in 1891), listing the then best methods by which to obtain fingerprints and preserve the records in a catalog, according to clear criteria, in order to enable the identification of individuals with certainty.

The reasons, according to Galton, which made this technique necessary, reflect the attitude of the British rulers to the "natives" in the colonies in the 19th century. Galton explains that decent and honest people, when signing a document, do not need to confirm their identity. In the colonies, however, such as India or China, all the natives look much alike (to the European eye). Moreover, they have no inhibition against perjury when taking an oath, and often forge documents of land ownership and other deeds. Therefore the governor of India at the time, a Mr. W. Herschel, ruled that a fingerprint would be required at the signing of every deed[1095].

Galton showed that each ridge in the fingerprint pattern could be considered an independent character (there was no correlation between the appearance of the ridges in different hands or persons). By sampling random different sections of the same finger pattern, he calculated that if the presence of 35 marked points on the pattern is recorded, the probability of getting exactly the same pattern for two fingers – whether of the same or different hands or persons – was extremely small – $(1/2)^{35}$. A positive identification was therefore practically certain.

Galton argued that fingerprints are valuable for proving identity – for example when verifying the authenticity of the signature of a deceased person on a document such as a will. The fingerprint pattern could be sent overseas when needed. Fingerprinting is still being used in police and forensic work throughout the world.

Like many of his contemporary intellectuals, Galton knew the Bible pretty well. As an example of the utility of fingerprints, he cites the story of Jezebel, wife of King Ahab, who was cursed by the prophet Elijah for slaying the prophets. The curse was that dogs would eat Jezebel "so that they

---

[1093] Ibid. p. 14.
[1094] Ibid. p. 16.
[1095] Galton 2006 {1892}, Finger Prints, p. 138.

will not say, this is Jezebel"; meaning that she could not be identified. When ordered to bury her [as a royal should be] they found only the skull and the palms of her hands and feet (Kings II, 9:37). Galton commented that the curse was not fulfilled: the palms of her hands could have led to certain identification!

22

The Biometricians: Karl Pearson

Karl Pearson (1857-1936) was a British physicist, philosopher, and mathematician, working in Cambridge. He was one of the founders of the science of biometry [applied statistics]. Pearson had a wide range of interests and was involved in various religious, social, and scientific projects. In 1884 he was appointed Professor of Applied Mathematics at University College, London, where he worked until his retirement in 1933, teaching – among other subjects – geometry. His book, "The Grammar of Science" (1892) is regarded as a classic text in the philosophy of science.

Together with Francis Galton (Chapter 21), Pearson advocated a statistical approach to evolution. In 1896 they founded the journal "Biometrica", which is still being published today. Like Galton, Pearson worried about the accumulation of heritable deleterious characters, which are protected by means of medical and social care, thus posing a threat to the future of the human race. After Galton's death in 1911, Pearson was appointed to the Chair of Eugenics – founded by a large endowment by Galton – and conducted research projects in genetics, anthropology, and psychology.

Pearson did not discuss genetics or evolution in his 1892 book, but in 1900 a second edition was published, in which two large chapters were added – entitled "Evolution"[1096]. The two chapters are dedicated to aspects of detecting natural selection by means of statistical methods.

Pearson's arguments were based on blending inheritance and the accumulation of small changes in a continuous distribution of characters. Although he claimed that a considerable number of quantitative measurements on heredity had been collected[1097], his argument is largely philosophical.

Advocating a Statistical Approach to Biology

Pearson insisted on a quantitative approach to biological science.

> Biologists... must throw aside merely verbal descriptions, and seek in future quantitative precision for their ideas[1098]. As we can predict little or nothing of the individual atom, so we can predict little or nothing about the individual vital unit (organism). We can only deal with statistics of average conduct[1099].

Pearson argued against the use of broad, qualitative terms in the description and interpretation of biological phenomena. His criticism was directed at the prevalent arguments, in the post-Darwinian era, of what natural selection can or cannot do – on the basis of logical and philosophical considerations. Such an approach, he contended, is not science but fiction:

> Every few months we find in one journal or another, a more-or-less brilliant hypothesis as to a novel factor of evolution. But how few are the instances in which this factor is accurately defined, or being defined, a quantitative measure of its efficiency is obtained[1100]!

[1096] In the same year Mendel's paper was re-discovered, which subsequently changed the entire scientific approach to heredity and evolution, but this is not reflected in Pearson's second edition.
[1097] Pearson 1900, p. 480.
[1098] Ibid. p. 374.
[1099] Ibid. p. 500.
[1100] Ibid. p. 373.

> It is imagination solving the universe, propounding a formula before the facts, which the formula is to describe, have been collected and classified.

Everything must be quantified, measured accurately, and tested:

> We now simply say: what is the numerical value of the variability of which you speak? Did you test the magnitude of the inheritance of that character? Till very definite answers are forthcoming to these questions, we are not... bound to pay much attention to those who are over-ready to "explain" not only organic but also social changes by a vague use of biological terms[1101].

Applied Statistics: Transmission of Characters between Generations

The practical application of statistical methods to biological data takes up much space in Pearson's two chapters on evolution. He describes in detail how to calculate the mean and the standard deviation, and the meaning of correlation and regression – two methods of which he made frequent use in the context of heredity[1102].

From a comparison of the frequency of a character in parents and in their offspring, Pearson argues, one can derive important conclusions about its heritability and whether or not natural selection has operated on that character between generations. Usually only fathers and sons were measured. If the sample included females, the female measurements were multiplied by a constant, converting them to male values ("reducing the females to male equivalents").

Pearson reasoned that the characters of an individual are affected not only by his parent's qualities, but also by the hereditary tendencies of all preceding generations:

> In the tenth generation, a man has 1,024 tenth-great-grandparents... it is the heavy weight of this mediocre ancestry, which causes the son of an exceptional father to regress towards the general population mean. It is the balance of this sturdy commonplaceness which enables the son of a degenerate father to escape the whole burden of the parental ill[1103].

Pearson advocated the extended use of mathematics in biology, in particular for the calculation of the inheritability of characters. If the distribution of character values in parents and offspring are known, the mathematician can determine several numerical constants which effectively describe the relative death rates of individuals in different classes of the frequency distribution,[1104] and so measure the selective pressure within populations between generations. Pearson considered that all the hereditary tendencies in Man can be reduced to a single parameter, gamma (γ). Combined with the average (midparent value) of a given character, gamma represents the hereditary value of the character.

> So long as the sexes are equipotent [have equal contributions to the child], blend their characters [at fertilization], and mate pangamously [= at random], the values of γ will be the same and all characters will be inherited at the same rate[1105].

[1101] Ibid. p. 372.

[1102] The extensive description of procedures which most biologists should consider elementary today indicates that they must have been little known by biologists at the turn of the century.

[1103] Ibid. p. 456. This follows Galton's Law of Regression (see Chapter 21 above).

[1104] Ibid. p. 409.

[1105] Ibid. p. 478.

Acquired Characters and Tradition in Human Populations

Some changes in human populations, in a long sequence of generations, can be acquired by mechanisms other than heredity. This set of mechanisms Pearson refers to as tradition[1106].

> The tradition of acquired modifications is clearly a factor of evolution in man. It is largely the means of differentiating civilized from uncivilized man. Habits of life, language, institutions, mechanical and other knowledge... react on the environment, upon food supply and relative fertility... The tradition of acquired modifications may give a progressive, but a comparatively unstable change to the higher types of life. It is a factor of evolution, but one which requires the action of selection to become of a permanent character[1107].

Erasmus Darwin, Lamarck, Charles Darwin, and Herbert Spencer had all assumed, erroneously and without proper validation, that acquired characters are inherited: Pearson declared that a satisfactory numerical demonstration of inheritance of such traits was still wanting[1108].

Natural Selection and Evolution

True to his insistence that everything in biology must be quantitatively measured, Pearson offered a quantitative definition of evolution:

> Thus mean, variability, correlation, determine the numerical specification of each form of life. And when we say that evolution is taking place, we mean that progressive changes are going on in one or all of the numerical values which fix the mean, variability and correlation of the system of organs and characters[1109].

According to Pearson, Darwin's Theory of Evolution involved three processes: *natural selection*, *inheritance of the characters*, and *sexual selection*. Pearson used the term sexual selection not in the strict Darwinian sense of selection by females of the most attractive males, but in a wider sense: sexual selection constitutes "all the differential mating due to taste, habits, or circumstances, which prevent a form of life from intercrossing", causing differential fertility[1110].

Rather than argue theoretically[1111], Pearson used the data-base of measurements of human characters collected by Galton to prove that sexual selection in that broad sense in fact exists. In addition, Pearson created a new data-base of the numbers of seeds in capsules of poppy (*Papaver*), grouped by the number of lines on the capsule (corresponding to the number of ovarian leaves). The distribution of seeds approached a normal distribution: capsules near the mean number of lines contained the most seeds, and capsules further away from the mean number of lines contained fewer and fewer. This meant that plants producing the mode ("type") number of ovarian leaves were the most fertile.

In Pearson's terminology, *fitness* refers to the general adaptation of the population to its existing conditions, as reflected in the mode of its distribution – high *fitness* may not only mean high *fertility*.

> If the type of maximum fertility is not identical with the type fittest to survive in a given environment, then only intensive selection can keep the community stable. If natural

[1106] This mechanism is termed "cultural evolution" in later literature.
[1107] Ibid. p. 380.
[1108] Ibid. p. 379.
[1109] Ibid. p. 403.
[1110] Ibid. p. 418.
[1111] Ibid. p. 426.

> selection be suspended, the most fertile [which may have lower fitness!] tends to multiply... it wants very few generations to carry the mode, the type, from the fit to the unfit[1112].

Natural selection works through differential fertility to change the mode. If a population is transferred to a new location where the environmental conditions are different, differential fertility will cause the "type" to change:

> The old modal center of fertility will alter with the new environment. Where the environment by natural selection produces a given type, with that type it ultimately associates the maximum fertility. Differentiation of type connotes differentiation of fertility. When two modes arise in a species, then arise two maxima of fertility[1113].

Eugenics and Racism

Expressions of concern for the future of the human race became frequent after the publication of "The Origin of Species". People believed that in ancient human populations, weak and deleterious variants must have been eliminated by natural selection, whereas modern medicine keeps them alive. In particular in 20th-century USA, means were suggested – and applied – in an attempt to reverse this trend.

Galton (Chapter 21) was one of those concerned, and he donated a large sum in his will to set up a Chair of Eugenics at University College, London. The first person to be awarded the Chair was Pearson, who became one of the more extreme supporters of eugenic ideas and practice in society.

> [Eugenics is] the new science concerned with the study of agencies under social control, that may improve the racial qualities of future generations, either physically or mentally [Pearson 1900, p. xii].

Pearson believed that mental and emotional characteristics of man are heritable and not environmentally induced, and that the poor living conditions and prevalence of alcoholism and crime in dwellers of city slums are the result of heritable properties of the poor. The high birth rate of the poorer and lower classes of society (whose social status is a result of their being less mentally-endowed) poses a threat to civilization. Something should be done to reduce the threat: on the positive side, the highly-endowed should be encouraged to contribute to society by breeding more offspring having their qualities:

> The Theory of Evolution is not merely a passive intellectual view of nature. It applies to man in the communities as it applies to all forms of life. From the standpoint of the patriot, no less than that of the evolutionist, differential fertility is momentous[1114]!
>
> We see how exceptional families, by careful marriages, can within a few generations obtain an exceptional stock, and how directly this suggests assortative mating as a moral duty for the highly endowed[1115]!

On the negative side, steps must be taken to restrict the reproductive rate of the lower classes, since there is no hope that they can be intellectually improved, except by infusion of better blood by marriage with the higher classes. But this is an undesirable step:

1112 Ibid. p. 466.
1113 Ibid.
1114 Pearson 1900, p. 467.
1115 Ibid. p. 486.

> The exceptionally degenerate isolated in the slums of modern cities can easily produce... a stock which no change of environment will permanently elevate, and which nothing but mixture with better blood will improve. But... we do not want to eliminate bad by watering it down with good, but by placing it under conditions where it is relatively or absolutely infertile[1116].

It was believed that the same attitude should apply in regard to the human races – the negroes are necessarily of lower value for civilization than the white race. Pearson put his position very strongly:

> It is a false view of human solidarity, a weak humanitarianism, not a true humanism, which regrets that a capable and stalwart race of white men should replace a dark-skinned tribe which can neither utilize his land for the full benefit of mankind, nor contribute its quota to the common stock of human knowledge.
>
> This sentence should not be taken to justify a brutalizing destruction of human life. At the same time, there is cause for human satisfaction in the replacement of the aborigines throughout America and Australia by white races of far higher civilization[1117].

[**Author's comment.** *The "Eugenic" ideas reflect prejudice and racial hatred. In particular, in view of the Nazi racial discrimination and the resulting holocaust of the Jewish people during World War II, today's social and politically-correct standards condemn the eugenic ideas and practices as racist and inhuman. But in the early 20th century the matter was discussed seriously by scientists who were truly concerned about the welfare of mankind. Entire symposia were devoted to this problem, with the participation of politicians, literary persons and university professors.*]

> It was not bad science, by the standard of the time that led to the scientific conception of hereditary mental differences between races: it was good science. By present standards, of course, the evidence, arguments and conclusions of Darwin, Huxley and Galton would be bad science[1118].

[1116] Ibid.
[1117] Pearson 1892, p. 310.
[1118] W. Provine, in Science as a Way of Knowing, 1985, p. 865.

The Fourth Circle

The Theory of Evolution in the 20th Century

The whole question of inheritance has assumed a new aspect, first on account of the work of De Vries in regard to the appearance of discontinuous variation in plants, and secondly on account of the remarkable discoveries of Gregor Mendel as to the laws of inheritance of discontinuous variation [Morgan 1903][1119].

The Theory of Evolution underwent some fundamental changes in the early 20th century. Symbolically, the turning point can be traced to the year 1900, although the central ideas and research in the early years of the 20th century had already originated in the 19th century.

In 1900, a forgotten paper by a Czech monk, Gregor Mendel (Chapter 23), was re-discovered by three botanists in three different countries. The paper dealt with the inheritance of flower and seed characters in peas, but provided a key to the understanding of "particulate" (rather than blending) inheritance. Mendel's theory provided the foundation for the new science of genetics. Later in the century, evolutionists incorporated genetics into the Theory of Evolution.

Ten years earlier the Dutch botanist, Hugo de Vries (Chapter 24) – one of the re-discoverers of Mendel's paper – had reported on the discovery of a new, true-breeding variant in a natural population of the "evening primrose". The characters of the variant were transmitted without "blending". De Vries termed the new form "mutation" and suggested that new species in evolution had originated suddenly as mutations, and not gradually through the accumulation of small changes, as the Darwinians maintained. The theory of mutations was accepted and supported by geneticists such as T. H. Morgan (Chapter 25) in the early years of the 20th century, but rejected by prominent evolutionists such as Ernst Mayr (Chapter 32).

From this turning point, two schools of evolutionists emerged. The "Biometricians" – notably Francis Galton (Chapter 21) and Karl Pearson (Chapter 22) – refused to believe in the sudden appearance of new species by mutation, and held to the Darwinian idea of gradual evolution of species by natural selection. They developed statistical methods and mathematical theories of selection in continuously distributed, metric characters, which were successfully applied by animal and plant breeders to improve their stocks.

The Mendelians (= geneticists) continued on the path established by Mendel and de Vries. They argued that small, "fluctuating" variation had no genetic basis at all: these were only the effects of minor environmental variations. New species appeared as mutations, which were environmentally tested for their ability to survive: natural selection removed those which were less fit. Intensive laboratory work by Morgan and his students (Chapter 25) discovered hundreds of new mutants which were truly transmitted between generations, with no blending.

The controversy over the inheritance or non-inheritance of acquired characters did not die with the complete rebuttal by Weismann (Chapter 18). A number of scientists, especially in America, kept the discussion alive by calling upon different kinds of data – from experiments with amphibians, learning behavior of mice, and arguments from education and social conduct in humans (Chapter 26). The controversy acquired a political twist when the supporters of Lysenko in the then Soviet Union adopted his Lamarckian ideas that the environment of the parents shapes the heritable characters of the next generations (Chapter 26). Most geneticists in Europe, and particularly the USA, dismissed these ideas as unproven if not entirely wrong, but the Soviet government stood behind Lysenko – with disastrous results for Russian science and economy.

[1119] Morgan 1903, p. 278.

The inheritance of discrete, rather than continuous, characters enabled the discussion of heredity not only as the transmission of characteristics in families from parents to children, but also their transfer between generations within populations during evolution. Shortly after the discovery of Mendel's paper, a mathematical theory published simultaneously in 1908 by G.H. Hardy in England and W. Weinberg in Germany spelled out the conditions for a population to arrive at and maintain a stable population composition (genetic equilibrium) (Chapter 27). The Hardy-Weinberg Law became the basis of an extensive quantitative theory of population genetics and evolution, to which the major contributors were R.A. Fisher (Chapter 28), J.B.S. Haldane (chapter 29), and Sewall Wright (Chapter 30). The merging of Mendelian genetics with the Darwinian Theory of Evolution led to "The Modern Synthesis" of evolution (Julian Huxley, 1942). The chief creators and proponents of the synthesis were Theodosius Dobzhansky (Chapter 31) and Ernst Mayr (Chapter 32), who refined and re-defined the concept of "species" in genetic terms, delimiting species by reproductive barriers ("the biological species concept").

Dobzhansky and his students applied the methodology of genetic crosses, developed in the laboratory, to studies of *Drosophila* in natural populations. Using discrete characters as genetic markers, they obtained quantitative measurements of genetic variation and estimated the rate of evolutionary change in real populations and derived a genetical theory of speciation in nature.

Evolutionary research was enhanced in the 20th century by a group of British biologists, headed by E.B. Ford. They established the sub-discipline of ecological genetics, aimed at the study of selection in nature. Perhaps the most important contribution of this group was their detailed study of industrial melanism, a phenomenon treated earlier mathematically by Haldane, and investigated in the field by H.B.D. Kettlewell. Similar attempts to measure natural selection in the field were carried out in snails, plants in contaminated soils, insects resistant to insecticides, and guppy fishes, in the latter half of the 20th century (Chapter 35).

An unusual challenge to the Theory of Evolution was presented by the prominent geneticist Richard Goldschmidt (Chapter 34). His studies of the gypsy moth in Japan and North America suggested that natural selection brings about adaptation to local conditions (micro-evolution), but does not lead to speciation: to affect macro-evolution, a different kind of mutation was needed – "systemic" mutation. Such mutation causes a large reshuffling of the genetic system and is mostly deleterious or lethal, but could also become the founder of a new species. Goldschmidt referred to carriers of such harmful macro-mutations as "hopeful monsters". He upset other evolutionary geneticists by challenging the accepted model of the genetic material and for many years was considered a heretic.

The introduction in 1966 of electrophoretic techniques for the study of genetic variation in natural populations led to an enormous increase in field studies, and led to the Neutrality Hypothesis of Motoo Kimura (chapter 36) – a controversial subject that was to prevail in evolutionary research for more than 20 years, until resolved (in favor of the neutralists) – with the introduction of the technology of DNA sequencing in modern molecular evolution.

This development is beyond the scope of current book, and the reader is referred to texts such as Graur (2016).

23

Gregor Mendel and the Origin of Genetics

Symbolically, the beginning of the 20th century marked the beginning of a new science – Genetics – which brought about a significant change in the methodological approach to the study of heredity and evolution: from logical and philosophical discussions to the investigation of real organisms in the laboratory and, later, in nature.

In 1900, the British botanist William Bateson[1120] discovered a neglected article published in 1865, entitled "Experiments in the hybridization of plants", by an Austrian monk, Gregor Mendel[1121]. In his paper Mendel described crosses of varieties of garden peas with different flower color and seed characters. To interpret his results Mendel provided a mathematical model, which predicted the frequency distribution of the characters' states in the offspring of a cross.

Bateson realized that Mendel had provided a key to the understanding of heredity of non-blending characters. He translated the paper into English and thereby greatly helped the dissemination of Mendel's theory.

Almost simultaneously, three other scientists discovered Mendel's paper and obtained results similar to Mendel's in their own research: H. de Vries (Chapter 24) in the Netherlands, Karl E. Correns in Germany, and Erich von Tschermak in Austria. Their independent results with different species of plants thus offered three independent confirmations of Mendel's theory.

The general applicability of Mendel's model was not appreciated at the time. His world fame came about when his paper was discovered, 16 years after his death. Mendel's theory currently forms the basis of the science of genetics and is taught in schools all over the world.

Why was Mendel's paper neglected for nearly 35 years? Contrary to widespread opinion, Mendel was not unknown in his own country. He had presented the results of his experiments at meetings of the Austrian "Society for Natural Science". Moreover, the journal in which the paper was published was sent by the Society to more than 100 libraries and universities in Europe, and was referred to by some writers before 1900. Hybridization had been experimented with for over 100 years in Europe, and several theories were suggested. Mendel's paper may have been regarded as just another report on the hybridization of plants[1122].

The Data and the Model

Mendel experimented with varieties of garden peas of different flower colors and seed shapes in the garden of his monastery. He crossed them and further bred the offspring of each cross. He applied simple considerations of probability to explain the results, then tested the predictions from the model in further experiments.

[1120] William Bateson – (1861-1926) was a devoted supporter of Darwin's theory. His book "Materials for the Study of Variation" (1894) provided data to support evolution by particulate inheritance. He coined the term "genetics" and later became the first Professor of Genetics at Cambridge.

[1121] Johann Mendel (he received the name Gregor when he was ordained) was a monk in a monastery in the Austrian town of Brünn (now Brno, in the Czech Republic). He taught Greek and mathematics in a school in his town although he did not have a formal teaching license. Later he was sent by his monastery to the University of Vienna, where he studied physics, mathematics, zoology, and botany for two years. He returned to his monastery and continued teaching these subjects. From 1868 until his death in 1884 he was the abbot there.

[1122] Sandler & Sandler (1986).

Mendel's choice of peas as research material was wise, since peas self-fertilize and pollination takes place in the still-closed flower bud. Controlled artificial pollination can thus be practiced easily – when the anthers are cut the chance contamination of the cross by visiting insects is negligible. Mendel concentrated on the inheritance of a single trait (or two) in each cross. He crossed varieties which differed in flower color or seed characters, collected the seeds produced by the hybrids (F_1), let the hybrid flowers self-fertilize to obtain second-generation seeds (F_2), planted them and grew the seedlings to maturity, and classified and counted them according to the character-states they expressed[1123].

Contrary to the then prevalent expectation, the characters carried by the male and female parents did not "blend" in the hybrids. Analyzing the proportions of plants expressing the character-states of interest, Mendel detected a pattern. In the cross of white with red-flowered plants, for example, all hybrid offspring had red flowers (and not pink). The white color reappeared in approximately 1/4 of the F_2 generation.

Mendel derived a simple model, expressed mathematically, which explained the pattern. In his model, Mendel suggested that the color of an individual plant is determined by a pair of factors, derived from the male and the female parent respectively (in the case of flower color, the red plant carried factors CC, and the white plant cc). In forming the gametes, the two factors separate (*segregate*), with each gamete carrying one factor. This is referred to as Mendel's First Law.

In forming the zygote, one factor is received from a male gamete and the other from a female gamete. The hybrids of a CC female and cc male are, accordingly, Cc. They expressed the red color because, Mendel suggested, C is stronger (*dominant*) than c (*recessive*). The hybrids, males and females alike, produce C and c gametes in equal proportions. The factors combine at random in forming the zygotes of the next generation. When the hybrids self-fertilize, independent assortment (Mendel's second law) resulted in 1/4 of the zygotes being CC, 1/2 being Cc, and 1/4 being cc.

These expected proportions can be obtained theoretically from the expansion of Newton's formula $(1/2 + 1/2)^2$ when the proportion of the factors C and c in the gametes are equal. In F_2, 1\4 of the plants should express the recessive character. A simple experimental test, performed by Mendel, confirmed that when the red-flowered F_1 plants are back-crossed to the white recessive parent, two phenotypes[1124] – red and white – are expressed in equal proportions among the offspring, as expected from the model.

Mendel crossed varieties which differed in pairs of characters – for example A, seed character, as well as B, flower color [dihybrid cross]. He assumed that each of the two characters was determined by two factors which segregated independently. Thus four kinds of gametes were expected to be produced by the male and female F_1 hybrids: AB, Ab, aB, and ab [lower-case letters represent the recessive character]. Independent assortment of the gametes was expected to result in 16 possible genotypes. Due to dominance, these should be expressed as four phenotypes in the ratio of 9 : 3 : 3 : 1. The actual results were similar to the expectations from the model (the numbers of one replicate reported by Mendel were 315:101:108: 32, quite close to the theoretical proportions[1125]).

R.A. Fisher's Criticism

The British statistician R.A. Fisher analyzed Mendel's published data and concluded that the results were too close to the theoretical expectation. Mendel had applied no tests of significance to his data. Fisher suggested that Mendel had not derived his model from the experimental results, but that he must have had a pretty accurate notion beforehand of what numbers he should expect. Mendel must have been familiar with Newton's formula and the properties of the binomial distribution from his studies of mathematics at Vienna. The results of the dihybrid cross served as a test of the accuracy

1123 See for example Sinnott, Dunn & Dobzhansky 1958, Chapter 3.

1124 The terms Genotype and Phenotype were suggested by Johannsen in 1911. Thus the phenotypically red flowers in the F_2 of the original cross were of two genotypes: CC and Cc, at an approximate ratio of 1: 2.

1125 See e.g. Sinnott, Dunn & Dobzhansky 1958, Chapter 3.

of the single-character crosses, and when he obtained results close to the expected 9:3:3:1 ratio, he decided to publish the model. Mendel was an experienced teacher, and chose to report in his papers only the numbers that fit the expectation – as a demonstration, omitting many "unnecessary" details[1126].

Fisher[1127] concluded that Mendel had carefully selected the varieties for crossing (only seven of 38 varieties were used). Characters of some varieties did not show dominance, and the F_1 offspring were of intermediate color, as if the characters had blended. Mendel had corresponded with a famous German botanist, K.W. von Nägeli, and tried to explain his results, but Nägeli was not convinced and suggested that Mendel should try another species, *Hieracium*, with which he himself worked. Mendel did so, but the results were confusing: he did not obtain the expected proportions of offspring (it later turned out that some of the seeds of *Hieracium* are produced parthenogenetically, without fertilization).

Reception of the Model

The biometricians Francis Galton (Chapter 21) and Karl Pearson (Chapter 22) believed that evolution proceeded through the accumulation of slight, minor variations in a continuous character distribution. Frank R. Weldon published in 1901 a counter-argument against the particulate nature of characters. William Bateson, who became an enthusiastic supporter of Mendel's model, responded by publishing a translation of Mendel's paper in 1902, claiming that Weldon had failed to understand Mendel's theory. Bateson argued against the biometricians, contending that, in real life, evolutionary changes appear discontinuous:

> We are taught that evolution is a very slow process, going forward by infinitesimal steps. To the horticulturist, it is rarely anything of the kind... it is going at a gallop. Whenever it can be shown that a variation comes discontinuously into being, it is no longer necessary to suppose that for its production long generations and accumulation of differences are needed...
>
> This supposition involved the most impossible hypothesis, that every intermediate form has successively been in its term the normal. Whenever there is discontinuity, the need for such a suggestion is wholly obviated[1128].

T.H. Morgan: What are Mendel's "Factors"[1129]?

The discovery of Mendel's paper stirred up enthusiasm among animal and plant breeders, eager to interpret the results of their crosses in the new light. The American geneticist, Thomas Hunt Morgan – himself a strong supporter of Mendelism (Chapter 24) – warned against a too simplistic and erroneous understanding of Mendel's model:

Mendel, Peas and Genes

> In the modern interpretation of Mendelism, *facts* are being transformed into *factors* at a rapid rate. If one factor will not explain the facts, then two are evoked; if two prove insufficient, three will sometimes work out...The superior jugglery sometimes necessary to account for the result, may blind us... to the

[1126] Fisher 1958, p. 9.
[1127] Fisher 1936.
[1128] Bateson 1900, cited by Moore, 1986b.
[1129] Morgan 1909 American Breeders Association Reports, 5: 365-368.

common-place that the results are often so excellently "explained" because the explanation was invented to explain them.

What are those factors? People assume that the characters are actually represented in the factors, and segregate in Meiosis into the gametes. We assume that [in the hybrid] the tall and the dwarf factors retire into separate cells after having lived together through countless generations of cells... The factors have become entities that may be shuffled like cards in a pack, but never become mixed[1130].

Morgan offers a more accurate interpretation:

The egg need not contain the character of the adult, nor need the sperm. Each contains a peculiar material, which in the course of development produces in some unknown way the characters of the adult. It follows that we are not justified in speaking of the materials in the germ cells as the same thing as the adult characters – until they develop[1131].

"Biometry" and "Mendelism" in Heredity

An important experiment, cited and described in many books on genetics, was carried out by the Danish botanist Wilhelm Johannsen in 1909-1911[1132]. Johannsen was interested in the inheritance of seed weight of beans (*Phaseolus*). He weighed individual seeds and grouped them into classes by average weight, then sowed seeds of each class separately – each "pure line" effectively from a single seed – and collected and analyzed the seeds of offspring of each line (fertilization in beans occurs in the closed flower and each line thus becomes closely inbred). He discovered that the pure lines produced seeds with mean weights corresponding to the parental seeds. Variation *among* lines was heritable and further selection on seed weight was possible by choosing suitable lines. Seeds *within* each "pure line", however, varied in weight, and this variation did not respond to further selection. Johannsen coined the terms *genotype* and *phenotype*: variation within groups was phenotypic, while among-group variation was genotypic.

The statistical methods developed by the "biometric" school provided important approaches to the understanding of quantitative characters, many of them of value for breeding plants and animals under domestication. Analyses of variance, regression, and correlation are widely in use today.

The "Mendelians", on the other hand, insisted that small individual differences in continuous characters ("fluctuating variation") were not heritable at all, but merely direct responses to a variable environment. Even geographic isolation cannot account for the origin of a new species, although it may account for the formation of geographical races[1133]. The Mendelians considered evolution in terms of discontinuous variation and particulate characters that could not blend. They insisted that new species arise discontinuously by mutation (Chapter 24).

In parallel with the detailed study of heritable characters in *Drosophila* (see Chapter 25) the Mendelians developed their own statistical and mathematical tools which matured into the quantitative basis of evolution. The "synthetic" approach was summarized in 1942 by Julian Huxley in his book "Evolution – the Modern Synthesis":

Genes are in many ways as unitary as atoms, although we cannot isolate single genes. They do not grade into each other, but they vary in their actions in accordance with their mutual relations[1134].

A gene unit is thus a section of the chromosome between two adjacent sites of potential breakage at crossing-over[1135].

[1130] Morgan 1909; American Breeders Association Reports, 5: 365-368.
[1131] Ibid.
[1132] e.g. Falk 2008.
[1133] Morgan 1903, p. 203.
[1134] Huxley, J. 1942, p. 48.
[1135] Ibid. p. 49.

24

Hugo de Vries and the Theory of Mutations

Hugo de Vries (1848-1935) was a professor of botany at the University of Amsterdam for 40 years, and a director of the Botanical Gardens. In 1886 he noticed an individual plant of the evening primrose, *Oenothera lamarckiana*, which differed from all other individuals at the site. When bred in the garden, the characteristics of the variant appeared unchanged in lines of its descendants. De Vries noticed that several different variants developed from the seeds of that original plant:

> I have discovered in the evening primrose... a strain which is producing them [variants] yearly in the wild state as well as in my garden[1136].

De Vries coined the term *"mutation"* for this and similar "spontaneously" appearing variants, and formulated a theory to account for the origin of species by mutation, rather than by the accumulation of small, continuous changes.

Intracellular Pangenesis

De Vries was interested in heredity. He reviewed the available information on cell division and chromosomes, and the ideas of their role in inheritance published by his contemporaries such as Nägeli and Boveri[1137] – as well as a similar review by Weismann (Chapter 20), of which he was aware. In a paper published later he briefly referred to Mendel's paper[1138] but did not go into detail.

De Vries was impressed by Darwin's Theory of Evolution, and especially by his book on Domestication. There he found Darwin's theory of *Pangenesis*[1139], which led him to think of heredity in terms of particulate characters. In a book entitled "Intracellular Pangenesis"[1140] De Vries suggested that each character is represented in the nucleus of a cell by a particulate entity, which he named "*pangen*" – in honor of Darwin's theory. The pangens of different characters were considered independent entities – in contrast to Darwin's "gemmules", which were thought to each represent all the characters of a given cell.

> These pangens do not each represent a morphological member of the organism, a cell or a part of a cell, but each a special hereditary character. These can be recognized by each being able to vary independently from the others. Their study opens a very promising field to experimental investigation.
>
> The pangens are not chemical molecules, but morphological structures, each built up of numerous molecules. They are the life-units, the characters of which can be explained in an historical way only[1141].

[1136] DeVries 1912, p. 17.
[1137] De Vries 1889, p. 53.
[1138] De Vries 1903, p. 255.
[1139] Darwin, Domestication, II, Chapter 7.
[1140] De Vries 1889.
[1141] Ibid. Conclusions.

De Vries thought that Darwin had complicated the concept of Pangenesis when he discussed the transport of the "gemmules" in the blood to the sex cells – an assumption which was shown by Galton to be incorrect (Chapter 21). The rejection of Darwin's "gemmules" masked the important, main concept of Pangenesis – the concept of particulate inheritance.

> The hypothesis, therefore, becomes one of **intracellular pangenesis.** To the smallest particles of which each represents one hereditary character, I shall give a new name and call them **pangens**, because with the designation of gemmule is associated the idea of transportation through the whole organism[1142].

> From the nucleus the material bearers of the hereditary characters are transported to the other organs of the protoplast. In the nucleus they are generally inactive, in the other organs of the protoplast they may become active. In the nucleus all characters are represented, in the protoplast of every cell only a limited number.

The History of the "Mutation" Theory

The original mutant of the evening primrose plant, as well as other variants in other plants, attracted much attention. The offspring of crosses of the mutants appeared in the ratios predicted from Mendel's laws of heredity. Since these mutants were viable and reproduced their kind without "blending", de Vries suggested that such discontinuous variants could, in fact, become the origin of new species.

De Vries's book "The Mutation Theory" was published in two volumes in German (1901-1903), and translated into English in 1910. De Vries was interested in the appearance of new species in evolution, not in understanding the mechanism of heredity. The objective of his book is expressed in the following words:

> The object of the present book is to show that species arise by saltation, and that individual saltations are occurrences which can be observed like any other physiological process. Forms which arise by a single saltation are distinguishable from one another as sharply and in as many ways as most of the so-called small species and as many as the closely-related species of the best histemotists [=systematists], including Linnaeus himself[1143].

When a new mutation appears, it may spread in the population – unless the change is lethal. De Vries thought that some of the categories which Linnaeus had classified as "polymorphic species" were actually collections of lines developed from mutants. De Vries recalls that many garden varieties of plants were originally developed from sudden "sports". Similarly, new species may develop from previously-existing ones, not gradually but by sudden leaps. Mutant individuals may multiply and spread, while the parental individuals remain unchanged[1144].

According to De Vries, many mutations have only a slight effect on the phenotype, and do not result in the formation of new species. He considered the distinction between "species" and "varieties" to be artificial: some "varieties" are very different from the parental type but are still not recognized as true species. The so-called "varieties" are those that differ from the type in less important characters than do "species". Varieties are derived in most cases from still existing types, through the loss of a certain attribute ("retrogressive" mutation) or acquisition of a lost character

---

[1142] Ibid. p. 7.

[1143] De Vries 1910.

[1144] This is contrary to the traditional Darwinian suggestion that species change gradually and diverge to become new species, and that when a species diverged into two, the parental species cease to exist.

("degressive" mutation). Only "progressive" mutations (causing the addition of a new character) actually form new species.

"Elementary Species"

De Vries defines "elementary species" as groups of varieties of which the original type is unknown or lost, and from whose characteristics is derived an hypothetical image of what their common ancestor might have been.

> Elementary species must have arisen by the production of new qualities. Each new acquisition constitutes the origin of a new elementary species. Elementary species and varieties are thus observed to be discontinuous and separated by definite gaps[1145].

An elementary species can be identified by controlled propagation in the garden, forming a pedigree culture. Any form which remains constant and distinct from its allies in the garden, preserving its distinctive characters from generation to generation, is considered an elementary species. New "elementary" species cannot be produced through the natural or artificial selection of small, inter-individual variation.

> That intra-specific selection may be regarded, as a cause of lasting and ever-increasing improvement… is assumed by biologists, who consider fluctuating variability as the main course of progression in the organic world. But the experience of breeders does not support this view, since the results of practice prove that selection according to a constant standard soon reaches a limit which is not capable of transgression[1146].

The "small" individual variants are incorporated into groups, the group means are the items selected by the breeder. This process must also operate in nature. It can lead to improvement of a crop variety, but not to the formation of new species.

> The struggle for existence goes on between individuals, and not between groups of brethren against groups of cousins. In every group the best adapted individuals will survive, and soon the breeding differences between the parents must vanish altogether. Manifestly they can, as a rule, have **no** lasting effect on the issue of the struggle for existence[1147].

Breeding and the Current Darwinian Selection Theory

Breeding has revealed that individual plants often continue to produce both mutant and normal offspring for many generations of self-reproduction. Such individuals are referred to as "double-races"[1148]. This is not expected according to the current Darwinian selection theory:

> According to the theory of natural selection, wild species can only retain useful or at least innocuous qualities, since all mutations in a wrong direction must perish sooner or later. Cultivated species on the other hand are known to be largely endowed with qualities which would be detrimental in the wild conditions[1149].

Botanists are familiar with the direct effects of the environment on the phenotype of a plant. If a lowland plant is cultivated at high altitude, it will show "alpine" growth characteristics. But,

---

[1145] De Vries 1910, p. 460.
[1146] De Vries 1904 [1912, p. 805].
[1147] Ibid. p. 825.
[1148] Ibid. p. 408.
[1149] Ibid. p. 445.

noted de Vries, there is no evidence that these new, environmentally-induced characteristics are hereditary, as the Lamarckians would claim. The "alpine" characters of the new form may become established not because the environmental change has become fixed, but because mutations in the opposite direction have been selected against[1150]. Relict plants of a previously moist environment, now growing in deserts in Ceylon [= Sri Lanka], did not evolve morphological adaptations to desert conditions, but instead a strategy of escaping the dry conditions: in the rainy season they germinate, grow very quickly, flower and produce seeds – which then remain dormant until the next rainy season. De Vries calls the ability to exist in both wet and dry conditions "double adaptations"[1151].

A mutation was investigated by De Vries while breeding the ornamental plant *Linaria* (Scrophulariaceae). The plant is perennial and is propagated vegetatively by stolons. A peloric variety (with radially-symmetrical flowers) was first described by one of Linnaeus's students and was known in many countries, indicating that the mutation must have recurred several times[1152]. The original plant produced both normal and mutant flowers, but the peloric form was not pollinated by bees (which were unable to enter the narrow flower tubes), and it did not seem to produce seeds in nature. De Vries pollinated the flowers artificially and so cultivated the peloric form in his garden as a true-breeding variety which produced only peloric flowers. The case of *Linaria* was the first mutation that occurred "under his very eyes". Of nearly 2,000 seeds that de Vries germinated and grew in his garden, only 16 were true-breeding pelorics:

The mutation took place at once. It was a sudden leap from the normal plant with very rare peloric flowers to a type exclusively peloric. No intermediate steps were observed… the whole plant departed absolutely from the old type of its progenitors[1153].

The Theory of Mutations

Darwin mentioned in his 1844 "draft" that, rarely, among millions of individuals of a species, individual variants occurred "spontaneously" which departed strongly from the normal. These were called "sports" by gardeners of the period. Darwin often referred to these variants as "monstrosities". In some cases such individuals were used to breed new races. Darwin gave no explanation as to how these sports suddenly appeared, but considered that perhaps this is how new species come into being. The critical review by Jenkin in 1867 ruled this out. In later editions of "The Origin of Species", Darwin argued that new species are formed through the gradual accumulation of small individual variations by natural selection.

De Vries objected to this theory. In a book published originally in Dutch in 1901-1903, he suggested instead a new theory of the appearance of new species: The Mutation Theory.

> The mutation theory is opposed to that conception of the theory of selection which is now prevalent. According to the latter view, the material for the origin of new species is afforded by ordinary or so-called individual variation. According to the Mutation Theory individual variation has nothing to do with the origin of species. This form of variation, as I hope to show, cannot even by the most rigid and sustained selection lead to a genuine overstepping of the limits of a species and still less to the origin of new and constant characters. The theory of mutations assumes that new species and varieties are produced from existing forms by sudden leaps. The parent type itself remains unchanged throughout this process, and may repeatedly give rise to new forms[1154].

[1150] Ibid. p. 449.
[1151] Ibid. p. 454. A familiar example among desert plants in Israel is "the rose of Jericho", *Anastatica hierochuntica*.
[1152] Ibid. pp. 464-467.
[1153] Ibid. p. 474.
[1154] Ibid. p. vii.

De Vries advocated in his book that the origin of new species should become a subject for experimental research. This was a true revolution in contemporary thinking.

> The origin of species has so far been the object of comparative study only. It is generally believed that this highly important phenomenon does not lend itself to direct observation, and much less, to experimental investigation... The object of the present book is to show that species arise by saltations, and that the individual saltations are occurrences that can be observed like any other physiological process... in this way we may hope to realize the possibility of elucidating, by experiment, the laws to which the origin of new species conform[1155].

The Mutationists

Thomas Hunt Morgan, a leading American geneticist (Chapter 25), described in 1903 the arguments of the mutationists, and underlined the advantages of their explanations over the traditional Darwinian theory[1156].

The Theory of Evolution, he wrote, is the best explanation of the known biological facts. But evolution is really two different phenomena: speciation and adaptation. These are not brought about by one and the same process, as Darwin proposed: new species are formed spontaneously, unrelated to the environment. Some of the new forms will die, others will succeed, survive, and reproduce. Natural selection will work on these survivors to bring about adaptation.

> De Vries defines the mutation theory as the conception that the characters of the organism are made up of elements that are sharply separated from each other. These elements can be combined in groups. Transitional forms like those that are so common in the external features of animals and plants do not exist between the elements themselves, any more that they do between the elements of the chemist[1157].

> Species have arisen from each other, not continuously but in steps. Each new step results from a new combination as compared with the old one, and the new forms are thereby completely and sharply separated from the species from which they have come (ibid.)

The mutation theory differs from the traditional Darwinian paradigm in the role assigned to natural selection. Natural selection does not create new species:

> If we suppose that new mutations and "definitely" inherited variations suddenly appear, some of which will find an environment to which they are more or less fitted, we can see how evolution may have gone on without assuming new species to have formed through a process of competition.

> [natural selection] is only a sieve, which decides what is to live and what is to die. Of course, with the single steps of evolution it has nothing to do. Only after a step is taken, the sieve acts, eliminating the unfit[1158].

> Natural selection may explain the survival of the fittest, but it cannot explain the arrival of the fittest[1159].

[1155] Ibid.

[1156] Morgan, "Evolution and Adaptation", 1903.

[1157] Ibid. p. 287.

[1158] De Vries 1912, p. 7.

[1159] Ibid. p. 462. The quotation is apparently from a paper by E.D. Cope, "The Energy of Evolution", American Naturalist 28: 205, 1894.

25

T.H. Morgan: Drosophila, Genetics, and Evolution

In the early 20th century genetic and evolutionary research flourished in the USA. An important figure in this development was the American-born geneticist, Thomas Hunt Morgan (1866-1945). Unlike the famous American evolutionists Dobzhansky and Mayr, who had immigrated to America after completing their education in Europe, Morgan had received his education at the University of Kentucky, followed by a doctorate from Johns Hopkins University in New York. In 1874 he traveled to the biological station in Naples, Italy, to work on embryology and marine invertebrates. Interacting with German biologists, he tended to support epigenetic ideas and became a critic of Darwin[1160].

Published shortly after the re-discovery of Mendel's paper and the publications by De Vries on discontinuous variation in plants, Morgan's important book[1161] offers a highly critical review of Darwin's Theory of Evolution, as expressed in his books "The Origin of Species" and "The Descent of Man", and is particularly critical of the theory of "Sexual Selection". This criticism can be considered as an example of the tendency of young scientists – especially in a new country – to reconsider and criticize established theories ("slaughtering sacred cows").

A major contribution of Morgan, then at Yale University, to science was the introduction in 1908 of the fruit fly, *Drosophila melanogaster*, as a research animal in his laboratory. This fly proved to be easy to maintain in the laboratory and very useful for research on inheritance, since it had only four chromosome pairs. A white-eyed mutant discovered in Morgan's laboratory was inherited in a "criss-cross" (sex-linked) pattern, and enabled the location of the mutation on the x-chromosome. Many more mutations were discovered in *Drosophila* in Morgan's laboratory in the years that followed – over 400 had been discovered by 1925 – many of which proved to be inherited according to Mendel's laws. The study of these mutations was followed by the analysis of linkage and mapping the genes on the chromosomes. The study of genetics became an independent and very fruitful branch of biology. For these discoveries Morgan was awarded the Nobel Prize in Medicine in 1933.

Morgan's book "Evolution and Genetics" (1925) was perhaps the first text on genetics, and marks the author's complete adoption of the Mendelian approach to evolution.

Chromosomes and Genetics

Although Weismann (Chapter 20) had suggested that hereditary information was carried in the chromosomes, it was W.S. Sutton[1162] who established the theory. Sutton studied grasshopper cytology and, in 1902, he noticed a similarity between the behavior of the chromosomes during meiosis and the theoretical expectation from Mendel's theory of character segregation. Sutton recognized that chromosomes have different shapes and can be distinguished individually. He observed that the

[1160] Benson 2001.

[1161] T.H. Morgan, Evolution and adaptation, 1903.

[1162] Walter S. Sutton (1877-1916), a New York farm boy educated at the University of Kansas, graduated in medicine. His MSc. thesis (1901) was on the chromosomes during spermatogenesis in a grasshopper. This material was the basis of his PhD. He pursued a successful career in surgery, and served in the Army during World War I in France. He died of acute appendicitis in 1916, aged 39.

chromosomes of the two parents are similar and attach themselves to each other, and then separate during cell division, similar to the theoretical "factors" in Mendel's model. He wrote in 1903:

> Thus the phenomena of germ-cell division and of heredity are seen to have the same essential features (viz. purity of units (chromosomes, characters), and the independent transmission of the same[1163].

Sutton also identified the "accessory (unpaired) chromosome" [now Y chromosome], which is found in the male meiosis but not in the female:

> Twenty-three is the number of chromosomes in the male cells [of the grasshopper] while twenty-two is the number I have found in the female cells, and thus we seem to find a confirmation of McClure's [his thesis supervisor] suggestion that the accessory chromosome is in some way concerned in the determination of sex[1164,1165].

Mendelian inheritance seems to have excited a wave of literature on heredity and animal breeding at the time. Morgan felt the need to warn against the enthusiastic attempts to naively explain away heredity of complex characters by simple models: he wrote sarcastically:

> If one factor will not explain the facts, then two are invoked. If two prove insufficient, three may sometimes work out. The superior jugglery sometimes necessary to account for the result, may blind us if taken too naively, to the commonplace that the results are often so excellently "explained" because the explanation was invented to explain them[1166].

Identifying the characters as physically represented by the "factors" in the germ cells, noted Morgan, is wrong! In a paper entitled "What are the factors in Mendelian explanations?" Morgan suggested a different interpretation:

> When we turn to the germ cells of the hybrid we… assume that the tall-factor and the dwarf-factor retire into separate cells after having lived together through countless generations of cells, without having produced any effects on each other. The assumption of separation of the factors in the gametes is a purely pre-formational idea. The factors have become entities that may be shuffled like cards in a pack… The egg need not contain the *character*s of the adult, nor need the sperm. Each contains a particular material, *which in the course of development* produces in some unknown way the character of the adult[1167].

In 1910 Morgan published a long essay on chromosomes as carriers of heredity. His papers[1168] signify his conversion to the mutation theory of heredity. In his description of the white-eyed mutant of *Drosophila* he wrote:

> A germ cell that produces white eyes differs from a germ cell that produces red eyes by one-factor difference. We think of this difference as having arisen through a factor in the red-eyed wild fly mutating to a factor for white[1169].

In 1925, Morgan strongly advocated the continuous study of the causes of genetic variation, but was still careful not to use "pre-formational" terms. He objected to the use of the term "lethal *genes*":

1163 Sutton 1903, Biological Bulletin 4: 231-251.
1164 Sutton 1902.
1165 Opler 2016: the Sutton-Bovery hypothesis.
1166 Morgan 1909.
1167 Ibid.
1168 Ibid. Morgan 1910.
1169 Morgan 1922.

We must find out what natural causes bring about variation in animals and plants, and we must also find out what kinds of variations are inherited, and how they are inherited[1170]. Death is an extremely complicated phenomenon. When the cause of death is simply referred to as a "lethal gene", this shows us clearly what a gene really is: we have clearly come to the conclusion that *a gene cannot be material*! A gene is a word, which enables a complicated happening to be briefly denominated[1171].

Morgan on Evolution

An entire chapter in "Evolution and Adaptation" is reserved for a critical review of the objections raised against the Theory of Evolution in the 19th century, and the difficulties in the theory noticed by Darwin himself. Morgan recognizes that the Theory of Evolution is composed of two different theories: the theory of Descent and the Theory of Transmutation, for which Darwin provided the mechanism of natural selection[1172]. The Theory of Descent can be verified.

> The history of life is written in the rocks. In the most ancient rocks we find fish. After them, we find, in sequence, amphibians, reptiles, birds and mammals. One of the last to show up is man. There should be no doubt that this order represents a series which starts from the most simple forms and culminates in the most complex[1173].

The theory assumes that a given group of organisms (a species) may produce one or more groups of offspring, which may differ in some characters from their predecessors. These offspring may replace the parental species, or they may coexist with it. This process is repeated, and new groups form new species. The similarity between these new groups is explainable by their common descent[1174].

> We accept the assumption that the Theory of Evolution is the best explanation of known facts. But we have to face two other questions: the formation of new species, and the significance of adaptation. These are two separate questions, and not one single question as Darwin's theory claims[1175]. In many cases the destruction comes in the form of a catastrophe to the individuals. So that the small differences in structure, whether advantageous or not, are utterly unavailable [to natural selection][1176].

Darwin had suggested that new species appear through the accumulation of small, favorable variations from a continuously-distributed population. However, noted Morgan, the theory of "transmutation" by natural selection is not to be accepted blindly. The assumption that heritable changes can accumulate from one generation to the next must first receive support before this assumption can be accepted. First and foremost, it must be determined whether the characters under study are heritable at all.

> But, within human history, not a single case is known of a transformation of one species into another. Perhaps the time was too short, or the probability of observing such a change is low – in particular if the change was too fast, or alternatively if the change was too slow, so that it will not be observed during the length of a human life[1177].

[1170] Morgan 1925, p. 15.
[1171] Ibid. p. 67.
[1172] As suggested by Haeckel in 1876 (Chapter 19).
[1173] Morgan 1903, p. 40.
[1174] Ibid. p. 31.
[1175] Morgan 1903.
[1176] Ibid. p. 120.
[1177] Ibid. p. 43.

> Even assuming that one or more individuals happen to possess a favorable variation, it by no means follows that natural selection would have free scope for the work of improvement, because the question of the inheritance of this variation and its accumulation and building up through successive generations, must be determined before we can be expected to give assent to this argument, so attractive when stated in an abstract and vague form[1178].

Morgan tended to accept de Vries's mutation theory. New species may arise from sudden, discontinuous "sports". The distinctions among existing species are sharp and clear, with no intermediate forms, while the environmental variation seems continuous. It is difficult to see how selection by the continuous environment may produce distinct species. The "fluctuating variation" is merely a response to the direct effect of the environment on the individuals. Selection may thus play no role in the formation of new species:

> Natural selection and heredity is actually at work, changing types. We have quantitative evidence of its effects in many directions. Yes, but no evidence that selection of this sort can do anything more than keep up the type to the upper limit attained in each generation by fluctuating variation[1179].
>
> Nature's test is survival. She makes new forms to bring them to this test through mutation, and does not remodel old forms through a process of natural selection[1180]... Selection does not do more than determine the survival of what is offered to it, and does not create anything new[1181].

Advantages of the Mutation Theory

The mutation theory escapes some of the greatest difficulties that the Darwinian Theory had encountered[1182].

> One of the greatest objections to the Darwinian Theory of Descent arose from the length of time it would require, if all evolution is to be explained on the theory of slow and nearly invisible changes. This difficulty is at once met and fully surmounted by the hypothesis of periodical but sudden and quite noticeable steps.

This is especially true as regards the formation of new species. The mutationists argued that new species arise suddenly in their final form by mutation, whether by multiple copy reproduction of a single mutant individual, or by the simultaneous occurrence of the same mutation in many individuals.

> From the point of view of the mutation theory, species are no longer looked upon as having been slowly built up through selection of individual variations, but the elementary species, at least, appeared at a single advance and fully formed... the most unique feature of those mutations is the constancy with which the new form is inherited[1183].

Whereas evolution and persistence of useless and even slightly injurious characters cannot be explained by the Darwinian theory of natural selection, the mutation theory offers a simple solution:

[1178] Ibid. p. 118.
[1179] Ibid. p. 274.
[1180] Ibid. p. 464.
[1181] Ibid. p. 460.
[1182] Ibid. pp. 297-8.
[1183] Ibid.

> If the organs appeared in the first place as mutations, and their presence was not injurious to the extent of interfering seriously with the existence and propagation of the new form, this new form may remain in existence. And if mutations continue in the same direction, the organs might become more perfect and highly developed[1184].

Morgan suggests that the process of *adaptation* does not necessarily have to be explained by natural selection. If a new form appears suddenly by mutation, regardless of the environmental conditions, it may be able to persist and spread provided it does not cause serious damage to the individual[1185]. This possibility could explain many difficult cases, such as the appearance of the 'soldier' caste in ant colonies:

> These have large, thick heads and large jaws. On the Darwinian theory this caste must have an important role to play, otherwise their presence as a distinct group cannot be accounted for... From the point of view of the mutation theory, their real value may be very small, but so long as their actual presence is not entirely fatal to the community, they may be endured[1186].

Another difficult case is that of the evolution of self-sterility in plants (heterostyly). Flowers of the same species differed in the relative lengths of the styles and the anthers. Only a limited number of "legitimate" crosses of the forms led to seed formation – self-fertilization was sterile – so favoring cross-fertilization[1187]. Darwin described this phenomenon in several plant families, and discussed its implications in detail[1188]. He had argued that such self-sterility could not have been directly favored by selection, and must have been incidental upon and correlated with some other character, such as the length of the stamens or the pistil, which ensured the legitimate fertilizations. To Morgan, this explanation sounded like an excuse, avoiding the difficulty. The mutation theory provides a more plausible solution:

> It is clearly apparent that the attempt to apply the theory of natural selection has been broken down, and it is a fortunate circumstance that the Lamarckian theory cannot here be brought to the rescue, as it is so often the case in Darwin's writings, when the theory of natural selection fails to give a sufficient explanation... If these two forms of the primrose should appear as mutations, and if, as is the case, they do not blend when crossed, but are equally inherited, they would both continue to exist as we find them today[1189].

[1184] Ibid. p. 372.
[1185] Ibid.
[1186] Ibid. p. 350.
[1187] Darwin's explanation stems from the assumption that the different forms of flowers are visited randomly by pollinators and have the same probability of being pollinated. However, the assumption may be incorrect. Experiments show that pollen-collecting hive bees on the distylous plant *Lythrum*, remain faithful to the tall-anther form and move only between tall-anther plants, effectively ensuring "legitimate" pollination. Bees collecting nectar, however, move randomly among the different forms of flowers. Eisikowitch & Ionescu 2007.
[1188] Darwin, 1986 (1877), "On the different forms of flowers...".
[1189] Morgan 1903, pp. 368-9.

26

The Resurrection of the Inheritance of Acquired Characters: from Weismann to Lysenko

The Dispute

The dispute regarding the inheritance of acquired characters could have ended with Weismann's publications at the end of the 19th century. T.H. Morgan wrote, in 1925, that although Lamarck and Darwin had often used "Use and Disuse" arguments to explain evolutionary phenomena, he was convinced by Weismann's arguments that such characters are not heritable:

> There are no measurements, so far as I know, to prove or disprove the claim that children of blacksmiths have stronger arms than other children, or that children of football players have bigger legs. Chinese women of high caste have had their feet bound and deformed for many generations, and now that the custom is being abandoned, the children do not appear to have feet different from those of other Chinamen. We do not observe the effects of the corsets of our grandmothers on the size of the waists of our children[1190].

But not all biologists were so convinced[1191]. Weismann himself noted that at the end of the 19th century, a group of neo-Lamarckians in the USA sought new kinds of evidence to support the inheritance of acquired characters, for example claiming that the differentiation of teeth in mammalian skulls was a result of accumulated pressure and abrasion of the teeth during mastication. The editor of the English edition of Weismann's book, the zoologist E.B. Poulson, noted in 1891 that this explanation did not make sense, as the teeth differentiate (in the calf) before they cut the gums, and therefore before they can be used for feeding.

In the 20th century the controversy moved from biology to other spheres –education, culture, and politics. Strong opposition to the categorical dismissal of the inheritance of acquired characters was raised by educators and social scientists, who were hoping that education and penal reform would bring about heritable changes towards a better society.

Studies of learning in animals provided a new source of support. At the Congress of Genetics in Edinburgh in 1923, Pavlov's laboratory in Moscow reported the results of experiments on the ability of mice to respond to signals. Mice were trained to run to a feeding place to obtain a piece of cheese upon the ringing of a bell. The report said that after 150 training runs, the mice would respond to the bell even without a reward ("conditioned reflex" was the term used). The offspring of the trained mice needed only 100 training runs to get to this stage, and subsequent generations only 50 and 5: the learned experience was transmitted to the next generation. Pavlov wrote that a future generation of these mice would acquire the ability to react to the bell without training at all. T.H. Morgan responded to this in 1925:

1190 Morgan, 1925, pp. 164-167.

1191 Julian Huxley (1942, pp. 457-466) reviewed many reports supporting Lamarckian explanations in the first 40 years of the 20th century.

> How simple would our educational questions become if our children at the sound of the school bell learned all their lessons in half the time their parents required! We might soon look forward to the day when the ringing of bells would endow our great-grandchildren with all the experiences of the generations that have preceded them[1192]. If we cannot inherit the effects of training of our parents, we escape at least the inheritance of their misfortunes[1193].

Kammerer: Education and Zoology

A strong supporter of these sentiments was the Viennese zoologist, Paul Kammerer [Kammerer's name became known mainly due to a notorious scientific scandal, which led to his suicide in 1926; see below]. Speaking before an assembly of educators and teachers on "the importance of heredity in education", Kammerer said that if Weismann was right, then whatever good characters the parents had acquired as children will die with them. But how different would be the outcome if Weismann was wrong!

> However, on the hypothesis of the inheritability of acquired characters – which seems to be closer to the truth – the individual's efforts are not wasted, they are not limited to his own life span, but enter into the life span of generations[1194].

As a young man, Kammerer had studied zoology and been employed as a curator of reptiles and amphibians at the Zoological Institute in Vienna. The Institute was provided with temperature-regulated growth-chambers in which animals could be maintained under constant climatic conditions, apparently an innovation which was not available in other institutions in Europe at the time. Kammerer appears to have been very successful at keeping creatures alive and breeding them, especially amphibians [he even named his daughter *Lacerta* [lizard].

Experimental Work

Although Kammerer is known mostly as a fraud and his work is generally ignored, his experiments [not his interpretation of them!] nonetheless seem to demonstrate a genuine effort to understand heredity.

Color Patterns in the Salamander: European salamanders dwell in moist habitats and return to freshwater about twice a year for reproduction. If the female is denied access to water, the young usually die. Kammerer reported that after a few failed events of reproduction away from water, some of the eggs – albeit very few – develop inside the female's abdominal cavity and are born not as tadpoles but as little salamanders. This is accompanied by a change in their color pattern from the normal black with irregular yellow spots (the natural "spotted" form), to a uniform black, uniform yellow, or the yellow spots arranged in rows on a black background ("striped").

Kammerer bred three strains of the salamander – the natural spotted form, collected in nature, and two striped strains: one collected in nature and the other selected in the laboratory. When he crossed the "natural" striped with the spotted female, the resulting offspring varied in color, as expected from Mendelian segregation. Crossing the laboratory striped strain with the normal spotted female yielded offspring intermediate in color. Kammerer interpreted these results to mean that the laboratory strain had acquired a different color pattern in the laboratory environment, and transmitted it to the next generation.

1192 Morgan, 1925, p. 159.
1193 Ibid. p. 178.
1194 Cited by A. Koestler, "The case of the midwife toad".

To further support his conclusion, Kammerer transferred ovaries from the spotted females to the abdominal cavity of striped females. When the ovaries were transplanted into a foster-mother of the "natural" striped strain, the offspring color pattern matched their *biological* spotted mother. When the striped foster-mother was of the laboratory strain, however, the offspring matched the color pattern of their "striped" foster mother. Kammerer considered this as evidence of a difference between the two striped strains (the salamanders reach reproduction at age 3, and he was unable pursue these experiments further].

Eyes in the Blind Amphibian: The blind amphibian, *Proteus*, is found in dark caves in Europe. Kammerer's very successful report, which brought him much fame in the USA, stated that large and functional eyes had appeared in the animals in captivity, under red light with bursts of white light. The finding was reported in the press and Kammerer was hailed as a "second Darwin" for his demonstration that the accumulation of small changes had led to a major adaptive change in the amphibian.

His opponents, however, claimed that the findings merely showed that this apparent possession of functional eyes was simply a response to light, with no genetic component at all, because it was modifiable by direct effect of the environment. Kammerer strongly rejected this objection:

> If we accept as hereditary only that which is unchangeable, and if changeable characteristics are not permitted to be hereditary, then just as in the Middle Ages, it is left to us only to expound the fixity of species. And with this not only the inheritance of acquired characters, but the whole theory of evolution is dogmatically done away with[1195].
>
> On the basis of the examples supplied here, we may assume that the inheritance of acquired characteristics has been proven, and this proof seems valid as regards the animal and vegetable kingdoms[1196].

The Case of the Midwife Toad: A Scientific Scandal

Among other organisms, Kammerer kept and bred an unusual species of toad, the midwife toad (*Alytes obstetricians*).

Most toads live near water or in shady habitats: a major part of their respiration is carried out through their moist skin. Although adult toads may wander quite far from a source of water, they must return to water to mate and reproduce. The male clings to the back of the female and releases his sperm when the female produces her eggs. During the mating season the males develop on their fingers small protuberances – usually black in color – to facilitate their grip on the female's moist and slippery back.

The midwife toad is different. Mating takes place on land, when the skin of the female is dry. The male wraps the fertilized egg-strips around his hind legs and carries them to the water, where the tadpoles develop. Accordingly, the males of this species do not develop the protuberances.

Kammerer kept his toads at a higher temperature than in their natural habitat, and forced them to mate in water. Most of the eggs did not survive, but he bred a second generation from the few "water eggs" that did. He claimed that the male offspring of his water-breeding toads also developed the protuberances, a character acquired due to the change in environment.

This claim was greatly disputed by the geneticists in England. William Bateson, the first Professor of Genetics at Cambridge and a strong supporter of the new genetics theory of Mendel (Chapter 23), questioned the claim and demanded to examine a specimen for himself. His request was denied, with the excuse being that these few toads were needed for breeding – and the protuberances

[1195] Kammerer 1924: p. 178.
[1196] Kammerer 1924. Ibid. p. 166.

appeared only in the breeding season. Bateson therefore traveled to Vienna and was received by the head of the Institute, but he was still not shown the toads. The dispute grew into a personal conflict between Kammerer and Bateson.

The First World War cut short the dispute, but it was resumed in 1919. Kammerer sent over microscope slides of sections through the protuberances on the fingers of the toads, but Bateson refused to examine them and declared that they had been retouched or taken from a different species. Kammerer was invited by the students at Cambridge (not the professors!) to give a talk, and he even brought a toad with him; but Bateson did not attend the lecture. Kammerer admits that he was deeply insulted by Bateson's refusal to accept his word that the results were genuine:

> W. Bateson ...did his best to maintain his contention that the nuptial pad [in *Alytes*] was no nuptial pad at all, but just a spot of black pigment... and finally went so far as to insinuate that the pad was brought about by retouching the picture... The botanist Bateson blamed me, the zoologist, for not knowing my own field of study!!

The war destroyed Kammerer's family finances [as well as destroying the Institute], and he had to earn his living by lecturing. He was twice invited to give a series of lectures in the USA, where he was declared a great biologist, second only to Darwin, because of his experiments with salamanders – especially the blind amphibian, *Proteus.*

His new fame led the Soviet government to invite Kammerer to Moscow to set up an institute like the one in Vienna, and in 1926 the Russian embassy in Vienna financed the transfer of all Kammerer's equipment to Moscow.

Kammerer finally sent a preserved specimen to England for examination. It did have black protuberances on the fingers, but upon examination the black pigment was shown to be black ink injected under the skin. An American herpetologist, G.K. Noble, in a letter in *Nature*, openly accused Kammerer of fraud. The story was greatly publicized and Kammerer, unable to face the humiliation, shot himself[1197].

Acquired Characters, Education, and Civilization

The concept of "acquired characters" was extended by educators to include social behavior and knowledge transmitted by education. In a book entitled "Evolution", published in 1925, the author commented on the inheritance of acquired characteristics:

> If the orthodox view, that acquired characters can never be transmitted, is correct, all our attempts at social improvement and education are vain, and the whole of civilization is doomed to crumble at no distant date, owing to the fact that all highly developed communities are principally recruited from their lowest ranks, the birth rate dwindling with each upward grade of development[1198].

Addressing a meeting of educators, Kammerer advanced the following argument:

> If acquired characteristics cannot be passed on, as most of our contemporaneous naturalists contend, then no true organic progress is possible. Man lives and suffers in vain. Whatever he might have acquired in the course of his life dies with him. His children and his children's children must ever start from the bottom[1199].

[1197] Koestler, The Case of the Midwife Toad.
[1198] Williams 1925, p. 57.
[1199] Kammerer 1924, p. 30.

Kammerer expressed his belief in a liberal view of Darwinism. In his book published in 1924, in a section entitled "Slaves of the Past or Captains of the Future?" he advocates that the effects of education and other social improvements are immensely important in human populations, and thus must be heritable in order to permanently improve society. In an outpouring of idealism, probably influenced by the Marxist theory and the Russian revolution, he wrote:

> Evolution is more than the fairest dream of the last century – the century of Lamarck, Goethe and Darwin. It is not unmerciful selection that shapes and perfects the machinery of life, nor dishonorable struggle for life which governs the world, but rather out of its own strength every creature strives upward toward light and the joy of life, burying only what is useless in the graves of selection[1200].

If natural selection is considered as a cause of progress, and the screening of populations as a creative force in human evolution,

> Darwinism becomes anti-Darwinism and the Theory of Evolution becomes regressive rather than progressive; the living world will be devoid of morality, and every case of cruelty will receive a halo of a natural process of selection, and not only be permitted – but actually required[1201]!

> Far above its interpretation as a theory of higher evolution, Darwinism is a doctrine of natural, world-embracing humanness. The belief in evolution is indispensable for the working-man's movement, which has disposed all other faiths[1202].

"Lamarckism" in the Soviet Union: The Heritage of Lysenko

The shortage of food in the Soviet Union in the years between the two world wars resulted in strong pressure to provide better-yielding field crops. A number of research institutions headed by famous geneticists were working on the genetic improvement of crops, but those involved insisted that they needed time for careful selection and breeding before releasing any new improved crops for agricultural use. Government officials despaired of this slow rate of progress.

Trofim D. Lysenko, a Ukrainian-born agronomist, suggested a fast way of improving agricultural yields. He believed that the "training" of plants under environmentally-induced harsh conditions would prepare them for survival in harsh winters, and the acquired properties would be transmitted to future plant generations. Lysenko supported his argument through application of a process (called vernalisation) which reduced the damage to wheat from spring frosts. Seeds were moistened to induce germination and then buried under snow (at 1°C) for several weeks to arrest growth until being sown in the field (the British geneticist J.B.S. Haldane, a devoted Communist in his youth, described the process enthusiastically in 1938). Lysenko suggested that the wheat plants thereby acquired resistance to cold, preparing them for the winter frosts, and if the same process is repeated, the trait will be inherited and cold resistance will increase. Following the experience of Michurin, a non-scientist but a successful apple breeder (famous for his use of grafting), the idea of training plants to develop useful properties was generalized into a principle, and was suggested for other fields of agriculture and horticulture. (Lysenko also recommended that rows of oak trees should be planted across Siberia as windbreaks. This should increase the temperature in parts of Siberia and enable crops to be grown there. The oak seeds should be sown in groups, because the individual seedlings will assist each other in withstanding the harsh environment).

[1200] Kammerer 1924, p. 258.
[1201] Kammerer 1924, p. 261.
[1202] Ibid. p. 264.

The Lysenko doctrine was rejected by the Soviet geneticists who headed the research institutions. They believed that Mendel's and Morgan's models of heredity were convincing, and should be followed. The Russian Communist Party officials however were delighted with Lysenko's suggestion that training is heritable. This had far-reaching effects.

> Some fanatic had come to the conclusion that the Marxist doctrine required the inheritance of acquired characters. The idea was that the proletarian, if given a chance, would acquire wonderful traits, and that his offspring would inherit them[1203].

Lysenko and his supporters used the press and party propaganda publications to raise public opinion against the "Mendelian" scientists who opposed Lysenko. In 1948, at the general assembly of the Academy of Agriculture, they succeeded in passing a resolution – by democratic majority rule and through the silencing of opponents – condemning "Mendelism, Weismannism, and Morganism" as wrong and reactionary, while Lysenko's (disguised as Michurin's) ideas were declared to be solid and the only truth. Following that resolution, all research on Mendelian genetics was banned. Research institutions were shut down and all traces of Mendel's and Morgan's genetics were systematically removed by the Ministry of Education from textbooks and curricula in schools and universities. Senior geneticists were accused of reactionary anti-revolutionary ideas, arrested, and sent to labor camps. The wheat expert Vavilov died in a concentration camp. The leading geneticist Dubinin was fired and his institution closed down. Dobzhansky used a stipend for training at Morgan's laboratory in New York and never returned to his homeland.

The dispute led to a rift between western and Russian geneticists[1204], and Lamarck's name – enlisted by the Lysenkoists to label their own ideas – became a derogatory term for opponents of the Theory of Evolution[1205]. Lysenko dominated Soviet genetics even after Stalin's death, and was removed from the academy of sciences in 1964, blamed for retarding the development of genetics in the USSR[1206].

[1203] Richard Goldschmidt, 1960, "In and Out of the Ivory Tower", p. 236.
[1204] Wolfe, A.J. 2016: J. hist. biol. 45: 389-414.
[1205] Hull 1984.
[1206] Joravsky 1971.

27

Genetics in Populations: An Introductory Overview

Variation in Natural Populations

The emergent science of genetics in the early 20th century required a new theoretical framework for the discussions of natural selection – based on non-blending characters. De Vries's suggestion of "elementary species" identifiable by experimental techniques, and his mutation theory (Chapter 24), required quantitative measurements.

One of the strongest objections to Darwin's Theory of Evolution was that too little heritable variation was available in natural populations, on which natural selection could work[1207]. Two kinds of variations were known at the time: minor differences among individuals in "biometrical" characters; and rare, occasional "sports" featuring large effects, mostly detrimental to the well-being of the animals and referred to by Darwin as "monstrosities"[1208].

The introduction of *Drosophila melanogaster* into Morgan's laboratory and subsequent genetic work, led to the discovery of many heritable mutations in characters such as eye color and wing shape. Many of these mutations had deleterious effects on their carriers, but "domesticated" mutant strains could be maintained indefinitely in the laboratory, and it was confirmed that they followed Mendel's laws of heredity.

The improvement of chromosomal techniques in the 20th century, and the introduction of electrophoresis of proteins for the study of real populations in nature – followed shortly by methods for the molecular analysis of DNA, enabled the quantitative measurement of genetic variation on a large scale, as well as enabling the prediction of future evolutionary changes in populations on the basis of real observations. A quantitative Theory of Evolution – *population genetics* – thus emerged.

Population Genetics in Nature

The next, revolutionary, step was taken by Theodosius Dobzhansky and his students: a broad-scale study of genetic variations of *Drosophila* in natural populations. Hundreds of thousands of flies were collected in nature, and lines derived from them were studied genetically in the laboratory[1209]. Cytological staining techniques revealed variations in chromosomal inversions – undetectable in adult flies but identifiable in the salivary glands of the larvae. Lines of *Drosophila* carrying these inversions were maintained and used for experiments. The frequencies of these inversions in natural populations were monitored, establishing a connection between genetics and the ecology of the flies. Other types of mutations were discovered in natural populations of the flies – fitness modifiers and even lethal genes were detected and mapped on wild chromosomes[1210].

[1207] Darwin admitted in 1844 (Chapter 8) that natural populations varied very little – and inferred, from the success of artificial selection, that hidden heritable variation must exist in nature.

[1208] A.R. Wallace, in his book "Darwinism", attempted to demonstrate that a great deal of variation in morphological characters indeed existed in populations of birds (Chapter 11).

[1209] See below Chapter 31.

[1210] See Lewontin 1974.

Another revolutionary step followed the introduction of electrophoretic analysis of proteins into population genetics[1211]. The number of genetic markers available for research in wild populations soared, as did the number of species other than *Drosophila* which could be investigated by these techniques[1212]. DNA sequence analysis at the end of the 20th century confirmed beyond doubt what had been suggested by Weismann 100 years earlier: except in rare cases, every individual is genetically different from all others. [The number of genotypes increases dramatically with the number of genes. With n loci and k alleles per locus, the number of possible genotypes is $[k(k+1)/2)^n]$. For n = 2 the number of genotypes is 9, but with n = 10 it rises to 59,049.] There is no shortage of genetic variation.

The Quantitative Theory of Evolution: Definitions

A crucial change in the approach to evolution occurred with the shift toward thinking in terms of changes in populations. A *population* is defined as a group of individuals of a single species, living in a specified area in a given period of time. The individuals are assumed to freely interbreed within a population. The theory generally deals with sexually-reproducing organisms.

The quantitative basis of evolutionary theory was established by three outstanding theoreticians: Ronald Fisher, J.B.S. Haldane, and Sewall Wright. Evolution was considered to occur as a result of a change in the frequencies of carriers of different genotypes in a population: new mutants could become extinct, or be driven to fixation by selection [as elaborated by Fisher[1213] and Haldane[1214]], or by random genetic drift [as suggested by Wright.[1215]].

An evolutionary process is defined as any process which causes a heritable and irreversible change in the frequency of genotypes in the population.

A genetic marker is some observable trait which should represent one or more loci, linked to this trait in some way (e.g., in an inversion). Genetic markers are essential for the study of changes in populations.

Genetic equilibrium is defined as a situation in which the frequencies of genotypes at the genetic–marker locus do not change from one generation to the next.

The Hardy-Weinberg Law

In 1908, G.H. Hardy, a Cambridge mathematician, and W. Weinberg, a German physician, independently derived the necessary conditions for a population to reach equilibrium frequencies and remain at a genetic equilibrium. Their model is known as the Hardy-Weinberg Law.

The model deals with a genetic marker with two alleles, say **A** and **a**. These alleles give rise to three diploid genotypes, **AA**, **Aa** and **aa,** at some arbitrary frequencies D, H, R (**D + H + R = 1.0).** If D, H, and R are known from observation, the frequencies of the two alleles can be calculated: **p** = D + 1/2 H; **q** = R + 1/2 H. (p + q = 1.0 by definition).

Hardy and Weinberg showed that, at equilibrium, the expected genotype frequencies should be $\mathbf{p^2}$ (AA); **2pq** (Aa); $\mathbf{q^2}$ (aa). [These frequencies can be calculated from Newton's binomial expansion, $\mathbf{(p + q)^2}$.]

Given five necessary conditions, equilibrium frequencies will be reached in one generation, and will not change so long as the five conditions continue to apply. The five conditions are: the population is infinitely large; mating is random; no mutations (at the marker locus); no migration; and no selection. Deviations from the expected equilibrium frequencies can result from a change in any of the five conditions, but it is possible to analyze these deviations and understand which of the

[1211] Lewontin & Hubby 1966; Hubby & Lewontin 1966.
[1212] e.g. Nevo 1984.
[1213] Chapter 28.
[1214] Chapter 29.
[1215] Chapter 30.

factors has not been met and why. The Hardy-Weinberg law is described and discussed in detail in many textbooks[1216].

A particularly useful graphic technique for illustrating equilibrium frequencies, and analyzing the departures from them, is the De Finetti diagram[1217]. This diagram is based on the geometrical rule that, *in an equilateral triangle, the height of the triangle is equal to the sum of the lengths of the three perpendiculars from a given point in the triangle to the three sides.* Thus, in a sample of organisms, the frequencies of the three genotypes – expressed as proportions D, H, R, are plotted as the perpendiculars from a point in an equilateral triangle. This point represents the composition of the population (several populations can be plotted in the same figure, to illustrate the differences in their composition, or a single population may be followed over time to illustrate the effects of selection). The expected frequencies of the genotypes at equilibrium – when allele frequencies change between q = 0 and q = 1, form a *parabola* in the diagram, with its maximum at p = q = 0.5, illustrating at a glance the deviation of the sample from equilibrium[1218].

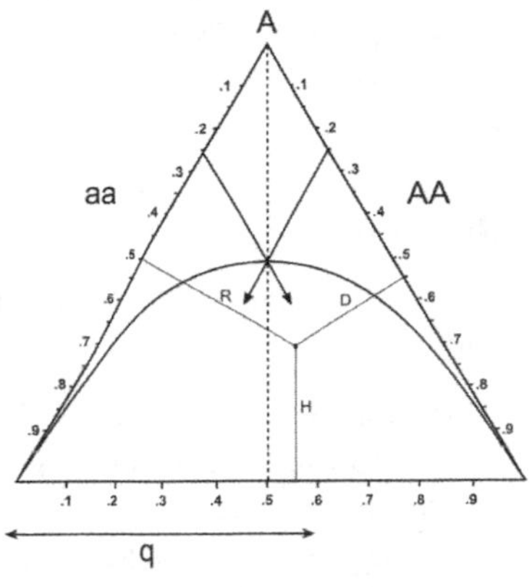

De Finetti diagram

Polygenic (Quantitative) Characters

The biometrical methods of Galton and Pearson[1219] found wide application in animal and plant breeding, and by the middle of the 20th century had developed into the new science of *quantitative genetics*, separate from the Theory of Evolution. The reason for this was that many characters of interest to animal breeders proved to have a continuous distribution when large samples of data were analyzed, and they responded to (artificial) selection just as Darwin and the biometricians had thought all characters should do. Weight, stature, milk yield in cows, yield of any crop, even discrete characters like egg production in domestic fowl or, in *Drosophila*, the number of facets in the fly's compound eyes, or the number of seeds in the capsules of *Papaver*, could be analyzed using statistical methods, such as the analysis of variance, regression, and correlation, based on the continuous normal distribution.

The theoretical incorporation of Mendelian genetics with classical biometric methodology in the Theory of Evolution became possible when it was realized that the simultaneous and cumulative effects of many Mendelian genes – each of them contributing a small and equal effect to the phenotype – form a normal distribution, with the additional proviso that the phenotype is affected not only by the genotype, but also by the environment[1220]. The mathematical models developed by Fisher, Haldane, and Wright could be applied to these "quantitative" (polygenic) characters. Quantitative inheritance thus became part of the unifying Theory of Evolution.

The numbers of genes involved in determination of a phenotypic character is generally unknown. Of greater interest for practical application is the relative importance of genetic (versus environmental) effects on the character, termed *heritability*[1221]. An estimate of heritability enables breeders to predict the response of their animals to selection for improving a desirable trait.

[1216] e.g. Li 1956; Wallace 1969; Spiess 1977; Wool 2006.
[1217] e.g. Wool 2006.
[1218] e.g. Wool 2006.
[1219] Chapters 21 and 22 above.
[1220] Falconer 1960. Symbolically, the total phenotypic variance V_p is the sum of the genetic (V_g) and environmental (V_e) effects.
[1221] Heritability is defined as the proportion of the additive genetic variance, V_a, in the total phenotypic variance V_p. Symbolically, Heritability $h^2 = V_a/V_p$.

Different characters are affected by the environment to a greater or lesser extent. For example, in seed samples of emmer wheat and wild barley, collected from different sites and sown in two common-gardens, the amount of rainfall (Ve) was the dominant component in total plant biomass, but the genetic differences among source populations (Vg) was predominant for seed (spicule) weight[1222].

Selection

The modern quantitative formulation of selection is based on the Hardy-Weinberg equilibrium model.

Definitions: Selection against a genotype implies that this genotype transmits fewer offspring to the next generation than the other genotypes. If the reproductive output of the most prolific genotype is taken as 1, the selected genotype will transmit **1 – s** offspring. **s** is referred to as the *selection coefficient*, ranging between 0 (no selection) and 1(lethality). (If **s** is assigned a negative value, then the selected genotype has an advantage – but this formulation is rarely used).

The quantity $\mathbf{w_i = 1 - s_i}$ is referred to as the *relative fitness* of genotype i. It is a measure of the reproductive output of that genotype relative to the most prolific one.

The population (or Darwinian) fitness is the average fitness of all genotypes in the population after selection, weighted by frequency.

The following Table summarizes the models for prediction of the results of different cases of selection.

Selection against:	Frequency of AA	Frequency of Aa	Frequency of aa	Fitness formula	W
No selection	p^2	$2pq$	q^2		
aa	1	1	$1 - s$	$p^2 + 2pq + (1 - s)q^2$	$= 1 - sq^2$
Aa	1	$1 - s$	1	$p^2 + 2pq(1 - s) + q^2$	$= 1 - 2pqs$
Advantage to Aa	$1 - s$	1	$1 - t$	$(1 - s)p^2 + 2pq + (1 - t)q^2$	$= 1 - sp^2 - tq^2$

Example: Results of selection against a recessive allele are calculated by selecting against the homozygous genotype **aa.** Following the equations in the Table, the average fitness after selection is $1 - sq^2$. The frequency of allele **a** after selection depends on both **s** and the initial allele frequency q. It is convenient to consider the limiting conditions: in the trivial case that s = 0 – no selection – q_1 is equal to q. When **s** = 1 the allele a is lethal. The equation reduces to

$$q_1 = 1/1 + q$$

substituting q_1 for q, we obtain for subsequent generations the series

$$q_2 = 1/(1 + 2q);\ q_3 = 1/(1 + 3q);\ \ldots\ q_n = 1/(1 + nq)$$

This series will tend to zero when n tends to infinity. The calculations show that *selection cannot effectively remove a deleterious recessive allele from the population – even when it is lethal.* (Many recessive deleterious alleles are known in human populations).

The expectations for other models of selection can be derived from the equations in the Table. They are elaborated upon in all textbooks on population genetics[1223].

[1222] Nevo 1978.

[1223] e.g., Li 1956; Spiess 1977; Wool 2006.

28

R.A. Fisher: The Fundamental Theorem of Natural Selection

Ronald A. Fisher (1890-1962) graduated from Cambridge University and was appointed statistician at the Rothamsted Experimental Agricultural Station. He developed statistical methods for analyzing agricultural and genetic experiments, in particular the Analysis of Variance – now commonly used – and wrote statistical manuals for their field application. His introduction of the principle of randomization into experimental design helped reduce bias and enabled isolation of the desired variable from the side effects of the experimental environment. After the retirement of Karl Pearson, Fisher accepted the Chair of Eugenics at Cambridge and taught genetics there for 13 years. His statistical methods were important in the new science of genetics, as they enabled the analysis of linkage. Fisher received a knighthood in 1952.

Natural Selection

Fisher's famous book, "The Genetical Theory of Natural Selection" (first published in 1930) was dedicated to the quantitative analysis of natural selection based on particulate, rather than blending, inheritance. He declared that the purpose of his book is:

> To state the principle of natural selection in the form of a rigorous mathematical theorem, by which the rate of improvement of any species of organism in relation to the environment is determined by its present condition[1224].

Fisher devotes considerable space in his book to describing the advantages of particulate inheritance for evolution. If characters blend, as believed earlier, an extraordinarily high mutation rate must be assumed in order to maintain large enough genetic variation for the operation of natural selection. But if inheritance is particulate, there is no need for a high mutation rate because the mutations are retained in the heterozygotes:

> The heterozygote is possessing variance in a latent form, so that instead of being lost it is really stored in a form from which it will reappear when the heterozygote genotypes are mated[1225].

Unlike other evolutionary theoreticians, Fisher treats natural selection from an ecological perspective: the reproductive rates of genotypes are part of his genetic calculations. The frequencies of genotypes change because their reproductive output is not the same. The growth rate of the entire population is the sum of the contributions of the different genotypes and is measurable as the specific rate of increase, **r**, "the Malthusian parameter". With one gene and two alleles, this can be expressed as $W = \Sigma(2\,p\,q\,a\,\alpha)$, where W is population reproductive success (= fitness), p and q are allele frequencies, **a** is the probability of substituting allele A by a, and α is the effect of this substitution on the growth rate of the population.

[1224] Fisher 1958, p. 22.
[1225] Fisher 1958, p. 10.

The mathematics of this formulation is hard to follow, belying the expressions like "clearly" and "it is easy to see that" that abound in the text[1226].

The central conclusion of Fisher's theory is that *natural selection will invariably increase the fitness of the population.* The speed of this increase is proportional to the genetic variance in the population:

> The rate of increase in fitness of any organism at any time is equal to its genetic variance in fitness at that time. [$dW/dt = \sigma^2_W$]

This formulation is known as the *Fundamental Theorem of Natural Selection.*

Natural Selection and Adaptation

Fisher, like Darwin, considered natural selection to be the driving force in adapting organisms to their environment. Being adapted to an environment, he wrote, is a relative state: an organism is considered adapted only if we can conceive of other organisms, slightly different from it, which are less well-adapted to the same environment – or, alternatively, other slightly different environments to which the organism is less well-adapted.

The process of adaptation depends on the magnitude of change required in the organism to make it better adapted. Fisher considers the change from the point of view of the individual organism: selection and evolution work on populations, but the changes occur by mutation in individuals, and mutations of large effect are often deleterious; the larger the change, the greater the probability that it will be deleterious[1227].

The environment itself is continuously changing. One of the principal dimensions of the environment that changes is the density of the population, which increases as a result of population growth (this is where ecological parameters enter Fisher's selection formulations).

> An increase in numbers of any organism will impair its environment in a manner analogous to, and more surely than, an increase in the numbers of its competitors. Probably more important than the changes in climate, will be evolutionary changes in associated organisms: as each organism increases in fitness, so will its enemies and competitors increase in fitness[1228].

Selection and the environment restrict the reproductive output of each genotype (w_i), but do not affect the potential growth rate of the population (**r**). The difference is expressed in the production of more individuals than are needed to maintain population size. Ideally, population size will remain stable if each adult female produces two fertile offspring at the end of a generation; but in reality, more than two offspring are needed to replace losses due to disease, predators, and parasites. A "Malthusian" geometric growth rate will materialize only if all these "surplus" adults reach maturity and reproduce.

[1226] Fisher may have been a mathematical genius, but he did not bother to explain his thinking to the less-endowed readers. Even professional statisticians had difficulty in following the details. "When I come to 'evidently', I know that it means two hours of hard work at least before I can understand why" [W. Gossett, statistician at the Guinness brewery, "student" of the famous t- test). "Of the books, I would like to recommend especially R.A. Fisher's "A Genetical Theory of Natural Selection" for its brilliant obscurity. After two or three months of investigation it will be found possible to understand some of Fisher's sentences." [Fred Hoyle, Cambridge astronomer and mathematician]. Both quotations cited by Bodmer (2003).

[1227] Fisher 1958, p. 50.

[1228] Ibid. p. 45.

> The production of offspring is only expressive in relation to an imaginary world, and the high geometrical rate of increase is only obtained by abolishing a real death rate, while maintaining a real rate of reproduction[1229].

A high death rate, Fisher wrote, is not proof of selection. The fact that a fish lays a million eggs of which only a few survive, does not mean that selection is taking place (but it would be interesting to investigate what is the physiological reason for a female fish investing so much energy in reproduction[1230].

Mutation, Variation, and Evolution

Natural selection, Fisher wrote, works to promote characters which are beneficial for the *individual*, improving its chances of survival and reproduction – regardless of whether these characters seem to be of value for the species as a whole.

Variants are produced by *mutation*. Mutations are rare events (detected at the rate of 10^{-6} or less per generation). Mutations which occur once only, in a single individual, have a high probability of being lost within one or very few generations[1231].

> The great majority of mutations in *Drosophila* are lethal, nearly all completely recessive. Of 221 mutations reported by Morgan, 208 were classified as recessive and 13 are dominant[1232].

Deleterious mutations have no evolutionary value, because deleterious mutations are constantly being removed by natural selection – although they may be important for the breeder and the geneticist. Advantageous mutations occur very rarely. Neutral mutations are more frequent. But the important variable is not the *mutation rate*, but whether a given mutation increases or decreases in frequency in the population (later referred to as the *replacement* rate).

> It is not only the frequency of a gene, but the reaction of the organism to it, which is at the mercy of natural selection[1233].

Any gene which increases in numbers – whether this increase is due to selective advantage, an increase in mutation rate, or to any other cause – will so react on the genetic constitution of the species to increase its selective advantage if this is increasing, or to retard it if it is decreasing[1234].

Dominance

One of Fisher's greatest contributions to the theory of population genetics and evolution was his understanding and analysis of the role of dominance.

> It should be emphasized at the outset, that dominance is an observational fact, involving a comparison of the somatic characters of three different genotypes, two homozygotes and the heterozygote formed by crossing them. If the heterozygote is found to be indistinguishable from one of the homozygotes, that homozygote is said to be completely dominant, and the

---

[1229] Ibid. p. 46.
[1230] Ibid. p. 52.
[1231] Ibid. p. 83.
[1232] Fisher 1931, p. 349.
[1233] Fisher 1958, p. 59.
[1234] Ibid. p. 103.

> other completely recessive[1235]. In many cases of multiple allelomorphism, where several distinguishable mutant genes have arisen from the same gene of the wild fly, the wild type is completely dominant to all its mutants[1236].

Dominance has a clear advantage, because recessive deleterious mutations will not be expressed in the phenotype of the heterozygote and will cause no damage[1237].

Fisher's concept of a wild type implies that, in most individuals, the two alleles at most loci are identical. Only a few are heterozygous for a mutant allele. Natural selection works to remove these mutant alleles from the population.

> It may be inferred that whereas genetic diversity may exist perhaps in hundreds of loci, yet in the majority of loci the normal condition is of genetic uniformity. Unless it were so, the concept of the wild type could be an indefinite one[1238]. In order to take full advantage of the possible occurrence of advantageous mutations, mutation rates must be generally so low that in the great majority of loci, the homologous genes are almost completely identical[1239]. The variations within a species are like the differences in color of different threads, which were crossed and re-crossed a thousand times in the manufacture of a homogeneous fabric.

Fisher suggested that the widespread "wild type" phenotype of the organism proves that it has a selective advantage, because none of the deleterious mutations it may carry are expressed. Selection will favor any genes which improve the coverage of the deleterious recessive mutations (modifier genes)[1240]. New mutations must become dominant before they become established in the population.

In poultry, which Fisher studied experimentally,[1241] the majority of "domestic" characters were partially or entirely dominant. Fisher suggested that the domestic fowl in Asia were often crossed and re-crossed with wild cocks, preventing the fixation of recessive characters[1242].

However, one should keep in mind that dominance may not always be advantageous for the species, because it protects the recessive deleterious mutations from selection, and their frequencies in the population may therefore increase:

> The acquisition of dominance to harmful mutations cannot be said to improve the species, for its consequence is that harmful genes are concealed and allowed to increase[1243].

Theoretically, selection favoring heterozygotes is considered an important mechanism for maintaining polymorphism in populations, because it may lead to a *stable* equilibrium.

> The stability of the frequencies of the different genes [in butterflies]... would find its simplest explanation if the heterozygotes could be postulated to be at a selective advantage compared to both of the alternatives[1244].

[1235] Fisher 1931, p. 348.
[1236] Ibid. p. 349.
[1237] Fisher 1958, pp. 53-54.
[1238] Fisher 1958, p. 138.
[1239] Ibid. p. 160.
[1240] Ibid. pp. 62-67.
[1241] Huxley, J., 1942, p. 81.
[1242] Fisher 1958, p. 357.
[1243] Fisher 1958, p. 74.
[1244] Ibid. p. 359. See details in Wool 2006.

The Fundamental Theorem

The calculated theoretical result, that selection always increases average fitness W, agrees with the expected effects of *natural* selection as envisioned by Darwin: selection is supposed to make the population better adapted to the environment, and a better adapted population should have higher fitness in that environment. [This tendency explains why W is sometimes referred to as "Darwinian Fitness", although it does not reflect adaptation but is a relative measure of reproductive success].

Fisher demonstrated that the rate of change of the average fitness W is related to the genetic variance σ^2 in the population:

$$dW/dt = \sigma^2_w$$

This expression is referred to as the Fundamental Theorem of Natural Selection. This is phrased verbally thus:

> The rate of increase in fitness of any organism at any time [ascribable to natural selection acting through changes in gene frequencies] is exactly equal to its genic variance in fitness at that time[1245].

Sexual Reproduction and Assortative Mating

In the later chapters of his book, Fisher deals with many evolutionary subjects – such as the evolution of the sex ratio, mimicry, and even social processes and social structure in Man. These subjects are discussed in general, non-mathematical and philosophical terms, imbued with the fundamental belief in natural selection as the only evolutionary force in nature.

The Advantages of Sexual Reproduction: The primitive method of reproduction in unicellular organisms must have been asexual: it is a continuation of the phenomena of growth and division of cells, which also enable the regeneration of lost or injured parts. Asexual organisms may have each given rise to a separate species: groups of individuals of common descent may have developed from single mutant individuals which had some reproductive advantage, so that their chance of survival was higher than the probability of extinction of single mutations.

The fact that the prevalent method of reproduction in *multicellular* organisms is sexual, indicates that this system of mating must have had an advantage over asexual reproduction. The advantage, according to Fisher, lies in that a sexually-reproducing line can respond much more quickly to changes in the environment.

> In consequence, an organism sexually reproduced can respond so much more rapidly to whatever selection is in action, that if placed in competition, on equal terms, with an asexual organism similar in all other respects, the latter would certainly replace the former[1246].

Assortative Mating and Sexual Selection: If the selection of mating partners is not random but assortative, the result may be the isolation of subgroups within a population. This may lead to sympatric speciation without geographic separation.

[1245] Edwards 1994.

[1246] Fisher 1958, p. 160. It should be emphasized that ecologically, and in the short term also evolutionarily, an asexual line can be expected to have an advantage over a sexual one: when new habitats are colonized, even a single asexual individual may found a new population (while a sexual individual needs one of each sex to reproduce), and the growth rate of an asexual population should be twice that of a sexual one, in which only half the individuals produce new offspring.

Fisher explains that the attraction of the sexes is not equally strong in different individuals. For this to have evolutionary significance two conditions must apply: 1) that the selection by the female of suitor A_i cancels the chance of mating with any other suitor A_j ($j \neq I$; and 2) That many potential suitors exist in the neighborhood, and the rejection of one suitor leads to courting by others. There should also exist a complex system of recognition cues among members of the same species in order to avoid the chance of mating between different species.

Fisher expresses support for Darwin's theory of sexual selection (which Morgan (1903) had rejected as unscientific)[1247]. In mammals during the breeding season, selection by the females of the victorious males may drive the evolution of big horns and sharp canines, characteristic of males only. Darwin, says Fisher, courageously suggested that *all* cases of sexual dimorphism were brought about by the same mechanism – including the color pattern dimorphism in birds like the peacock and the pheasant. This generalization is controversial, and many scientists have claimed that each case should be examined separately to determine whether females do indeed prefer the most colorful males[1248]. If female selection of mates is based on heritable mating behavior incorporated in the genetics of every species, then female attraction to a certain color or pattern of the male may bring about an evolutionary change in the color of males, provided that the color or pattern affects in some way the fitness of the males:

> The importance of this situation lies in the fact that the further development of the plumage character will still proceed, by reason of the advantage gained in sexual selection, even if it had passed the point in development at which its advantage in natural selection had ceased[1249].

If the two processes – improvement of the bright color or pattern in the male and increased preference of the females for the brightest or most elaborately-patterned males – occur together simultaneously, improvement of both characters may proceed at an accelerating speed (a "runaway process").

> In any case where sexual selection can bring about a great reproductive advantage – a situation which in fact exists in some polygamous birds – there is the possibility of an accelerating process, which in advanced stages may proceed very quickly, and if it is not suppressed may bring about impressive results[1250].

Natural Selection and the Sex Ratio

The fact that males and females are born in approximately equal numbers in human populations was being pondered on already in the 18th century. Dr. John Arbuthnot, physician to Queen Anne and a member of the Royal Society, published as early as 1710 a research paper based on 82 years of birth records in London hospitals. He used the Newtonian binomial formula $(p + q)^n$ (gambling results with two dice, assuming the results to be random) to show that the probability of getting *exactly* equal numbers in a family diminishes with increasing sample size. His records showed that *more males* are born than expected by chance, and he explained that this is due to the wisdom of the Creator:

> We must observe that the external accidents to which males are subject (who must seek their food with danger) make a havoc of them, and that this loss exceeds by far that of the other sex occasioned by disease incident to it as experience convinces us. To repair the loss,

[1247] See above, Chapter 25.
[1248] Fisher 1958, p. 149. See the objections by A.R. Wallace, Chapter 16 above.
[1249] Ibid. p. 152.
[1250] Ibid.

> Provident Nature, by the disposal of its wise Creator, brings forth more males than females. And that in an almost constant proportion… There seems to be no more probable cause to be assigned in physics to this equality of the births, than that in our first parent seed were at first formed an equal number of both sexes[1251].

Charles Darwin, in "The Descent of Man", also reflected on the sex ratio. Surveying the available evidence on the sex ratio in many (mostly domesticated) species of animals, Darwin came to the conclusion that, in most cases, the sex ratio did not differ from 1:1. Fisher cites Darwin's statement verbatim:

> I once thought that the tendency to produce the two sexes in equal numbers should be advantageous, and that natural selection has brought it about. But now it seems to me that the problem is so complicated, that it is safer to defer its solution to the future[1252].

Fisher commented that Darwin's suggestion, that natural selection favored the production of the sexes in equal numbers, is perfectly acceptable. In Fisher's interpretation, the reproductive effort in every generation must be equal in the two sexes. If the investment necessary for producing a male is less than that for producing a female, then some females may tend to produce a larger number of males, and the sex ratio at birth may change creating a surplus of males – which will be balanced later by higher male mortality.

On Mimicry

Fisher offered a theoretical analysis of the genetic aspects of the similarity of certain species of butterflies (mimics) to other, model, species which are inedible and are avoided by predators. Two possible scenarios had been were presented to explain the evolution of mimicry. The first – "Batesian" mimicry – was suggested by Henry Walter Bates[1253], and assumes that the similarity in pattern and color of the mimic and the model misleads the potential predators into refraining from attacking both the mimic and model. The recognition of the model's pattern is heritable in the predator. "Batesian" mimicry will work if the model is common and the mimic rare. Natural selection will work to increase the similarity in both the mimic and the model, because the closer the similarity, the better both are protected. The second explanation suggests that ["Müllerian"] mimics of different species tend to have similar conspicuous warning coloration – red, yellow, and black – as well as other characteristics in common. All these species may be in some degree inedible, and actually function as both models and mimics. Müller noticed that birds learned individually from experience – by trial and error – to identify the pattern as indicating an inedible prey.

After several attempts to compare the Batesian model and the Müllerian model mathematically, Fisher concluded that he could not decide which one offers a better explanation of the phenomenon.

On Human Society: the Biological Paradox

An extensive chapter in Fisher's book is dedicated to processes in human populations. He has this to say on the social structure:

> Biologically, [the different occupations within populations] are of importance in insensibly controlling mate selection through the influence of prevailing opinion, mutual interests

[1251] Arbuthnot 1910.

[1252] Fisher 1958, pp. 158-159. Darwin 1874, p. 277.

[1253] See Chapter 12.

and the opportunities for social intercourse which they afford. Social classes thus become genetically differentiated, like local varieties of a species[1254].

Whenever, then, the socially lower occupations are the most fertile, we must face the paradox that the biologically successful members of our society are to be found principally among the social failures, and equally that classes of persons who are prosperous and socially successful are, on the whole, the biological failures, the misfits of the struggle for existence. The struggle for existence (within such societies) is the inverse of the struggle for property and power[1255].

Smoking and Lung Cancer

In his later years, Fisher became involved in a public controversy over the relationship between cigarette smoking and lung cancer. In the middle of the 20th century the habit of smoking became widespread in Europe, stimulated by the mass production of cigarettes (the smoking of pipes, and of imported cigars, was a much more restricted practice, limited to the wealthy). In England, the consumption of cigarettes increased 100-fold between 1900 and 1965, with a parallel increase in the reported incidence of lung cancer. A public movement in England, motivated by both moral and social arguments (e.g. that "cheap" women smoking in public were a disgrace), blamed smoking as the cause of the rise in mortality from cancer. The cigarette manufacturers hired Fisher as a statistical advisor. He examined the data and declared that it was based on correlation; and that, statistically, correlation does not necessarily indicate a cause-and-effect relationship: there was no evidence of smoking causing cancer, for both smoking and cancer could result from an external factor affecting both variables, and experiments with humans to test the effect of smoking were of course impossible (from a statistical point of view he was indeed right).

Despite the position of the statisticians, a committee of experts set up by the American Ministry of Agriculture ruled in 1964 that smoking was the cause of cancer. Severe restrictions were imposed on the cigarette manufacturers, and the sale of cigarettes dropped considerably – with a subsequent reduction in the incidence of lung cancer[1256].

[1254] Fisher 1958, p. 244.
[1255] Ibid. p. 241.
[1256] Moses 2008 (in Hebrew).

29

J.B.S. Haldane: Science in Everyday Life and Genetics in Populations

John B. Sanderson Haldane (1892-1964) was an eccentric figure among scientists. He had participated in applied scientific research since childhood. His father, John Scott Haldane, was an eminent physiologist at Oxford[1257], and as a boy Haldane had accompanied his father in his research on methane accumulation in old coal mines[1258]. Graduating with distinction in mathematics from Eaton and Oxford, he served as an officer in the British army in France during World War I, and then taught mathematics in Cambridge and London. He was an enthusiastic Marxist for many years, but left the Communist Party when Lysenko took over responsibility for science in the Soviet Union[1259]. In 1957 he moved to India, became a citizen there, and headed the Department of Biometry and Genetics in Calcutta.

Haldane's research interests spanned a wide range of both applied and theoretical subjects. He participated in research on toxic gases for chemical warfare, and in developing rescue apparatus for crews in submerged submarines. He did not hesitate to test the novel methods on himself as a research animal, exposing himself to life-threatening atmospheric changes[1260]. Haldane spoke lightly of this kind of research:

> Experiments of this sort are safe if you calculate the quantities correctly and make no mistakes – and if you do make a mistake, then there will be one less bad biochemist in the world.

Haldane published theoretical calculations of genetics in populations, as well as an idealistic forecast of the bright future that science will provide for humanity [see below]. He published many books and hundreds of articles. Among other subjects, he studied the effects of body size on the anatomical and physiological characteristics of animals. The relationships he discovered are referred to as "Haldane's principle". A collection of his papers on the subject was published under the title "On Being The Right Size"[1261].

[1257] Among his other achievements, Haldane Sr. developed the decompression tables which are used to this day by divers when they surface from deep-sea diving, to avoid the decompression sickness (the "bends") of gas bubbles developing in the blood.

[1258] He was asked to stand up in suspect places in the mine and recite aloud a Shakespeare poem until he fainted and collapsed: when he fell, he could again breathe fresh air since methane is lighter than air.

[1259] See above, Chapter 26.

[1260] Haldane remained for 14 hours in a sealed chamber, recording the negative effect of the reduction of oxygen and accumulation of CO_2 to dangerous levels before the chamber was automatically opened. To prove that urea is not toxic to humans, he ate about 100 grams of urea (1/4 pound) of it [urea is formed in the liver from metabolic ammonia, and excreted in the urine]. Similarly, to examine the effect of blood pH on respiration rate, he increased his blood pH by eating 40 grams of sodium bicarbonate (his blood pH went up as expected, and his breathing rate slowed down). He then drank a low concentration of ammonium chloride to reduce the blood pH [the material dissolved in the blood and left residues of hydrochloric acid, which caused his respiration rate to increase dramatically – forcing him to pant for several days!].

[1261] Haldane 1927.

Haldane gained fame as a popular writer, publishing articles on the application of science in *The Worker* – a newspaper of the British working classes – later collected into a volume entitled "Science and Everyday Life"[1262].

"On Being The Right Size"

> It is easy to show that a hare could not be as large as a hippopotamus, or a whale as small as a herring. For every type of animal there is a most convenient size, and a large change in size inevitably carries with it a change of form[1263].

Haldane's argument is based on the simple principle, that a ten-fold change in linear dimension of a vertebrate animal means a 1,000-fold change in volume (and therefore weight), but only a 100-fold change in the cross-section of the bones. Therefore, the bones will have to withstand a ten-fold heavier load per square inch.

A similar size-limitation is imposed by gravity. A small animal such as a mouse can fall from a height with no damage to its skeleton, while a rat or a man falling from a similar height will be killed. This is because the resistance of the air to movement is proportional to the surface area of the falling object. A 10-fold smaller size means a 100-fold smaller surface area – and the air resistance to the fall is thus 10 times greater. This is why a small insect is almost unaffected by gravity and can walk upside-down on the ceiling.

Large homeothermic vertebrates have other problems that limit their size, such as the need to obtain enough oxygen (which is translated to lung or gill size); the need to pump blood to greater distances in their bodies than in small animals; and the need to consume much more food to sustain them. On the other hand, large animals survive better in a cold environment, because their smaller surface-to-volume ratio enables them to maintain their body temperature better than small animals.

Similar considerations apply to birds. In order to enable flight, birds cannot be as heavy as mammals of the same size (the hollow bones make the difference); heavier weight must be limited by wing area – which explains why large birds soar, rather than flap their wings in flight.

This "Haldane's principle" seems to be applicable to the entire animal world.

Daedalus

Haldane, the idealist and true believer in Marxism, gave a lecture at Cambridge in 1923, entitled "Daedalus[1264] – or science and the future". In his lecture, Haldane described his vision of an ideal world. Food would be synthesized in factories, in which microorganisms would be used to convert cellulose to sugar. This industry would make agriculture unnecessary: the factory worker would replace the farmer. The bulk of the human population would move to the cities, with a great improvement in living conditions. Haldane's imagination carried him very far indeed:

> Within the next century, sugar and starch will be as cheap as sawdust. Many of our foodstuffs, including the proteins, we shall probably build up from simple sources, such as coal and atmospheric nitrogen. This will mean that agriculture will become a luxury, and that mankind will be completely urbanized. Personally I do not regret the disappearance of the agricultural worker in favor of the factory worker, who seems to me a higher type of person from the most points of view… Synthetic food will substitute the flower

[1262] Haldane 1940a.

[1263] "Possible worlds and other essays", pp. 18-26.

[1264] Haldane 1923. In Greek mythology, Daedalus was a sculptor who carved the beautiful statues of the gods. Daedalus serves as a symbol of applied science, because he designed and built the wings which enabled him and his son Icarus to escape from the Minotaur (Icarus flew too close to the sun, the wax used in the structure of the wings melted, and he fell into the sea and drowned).

> garden and the factory for the dunghill and the slaughterhouse, and make the city at last self-sufficient[1265].

Haldane's vision of energy production is even more futuristic. Within 400 years, he predicted, England will be covered with a grid of wind-driven generators which would supply electricity to central power stations. These stations would use the surplus energy for electrolysis of water, and the resulting hydrogen and oxygen would be stored – in cold liquid form – in giant insulated tanks and provide cheap energy for uses like the operation of ships and airplanes, and for industry.

As an example of the enormous positive effects of science on society, Haldane refers to the great improvement in the health and sanitary conditions of the population, already brought about by the science of medicine:

> Although the living conditions in the cities are as bad as they are, there is not even one slum in the country in which the infant death rate is even one third of what was the death rate in royal families in the Middle Ages.

Haldane predicted that the science of genetics would develop methods to control newborn genotypes, which would improve the breeding of desirable plant and animal varieties, as well as of humans: this would be achieved by artificial insemination in the test tube[1266]. Haldane wrote that this idea would probably elicit strong objections from religious circles – but that is the fate of every new scientific finding:

> If every physical and chemical invention is a blasphemy, every biological invention is a perversion... the biological invention then tends to begin as a perversion and end as a ritual, supported by unquestioned beliefs and prejudices[1267].

On the other hand, Haldane was very critical of the Eugenicists[1268] due to their blaming the impoverished classes in society for carrying genetic defects and endangering the fate of humanity. The Eugenicists also advocated the restriction of reproduction in the lower classes. In their attempt to create a better race of man, Haldane wrote, the eugenicists succeeded in arousing opposition and hatred:

> A number of earnest persons, having discovered the existence of biology, attempted to apply it in the then crude condition to the production of a race of supermen, and in certain countries managed to carry a good deal of legislation. They appeared to have managed to prevent the transmission of a good deal of syphilis, insanity and the like, and they certainly succeeded in producing the most violent opposition and hatred among the classes whom they... regarded as undesirable parents[1269].

On Genetical Processes in Populations

One of Haldane's most important contributions to the genetical Theory of Evolution was published in 1924. He listed a number of necessary preconditions before one could claim that a given trait may be under the control of natural selection: 1) determination of the mode of inheritance of the trait;

[1265] Haldane 1923, Daedalus, p. 5.

[1266] No one could have foreseen in 1923 that this bizarre idea would become quite a common medical procedure as the 20th century was nearing its end.

[1267] Daedalus, p. 6.

[1268] See Chapters 21, 22, 26.

[1269] Haldane 1923, Daedalus, p. 7.

2) knowledge of the mating system; 3) determination of both the intensity of selection; and 4) the frequency of carriers of the trait in the population.

Haldane then worked out detailed mathematical formulas to predict the frequency of dominant alleles, when exposed to different intensities of selection when mating was random, as required for equilibrium[1270]. Haldane's calculations are still listed to date in textbooks on population genetics.

Haldane was first interested in understanding selection favoring a dominant allele with positive effects on the individual. The selective intensities in his initial calculations were small – in the order of magnitude of 0.001 – since Darwin had insisted that any advantage of one individual over others, however small, will allow that individual to multiply and spread in the population. Haldane showed that, at such selection intensities, thousands of generations are required for the fixation of an advantageous mutation – regardless of whether it is dominant or recessive. He also demonstrated that selection against a recessive trait is very inefficient, and considered that recessive, advantageous mutations with a large positive value are rare. He derived this from the fact that of the many mutations causing melanism in moths, some of which turned out to be advantageous – not one was recessive. He concluded,

> It seems therefore very doubtful whether natural selection in randomly-mating populations can cause the spread of autosomal recessive characters, unless they are extraordinarily valuable to their possessors[1271].

Industrial Melanism

An increase in the frequency of melanic (dark-colored) moths in locations where moths of lighter color had previously been predominant, was noted in British industrial areas early in the 19th century. Haldane analyzed the conditions under which the melanic moth mutants had spread in England, and calculated that from the first appearance of the first melanic mutant in Manchester in 1848, until the time of total takeover (fixation) of the melanics before 1901, a period of only 50 years had elapsed. To become the most prevalent form in the population within so short a time – for a moth with but one generation per year – the mutant must have had a selective advantage of 50% over the native, typical form – a selection intensity that was several orders of magnitude higher than the values then considered characteristic of natural selection. At the time, in 1924, Haldane did not specify what gave the mutant that advantage[1272].

In his book "The Causes of Evolution"[1273], Haldane gave an example of changes in frequencies of moths in a wood in Yorkshire. A storm in 1885 had felled the pine trees in half of the wood and they were replaced by birch. In 1907 a difference was observed in the color frequencies of moths in the two halves of the wood. In the replanted area, 15% of the moths (*Oporabia automnata*) were dark-colored and 85% light-colored. In the nearby original, pine-dominated, half of the wood, 95% of the moths were dark and only 4-5% light-colored. Haldane suggested that the difference was caused by selective predation of the moths by birds and bats: dark moths were protected by the dark color of the bark of pine trees, but exposed to predators on the light-colored birch bark. This mechanism was confirmed by the finding of moth wing remains on the forest floor[1274].

In another important paper Haldane discussed the stabilizing effects of selection in natural populations. He demonstrated that when selection intensities are low – as Darwin and his followers assumed – it may not be possible to detect the effects of selection on allele frequencies, because

---

[1270] Chapter 27. See Li, 1966, for more details.

[1271] Haldane 1924

[1272] When the phenomenon called "industrial melanism" was analyzed in detail, predation by birds was shown to be effective and an accepted explanation for the spread of the melanic moths, (Chapter 30. Ford 1965, Kettlewell 1958,1961).

[1273] Haldane 1932.

[1274] Ibid.

estimates can vary widely and the time required for a noticeable change is long. Therefore, from the point of view of a human observer, species in nature can be considered stable and unchanging:

> We may rarely hope to notice evolutionary change within a human lifetime. From the standpoint of a human observer, species may be regarded as almost at equilibrium. It may be possible to observe evolution by natural selection in a species which is adapting itself to a new environment[1275].

Non-Random Mating: Sib-Mating and Inbreeding

Haldane was interested in uncovering the effects of deviations from equilibrium due to non-random mating. He worked out mathematical expressions for predicting the allele frequencies for cases that were attracting attention at the time, like the Rh alleles in Man – a case of selection against heterozygous offspring born to homozygous recessive mothers. Later he worked out an elaborate system of recurrent equations for predicting allele frequencies in cases of sib-mating – a mating system frequently employed in the breeding of animals; and showed that under non-random mating systems, recessive alleles may spread very fast. His conclusions have since been verified, and form today a part of the population genetics theory in all textbooks on evolution.

To predict the frequencies of alleles and genotypes, the actual system of mating that exists in the population must be specified. The range of possible mating systems between random mating and selfing, even in the simple one gene-two allele model, is infinite, and each system has different consequences. Haldane's solution was to take selfing as an extreme example. Once this system is understood, the less extreme systems could be expected to be qualitatively similar but milder.

Haldane noted that selfing may be advantageous for the species: if an advantageous recessive mutation occurs, it will spread more quickly in a selfing than in a random-mating population. This may be one reason why inbreeding is widespread in nature, despite its known deleterious effects[1276].

The theoretical analysis of selfing reveals that the result is an increase in the proportions of homozygous individuals, and a corresponding decrease in heterozygous individuals, which are halved with each successive generation[1277].

$$H_n = 1/2\, H_{n-1} = (1/2)^n H_0$$

This expected result is qualitatively valid for all systems of inbreeding. Haldane also worked out mathematically the expectations for a less extreme mating system – brother-sister (sib)-mating – which is commonly practiced in animal breeding. Inbreeding can be expected to increase the frequency of homozygous carriers of deleterious, recessive mutations, with harmful effects[1278]. Darwin was aware of this undesirable effect.

[1275] Haldane 1937, p. 337.

[1276] Asexual reproduction and selfing may be ecologically advantageous when new populations are formed after colonization of a new habitat. In contrast to sexually-reproducing organisms, every single colonizing individual may reproduce and establish a population without the need to find a mate. A selfing population may grow at double the rate of a sexually-reproducing one in which only the females bear young.

[1277] See textbooks e.g. Wool 2006.

[1278] Darwin worried about the possible deleterious effects of cousin-mating since he had married his cousin, Emma Wedgewood. Darwin's son George, a professor of astronomy and statistics at Cambridge University, conducted a survey of inmates in mental institutions in England, and a sensational mistaken report of his results claimed that about half of the mentally-disturbed and stupid inmates were the offspring of such marriages. George Darwin quickly corrected the report in a letter to "Nature" (in fact only 3.5% of the inmates were the offspring of related parents).

Evolution

In 1929, Haldane, contemplating about the origin of life, suggested that aggregates had been repeatedly formed and dissolved in the "primeval soup" until a combination of molecules with a capacity for self-replication appeared, which gave them an advantage.

> The world is full of mysteries. Life is one. It is not the concern of a Theory of Evolution to explain these mysteries. Such a theory tries to explain events that happened in the past, in terms of laws and rules which are known to be true in the present time, assuming that the past was no more, but no less, mysterious than the present.

Haldane presented his ideas on evolution in his book "The Causes of Evolution"[1279]. The book is written in a simple style, intended for the general public. Like Haeckel in 1876, Haldane too considered evolution as manifesting two theories: The Theory of Descent and the Theory of Natural Selection.

Haldane deals with variation in natural populations, using examples from research on plants, in particular plant cytology. In his discussion of species and speciation, he uses examples from crossing experiments in plants resulting in polyploidy, with consequent reproductive barriers with the parental species. He suggests that this process, which was known to be effective in artificial plant propagation, may have been a mechanism of speciation in natural plant populations.

In his discussion of natural selection (Chapter IV) and of fitness (Chapter V), Haldane avoided introducing mathematical formulations – which would not have been comprehended by the general public. However, he considered mathematics as a central method for the study of evolution. He added a supplement to the book, presenting his mathematical formulations and those of Fisher and Sewall Wright,[1280] and concluded with the following prediction:

> The absorption of mathematics in biology is just beginning, but will continue (unless the history of science shown a false trend) and the research summarised here will become the beginning of a new branch of applied mathematics[1281].

Trivia

Haldane was very famous for his barbed sentences, like the following:

> If one could conclude as to the nature of the Creator from a study of his creations, it would appear that God had a special fondness for stars and beetles.

He objected to the distinction between "pure" (basic) and "applied" science – so often emphasized today:

> What is commonly called "pure science" is really long-range science – that is to say, science which will not find a practical application for some years to come[1282].

The Darwinians measure fitness in terms of the number of offspring. This makes the poor in human societies – the winners in the struggle for life:

[1279] Haldane 1993 (1932).

[1280] Ibid. see Chapter 30. Leigh1993.

[1281] Haldane's foresight was amply proven right. In the years that followed, the mathematical theory of population genetics and evolution advanced much faster than the biological evidence, leaving many biologists behind. As examples, see Crow and Kimura (1970) and Leigh (1993).

[1282] Haldane 1940, p. 270.

> From a Darwinian point of view, the poor are fitter than the rich. The capitalist may win the struggle for cash, but the poor are winning the struggle for life[1283].

On altruism:

> Would I lay down my life to save my brother? No, but I would to save two brothers or eight cousins. [The genetic relatedness with a brother is 1/2, with a cousin 1/8.]

[1283] Ibid. p. 115.

30

Sewall Wright: Chance and Evolution in Small Populations

Sewall Wright (1889-1988) received his doctorate in zoology from Harvard University (1915) and was employed for ten years by the U.S. Department of Agriculture. He then joined the faculty of the University of Chicago and worked there for about 30 years, before moving to Madison, Wisconsin, where he continued to work and publish scientific papers until well after his retirement.

Wright's early publications dealt with the inheritance of coat-color patterns in *Cavia* ("guinea pigs").

> Sewall Wright's (1925) is… a very thorough study of albino series of multiple allelomorphs in the guinea pig. Using five allelomorphs of this series, Wright bred the five homozygous and the ten heterozygous types… in sufficient numbers to study both the average depth of pigmentation of the red and black parts of the animal and its variability between different individuals of the same genotype[1284].

Although his early work concerned animal breeding – his contributions to this field were mathematical. A series of Wright's theoretical papers on the inheritance of characters in non-random mating systems became the foundation of animal breeding theory[1285]. The fifth paper in the series contains a summary of the four preceding papers.

Wright entered the field of evolution later than Fisher and Haldane, but he outlived them and continued his scientific work during most of the 20th century. His most important contributions to the Theory of Evolution concerned the expected genetic differentiation and speciation in small populations. His papers in this field were collected into a volume entitled "Evolution in Mendelian Populations"[1286], followed by four volumes of a book of the same name.

Wright suggested that evolution should proceed faster in small, peripheral and isolated populations than in large, variable ones. His theory of the role of chance events in evolution (genetic drift and random fixation) was contrary to the main stream of thought since Darwin. Wright's theory was first referred to as "the Sewall Wright effect" and was considered of marginal importance at the time. Later in the 20th century, Wright's theory became firmly established as a central process in evolution, and with the advent of electrophoresis and molecular techniques of measuring variation, this theory dominated in evolutionary thinking.

Systems of Mating

Wright noted that when mating is not random, three factors affect the genotype frequencies in the population: the level of relatedness between the parents (inbreeding), unequal mating frequencies of parents (assortative mating), and selection. He studied the theoretical consequences of the three factors in detail, and showed that inbreeding and assortative mating may alter the *genotype* frequencies in the population, without having the slightest effect on the *allele* frequencies. Consequently, the

[1284] Fisher 1931, The Evolution of Dominance, p. 349.
[1285] Wright, S. "Systems of Mating", 1921 ff.
[1286] Wright 1931; 1984.

original genotype frequencies may be restored by a single generation of random mating. This is not the case if selection is operating: selection may alter allele frequencies irreversibly.

> With random mating, inbreeding, or assortative mating, the relative frequencies of the different genetic factors in the original population remain constant. Resumption of random mating leads to the restoration of the composition of the population[1287]. Whether due to differences of death rate, mating rate, or fecundity, [selection] modifies the relative frequencies of the factors [=alleles] and alone so effects a permanent change in the average composition of the population.

For example, in selfing – the most extreme inbreeding situation – the proportion of heterozygotes is halved every generation, and the proportion of homozygotes correspondingly increases. However, the allele frequencies do not change[1288].

The Inbreeding Coefficient and Inbreeding Depression

Wright suggested that the level of inbreeding in a given population with different levels of kinship can be quantitatively estimated. Suppose we could assign to a gamete carrying alleles A, the value 1, and a gamete carrying a, the value 0. Sampling a large number of pairs of gametes before fertilization, we could then calculate the *correlation coefficient* **r** between pairs of gametes forming the zygotes. If mating was random, the correlation between gametes should be close to zero; but if the parents were related to one another, the probability of gametes carrying identical alleles (A uniting with A and a with a), should be greater than zero and the correlation should have a positive value. Wright termed the correlation between gametes the *inbreeding coefficient*, and used the symbol F for it. F increases with closer relationships between the parents, and also depends on population size[1289].

Wright showed that F could be incorporated into the calculation of expected equilibrium frequencies, and that the Hardy-Weinberg law is a special case of a more general equilibrium formulation that includes non-random mating: in the generalized form the equilibrium proportions of $AA = p^2 + pqF$, proportion of $Aa = 2pq - 2pqF$, and proportion of $aa = q^2 + pqF$. In the special case that mating is random, $F = 0$ and the formulas are reduced to the familiar HW frequencies[1290].

Wright showed that it is possible to estimate the inbreeding coefficient from the genotype frequencies sampled from a population at any generation. At equilibrium with random mating, the expected frequency of heterozygotes is $H = 2pq$, and therefore $H^2 = 4p^2q^2$ [or 4DR]. When the population is inbred, the expected H should be lower than 2pq, and the homozygous genotype frequencies $D \neq p^2$ and $R \neq q^2$. F can be estimated from the expression

$$F = (4DR - H^2)/(2D + H)(2R + H)$$

Or

$$F = (4DR - H^2)/(4DR - H^2 + 2H)$$

This formulation was very useful for animal breeders since the relatedness of inbred parents is often accompanied by undesirable effects on fertility and viability, and an increase in the frequency of deleterious phenotypes (a phenomenon referred to as *inbreeding depression*). Darwin and his contemporaries were well aware of this phenomenon. Selection for any desirable trait inevitably involves some degree of inbreeding, because only a sample of individuals from the available animals are selected as parents for the next generation (unless special care is taken and the population is intermittently deliberately outbred). Using Wright's formulas, the breeder could monitor the level of inbreeding in his stock and take the necessary steps to avoid inbreeding depression.

[1287] Wright 1921-Systems of mating, p. 167.

[1288] e.g. Wool 2006.

[1289] See Li,1966 for details.

[1290] See Chapter 27.

The damage from inbreeding results, in the short term, from the accumulation of deleterious recessive homozygotes. Darwin was worried about these consequences, because consanguineous marriages were common in British society and he himself had married his cousin. On an evolutionary time scale, the damage from inbreeding is mainly a reduction in both the gene pool and the heterozygosity levels, which limits variation and therefore reduces the ability of the population to respond to natural selection when conditions change.

Path Coefficients

Wright realized that not only the genetics of the animals, but also the environmental conditions affect the characters which the breeder wishes to improve – and that these factors often work not independently but in combination.

> In the biological sciences, especially, one often has to deal with a group of characteristics or conditions which are correlated, because of a complex of interacting, uncontrollable, and often obscure causes[1291].

In order to estimate the effects of individual interacting causes, Wright developed the method of Path Coefficients. His purpose in developing this method was to enable the breeder to direct the characters of his stock to his desired goal:

> One of the foremost considerations in the mind of the breeder is to obtain such control over the heredity of his stock, that the characteristics of the progeny can be predicted from those of the parents[1292].

The method enables estimation of the relative importance of each of the factors affecting the target characteristic, when other factors are kept constant. The method is based on the calculation of correlation and regression statistics[1293]. The correlations among all known variables are calculated from empirical data, and the effect of each factor is estimated as the square of the regression coefficient b^2 – if the factors are not independent, the square of the correlation coefficient r^2 is used[1294]. The expected sum of the squared factors is 1.0, and if the observed total sum is smaller than 1.0, the difference is an estimate of the effect of factors which are not included in the model. Wright used this method in his early study of the inheritance of litter size and coat-color pigmentation patterns in guinea-pigs (*Cavia*)[1295].

Wright noted that breeders strive to obtain homozygous stock for a desirable trait, but this often has disadvantageous consequences in terms of fitness:

> Unfortunately, a high percentage of homozygosis is also apt to go with a reduction in fecundity, growth and vigor. These effects appear to be due to the greater frequency with which mutations are recessive than dominant, and to the greater likelihood that any mutation will likely be harmful rather than advantageous[1296].

Wright recommended various mating schemes that might achieve the desired outcome while minimizing the risk of negative effects on fitness – for example, mating of the same male with several females that are half-sisters; combining selection with a hierarchy of mating; and the

[1291] Wright 1921, p. 557.
[1292] Wright 1921, p. 169.
[1293] e.g. Sokal & Rohlf 1995.
[1294] See Li, 1966 for a detailed description.
[1295] Wright 1921, pp. 560-569.
[1296] Ibid. p. 171.

crossing of parallel lines from the same stock. He examined the theoretical consequences of each suggested mating system from which the breeder might choose.

Evolution in Small Populations

Wright's greatest contribution to the Theory of Evolution was his suggestion, in 1931, that evolution should take place faster in small, isolated, or marginal populations than in large and variable populations, as Darwin and most of the evolutionists after him had believed.

Wright suggested that most natural populations are in fact small and more or less isolated from each other. Isolation of a small population, which by chance happens to contain a unique gene pool, may enhance the differentiation of new species by limiting gene flow, without the need for a selective advantage to its component genotypes and without resort to natural selection as a driving force for speciation.

Wright suggested that it is unlikely that evolution should proceed by the fixation of favorable mutations, as the classical Darwinian model maintained:

> The great majority of mutations are either definitely injurious to the organism or produce such small effects as to be seemingly negligible.
>
> Since the time of Lamarck, a school of biologists have held that the primary changes in hereditary constitution must be adaptive in direction in order to account for the evolutionary advance. Unfortunately the results of experimental study have given no support for this view[1297].

Rather, Wright suggested that *chance* plays a major role in evolution. His model involves two processes, which are termed *genetic drift* and the consequent *random fixation*. Both processes have stronger effects the smaller the population. This has been amply described in textbooks[1298].

Genetic Drift and Random Fixation

Imagine a population of only ten individuals of three genotypes: five AA, four Aa, and one aa. Suppose the only aa individual is accidentally killed by a falling rock. This chance event changes the allele frequency of a in the population from 0.3 to 0.22, and the genotype frequencies from 0.49; 0.42; 0.09 to 0.61; 0.34; 0.05. This change will affect the offspring and is irreversible.

Genetic drift can be illustrated by a simple simulation experiment. A marble is rolling down a slanted board, across which are arranged rows of equally-spaced nails. The width of the board represent the allele frequency of a, from 0 to 1. The position of the marble relative to the width of the board represents the allele frequencies at any given point in time. In the margins of the board – at frequencies 0.0 and 1.0 – there are deep grooves which direct a marble reaching them to roll down out of the board.

The marble is placed initially at the top of the board at frequency 0.5, and rolls down hitting the nails. Its speed and direction at any point in time are determined by many factors, simulating random movement. The temporal change of position of the marble is an illustration of *genetic drift*. Sooner or later, the marble must hit, by chance, one of the two marginal grooves. This simulates *random fixation,* since the population reaching either allele frequency 0.0 or 1.0 has lost the other allele and will remain monomorphic (pending a rare, new mutation)[1299].

[1297] Wright 1921, pp. 141-142.

[1298] A class experiment for illustrating genetic drift and random fixation was suggested by Wallace, 1969.

[1299] See, for example, Wool 2006 and Wool & Mendlinger 1950.

The Genetic Fate of Small Populations

Wright showed mathematically that if small populations are followed through time, all of them must eventually reach fixation. The marble simulation described above can be repeated with many populations, all starting at allele frequency 0.5 and each monitored until fixation. If time is arranged in generations, in the first generation the distribution of allele frequencies of the population will be normal, with a peak at the mean of 0.5. But as time proceeds, more and more populations will become fixed at either 0.0 or 1.0 and the distribution will become U-shaped. This process has been simulated experimentally with *Drosophila* and with flour beetles[1300]. In the absence of selective advantage to any allele, the distribution will be symmetrical, with an equal number of populations fixed at either end.

Wright showed further that the *rate* of fixation depends on *population size*. In populations of size N, 1/2N of them will become fixed at each generation. When N is large, the process will be slow and may not be detectable. Infinitely large populations will not change at all by chance events[1301].

Shifting Balance Theory: According to Fisher's "Fundamental Theorem", natural selection invariably increases fitness – and will bring the population to a genetic composition equilibrium at the nearest fitness peak achievable at the present point in time – but not necessarily to the highest possible peak. As an illustration, if genotypes at two loci are plotted – such that the fitness is a third variable, vertically from the bi-plot surface – a sort of topographical map[1302] is obtained with peaks and valleys. From a starting genetic composition, selection will move the population towards the nearer peak, but once there, will not be able to move it to the next, higher peak, since it cannot cross a "saddle" with lower fitness. A random event – such as a population bottleneck – resulting in a small population with a different allele frequency, or an environmental change, which alters ("shifts") the selective values of present genetic combinations (turning "peaks" into "valleys") – may be necessary to achieve the next move towards higher fitness.

Isolation by Distance

The dispersal of organisms in search of mates depends on the mobility of the organisms and on population density: even the dispersal of pollen in wind-pollinated plants is not unlimited[1303]. The probability of mating of two individuals decreases with the distance between them. Wright termed this phenomenon *isolation by distance*[1304].

He defined the distance in space in which an individual wanders randomly to find a mate, as the *neighborhood*. He showed that when individuals in the search for mates are limited to a neighborhood, an increase in homozygosity and decrease in heterozygosity in the population is expected – just as in the case in an inbreeding population – although mating within the neighborhood is random.

Neighborhood size strongly depends on the mobility of the organism in question, and low mobility can lead to genetic differentiation over small distances. Selander and Kaufman[1305] found great differences in allele frequencies – detected elecrophoretically – in snail populations taken from adjacent city blocks.

[1300] Kerr &Wright 1954; Wool 1987.

[1301] "That non-adaptive differentiation will occur in small populations owing to "drift" ("Sewall Wright' effect"), or chance fixation of some new mutation… is one of the most important results of mathematical analysis applied to the facts of neo-Mendelism" (Huxley, J., The Modern Synthesis, 1942, p. 200).

[1302] "Adaptive landscape". Wallace 1969.

[1303] Recent examples: Dow and Ashley 1998; Imbert and LeFebre 2003.

[1304] Cavalli-Sforza (cited in Wallace 1969) reported that the probability of marriage between the inhabitants of villages in northern Italy decreased linearly with the square-root of the distance between their natal villages. The situation is perhaps not the same today due to the increased mobility of people, but may describe the Italian rural population in the 19th century. Wallace (1969) suggested that most natural populations are small, and can be described genetically as a fakir's bed of nails (Wallace 1969).

[1305] Selander & Kaufman 1975.

Support for Wright's isolation by distance model was later obtained in a computer simulation study[1306].

Gene Flow Among Populations: The Use of Wright's F Statistics

To measure the rate of migration among populations, we need to know not only the number of individuals exchanged between any two populations, but also whether the immigrants have contributed genetically to the recipient population. The transmission of alleles by the immigrants to the recipient population is referred to as *gene flow*. The effects of migration may be continuous and extended over many generations. Consequently, ecological measurements using release-and-recapture techniques are not sufficient.

A population may become further subdivided into smaller subunits for a variety of reasons, such as mobility and inter-individual distance. Wahlund[1307] demonstrated theoretically that if allele frequencies p_i, q_i were different among i subunits, in a population with mean *genotype* frequencies D, H, R, the equilibrium frequencies will deviate from the Hardy-Weinberg expectation by an excess of homozygotes and deficiency of heterozygotes as in an inbred population,

$$D = p^2 + \sigma^2$$
$$H = 2pq - 2\sigma$$
$$R = q^2 + \sigma^2$$

where σ^2 is the variance of p among the subunits[1308]. Wright developed a method for measuring the rate of migration among sub-populations. The method remained a theoretical exercise for decades, but found extensive use in the last quarter of the 20th century, when a large number of genetic markers became available for many populations of hundreds of organisms – first from electophoretic studies and later from the use of molecular DNA analysis.

To estimate gene flow, one needs to know the population size N of each subunit, and the migration rate **m** (the proportion of migrants from population B in population A). Such data are not available for any population in nature. Wright's method circumvents these difficulties by estimating the product, **Nm**, independently of the ecological variables. **Nm** is an estimate of the average number of migrants (that contributed to the gene pool) exchanged between the pair of populations per generation.

If the subunits have been isolated, they are expected to be inbred, and the inbreeding coefficient F – measured within each subunit – is expected to increase with time. With gene flow, F *within* subunits should be reduced.

Wright showed that the total F for a group of subunits (F_{it}) can be partitioned into two components: within subunits (F_{is}) and among subunits (F_{st}), such that $F_{it} = F_{is} + F_{st}$.

F_{st} is a measure of the variation among subunits: when gene flow is more extensive, this component becomes larger. This component can be estimated from data on heterozygosity in the population – obtained from studies of genetic markers as follows:

The expected heterozygosity [$H = 1 - \Sigma p^2$] can be estimated from allele frequency data in the individual subunits. The average of these estimates is labeled H_s. Heterozygosity in the total population (H_t) is estimated from the pooled allele frequencies. The *difference* between the two estimates reflects the genetic differentiation among subunits – and is measured as $F_{st} = (H_s - H_t)/H_t$

Wright showed that F_{st} is related to **Nm**:

$$F_{st} = 1/(4\,\mathbf{Nm} + 1), \quad \text{or} \quad \mathbf{Nm} = (1 - F_{st})/4F_{st}$$

[1306] Rohlf & Schnell 1971; see Wool 2006.

[1307] Wahlund 1928.

[1308] See Li 1966.

He further showed that if **Nm** is smaller than 1 (less than one individual exchanged per generation between two populations), they are effectively isolated and the differences between them could be the result of genetic drift and /or selection. But if **Nm** is greater than 1, gene flow may cancel this differentiation and homogenize the subunits into a single population.

Wright's method of F statistics was extensively used in the 20th century[1309].

[1309] e.g. Costa & Ross 1994; Schitzhuisen & Lombaert 1994; Martinez et al. 2005.

31

Theodosius Dobzhansky

"Nothing in biology makes sense except in the light of evolution"[1310]

Dobzhansky (1900-1975) was one of the most outstanding and productive scientists in population genetics and evolution in the 20th century. Born in the Ukraine, he received his first academic degree at the University of Kiev, studying the anatomy of ladybirds (Coccinellidae). He then became interested in *Drosophila* strains, which were being sent to Moscow from Morgan's laboratory in the USA. Dobzhansky received an international scholarship for a period of training in Morgan's laboratory at Columbia University, traveled to New York, and remained in the USA for the rest of his life.

Dobzhansky engaged in cytological investigations of *Drosophila* chromosomes at CalTech, including the induction of mutations by x-ray, and was appointed a professor there in 1936. In 1940 he moved back to New York and worked at Columbia University until 1962, and then at Rockefeller University until he retired.

Dobzhansky was the first to define the concept of species genetically[1311]. He published textbooks on genetics and collaborated with E. Mayr, G.G. Simpson, S. Wright, and other evolutionists[1312]. He published hundreds of research articles and edited the journal series "Evolutionary Biology" for several years. Among his more important books are "Genetics of the Evolutionary Process" (1970), "Evolution, Genetics, and Man" (1955), and "Mankind Evolving" (1962).

Dobzhansky was not a theoretician in the mathematical sense, but a broad-based biologist with an evolutionary perspective. His popular writing style exposed the general public to scientific research on evolutionary problems. In his books he facilitated the merging of ideas on natural selection with the Mendelian theory of genetics, resulting in a comprehensive Theory of Evolution, from primitive origins to the evolution of Man:

> Nobody has seen that the earth is a sphere[1313] or that it revolves around the sun, rather than vice versa. Nobody has caught a glimpse of atoms or of things within atoms... The least that can be said is that in our activities we take the earth to be a sphere and treat atoms as though they were facts. For similar reasons, it is not a matter of personal taste whether or not we "believe" in evolution. The evidence for evolution is compelling[1314].

Natural Populations of *Drosophila*

Dobzhansky recognized the advantages of using *Drosophila* as a research organism. Several species of *Drosophila* occur near human habitations and are attracted to bait. This facilitated the study of natural populations – applying the methodology developed in Morgan's lab to field-collected flies. Natural populations of *Drosophila* were shown to comprise a mixture of different species

[1310] Th. Dobzhansky, cited in Ayala & Kiger, 1980, p. 1.
[1311] Dobzhansky 1937, "Genetics and the Origin of Species".
[1312] Chapters 32, 29, respectively.
[1313] Nobody except a visionary like Jules Verne could have foreseen that 20 years after these words were written, the space missions would enable Man to actually see that the earth is indeed spherical.
[1314] Dobzhansky 1955 (1967)p. 319.

and strains. Hundreds of thousands of flies were sampled in natural populations and screened to measure the frequency of mutations in nature, and a mass of data was accumulated for the first time on the genetic structure of natural populations of these flies.

In the early studies, morphological mutations were studied based on techniques developed in Morgan's lab. Female flies captured in nature were isolated and their progeny were studied. Captured males were mated in the laboratory to females from a synthetic laboratory strain that carried conspicuous markers on each of the chromosomes. In this way new morphological mutants could not only be detected, but also allocated to the appropriate linkage group and mapped according to the number of recombinants (using any two markers on a chromosome, 1 Morgan = 1% recombinants).

Later, a new kind of genetic marker became available: chromosome aberrations (inversions). *Drosophila* larvae have giant chromosomes in their salivary glands. Different inversions in these chromosomes cause characteristic loops, which can be identified in microscope preparations of the larvae. Moreover, no crossing-over was observed between markers included in an inversion loop: all the genes within the loop are transmitted as a unit, making the recognizable loop itself a reliable genetic marker for a distinct – if unknown – group of genes. Females of each species were individually transferred to vials with food for oviposition, and the larvae were screened cytologically for inversions of the giant chromosomes. Pure-bred lines carrying individual inversions were selected in *D. paulistorum* and *D. pseudoobscura* for further studies.

Further research by Dobzhansky and his students on the frequencies of different inversions in natural populations revealed their possible relationships with environmental factors such as temperature and elevation. Some inversions showed seasonal fluctuations in abundance, supporting the suggestion that the genes included in the inversions may be affected by natural selection at different times of the year.

The viability and reproductive rate of strains homozygous for different inversions were tested in the laboratory. Pure strains could be maintained indefinitely in the laboratory, but when carriers of two inversions were placed together, initially in equal numbers, differences in survival (fitness) rates were discovered between them.

One of the experiments carried out by Dobzhansky's team demonstrated the effects of genetic drift. Two homozygous strains carrying chromosomal inversions, PP and AR – both collected in natural populations – were used to start ten "small" populations – each started with 20 flies – and ten "large" populations, each started with 4000 flies. Initially the flies were introduced in equal numbers of PP and AR. The cultures were maintained in the same environment and the frequencies of the inversions were monitored for 18 months, with the food replenished regularly. The frequency of carriers of PP declined in all populations – showing some selective disadvantage to the carriers of this inversion. However, the *variance between the lines* was considerably larger in the small than in the large populations: the frequencies of PP in small populations varied much more than in the large ones. This supported the claim that "the outcome of selection in the experimental populations is conditioned by random genetic drift"[1315].

Non-Random Mating in Natural Populations

Many studies by Dobzhansky and his students concerned the question of whether or not mating in natural populations was random (the assumption of random mating is an essential component of the Hardy-Weinberg equilibrium. Chapter 27). Are natural populations at equilibrium? If mating was random, gene flow should homogenize the populations. How do populations, "races", or "incipient species" co-exist with different frequencies of genetic (chromosomal) markers?

[1315] Dobzhansky 1957.

Drosophila paulistorum was considered a single species throughout its geographical range, but was divided into six "geographical races"[1316]. Mating frequencies were tested within and between replicate populations of the six races. The results showed that individuals, when given a choice, tended to mate preferentially with mates from the same race. There was only a weak correlation between the mating preference and the fecundity or the rate of offspring sterility in crosses between races. These experiments suggested that mating behavior could be a factor affecting possible speciation in *D. paulistorum*.

Experiments were also conducted with another species, *D. pseudoobscura*, using populations that were either monomorphic for one of two inversions, AR or CH, or polymorphic with both inversions present. Polymorphic populations tended to increase in size faster than monomorphic ones (i.e., produce more offspring). Observations on mating behavior in laboratory cages showed, surprisingly, that mate choice depended on genotype frequencies in the mating cage. When the adult carriers of CH and AR were introduced in equal numbers, mating was not different from random; but when male numbers were unequal, the males of the *minority* strain tended to mate more frequently than expected from their proportion in the mating chamber. This phenomenon attracted great attention and interest at the time, because it could in theory lead to a stable genetic equilibrium in a population: If one genotype becomes rare for any reason (selection disadvantage or chance) the advantage it gains from mating may restore the balance. The *rare-male advantage* phenomenon was reported in several species of Drosophila and also in several other insects[1317].

How did the flies of any genotype in the mating chamber perceive that they were rare? Twenty or 25 pairs of flies were introduced into the mating chamber for each trial, in some replicates with equal proportions of the inversion carriers, and in others with one of the inversions (in turn) being rare (5 versus 20 pairs). The identity of the strains of each mating pair was recorded. [The tested strains were marked experimentally by paint or wing notches for identification by the observers. Special care had to be taken, to prevent the marking from affecting mate selection]. In some tests, the mating chamber was placed on top of a similar cage containing many flies of the "rare" type (source). When the two cages were separated by a screen, the rare males had no advantage: they could sense the presence of their own type in the other cage. Finally, the two cages were placed widely apart but connected by a long tube. A gentle air current was blown through a directional pump. When the air was blown from the source *into* the mating cage, the rare males' advantage disappeared. When the air direction was reversed, however, the rare males had a mating advantage. The mating flies acquired the information by olfactory means – the intensity of the odor.

Genetic Polymorphism in Natural Populations of Drosophila

In the 1960s, electrophoretic methods were introduced for the study of genetic variation in natural populations[1318] and quickly replaced the chromosomal techniques. *Drosophila* was naturally one of the first organisms to which the new technique was applied.

One of the early uses of the new technique was a survey of electrophoretic and inversion variation intended to compare the genetic polymorphism in island populations of *Drosophila* with the South-American mainland populations[1319]. Inversion polymorphism was much lower in the island populations and different islands harbored different inversions. This was consistent with the possibility that the population on each island had been founded by a few migrants (founder effects and genetic drift). By contrast, electrophoretic patterns were the same in both mainland and island populations. This was interpreted to mean that natural selection affected the electrophoretic pattern

[1316] Dobzhansky et al. 1962, 1972, 1976.
[1317] Spiess 1977.
[1318] Lewontin & Hubby 1966, Hubby & Lewontin 1966.
[1319] Ayala et al. 1971.

and the same isozymes were selected in all the sampled locations – providing support for the claim that electophoretic variation is not "neutral".

By sampling the same populations at fixed intervals, it was possible to track seasonal changes in electrophoretic marker frequencies at several locations over a number of years. The investigators interpreted the observed increases and decreases of marker frequencies as results of seasonal changes in the direction and intensity of natural selection.

Species and Speciation: The studies of natural *Drosophila* populations by Dobzhansky and his associates led to general conclusions regarding speciation. Convinced that speciation in evolution must be a genetic process, Dobzhansky sought to define the species in genetic terms. Like Ernst Mayr (see next chapter), he too thought that reproductive isolation is essential for the formation of a new species, and therefore that this should be the criterion for recognition of species status. Dobzhansky seems to have been the first to offer genetic definitions of the term "species":

> Species are groups of populations, the gene exchange between which is limited or prevented by one, or a combination of several, reproductive isolation mechanisms[1320].

> Species ...are genetically closed systems, since they exchange genes rarely or not at all[1321]. A biological species is an inclusive Mendelian population. It is integrated by bonds of sexual reproduction and parentage[1322].

If a species is a coherent genetic system, then in order to speciate, a reproductive barrier must develop in the distribution area of the parent species and separate a part of it from the rest. Reproductive barriers may be geographic, ecological, or genetic, like hybrid sterility[1323].

> Considered physiologically, different isolating mechanisms have scarcely a common denominator. Yet genetically, their effect is the same: that is, limitation or prevention of gene exchange between populations of different species[1324].

Dobzhansky accepted the Darwinian gradualist mode of evolution and speciation by means of slow accumulation of small changes.

> Genetic differences between species are compounded gradually, of many genetic and chromosomal alterations, each change having arisen by mutation[1325]. Species evolve from races by the accumulation of genetic changes[1326]. [but] to state that races are incipient species, is not tantamount to saying that every race is a future species. To become species, races must evolve reproductive isolation[1327].

> When races become sympatric they tend to merge ...into a single, variable population. In contrast to races, species are able to maintain their genetic integrity despite sympatric coexistence[1328] (ibid. p. 356).

[1320] Dobzhansky 1937 (1951) p. 262.
[1321] Dobzhansky 1955, p. 165.
[1322] Dobzhansky 1970, p. 354.
[1323] Ibid. p. 314.
[1324] Ibid. p. 313.
[1325] Ibid. p. 361.
[1326] Ibid. p. 367.
[1327] Ibid. p. 376.
[1328] Ibid. p. 356.

Human Evolution

Dobzhansky wrote at length about human evolution and current processes in human populations. Two of his books address these issues specifically: "Evolution, Genetics, and Man" (1955) and "Mankind Evolving" (1962).

> Biologists have been so pre-occupied with proving that Man is a product of organic evolution, that they have scarcely noticed that man is an extraordinary and unique product of this evolution[1329].

> The leading forces of human evolution are intelligence, ability to use linguistic symbols, and culture which man has developed[1330].

> The question may again be raised as to whether upright stance, tools, constant sexual receptivity of females, symbolic languages, monogamous family, changes in food habits, or relaxation of male aggressiveness came first... What we are dealing with is a whole new evolutionary pattern, a transition to a novel way of life, which is human rather than animal[1331].

Dobzhansky describes the scientific data and opinions from a genetic point of view, but his books are impregnated with the author's personal attitude – in particular as regards human societies. He reviews the information on Man's nearest phylogenetic relatives, the primates, and states, with detailed tables in support, that the anatomical differences between Man and the primates are quantitative, not qualitative[1332]. He describes the very few remains of pre-historic Man that were known at the time, emphasizing the disagreements among anthropologists over the interpretation of these finds and their phylogenetic relationships to modern Man. He describes, with illustrations, the possible reconstruction of *Picanthropus*, *Australopithecus*, and the Neanderthal man, but is unable to decide which of them was an ancestor of *Homo sapiens*. Although different names had been given to different paleontological finds of human remains – such as Java man, Heidelberg man [which was later established as a fraud], and Peking man – Dobzhansky insists that there is no evidence that different species of man ever coexisted in the past:

> Human evolution never led to the differentiation of a single species into a group of derived species, some of which might have become lost and others survived. Mankind was and is a species which evolved as a body[1333].

One of the most important trends leading to human evolution must have been the increase in brain size in human ancestors, because "brain power confers enormous adaptive advantages to its possessors"[1334]. The development of the brain was of major importance in human evolution, because it enabled *cultural evolution* – the transmission of information by learning, not through sexual union: while genetic transmission is possible only from parents to offspring, cultural transmission has a much wider scope and may occur between unrelated individuals, between generations, between populations in different continents, and between nationalities. Therefore, cultural evolution is much more efficient and proceeds much more rapidly than biological evolution.

[1329] Dobzhansky 1955, p. 320.
[1330] Ibid.
[1331] Dobzhansky 1962, p. 199.
[1332] Ibid. pp. 164-178.
[1333] Dobzhansky 1955, p. 333.
[1334] Dobzhansky 1955, p. 334.

Dobzhansky on "Social Darwinism"

Dobzhansky objected strongly to the opinion of the "social Darwinists", which was popular in the USA in the first third of the 20th century. These Darwinists believed that the struggle for existence and competition activate human society and that these are characteristic not only of the interactions among individuals within societies, but also between classes, races, and nations[1335]. They saw the economic success of individuals as a demonstration of the principle of natural selection ("survival of the fittest") as advocated by Herbert Spencer. One supporter wrote:

> Millionaires are products of natural selection, acting on the whole body of men to pick up those who can meet the requirements of a certain work to be done[1336].

> When applied to the life of man in society, [they] suggested that nature will provide that the best competitors in a competitive situation would win, and that this process would lead to continuing improvement… It would be a big mistake to let our sentiments interfere with nature by helping the poor, the weak, and the generally unfit[1337].

Dobzhansky totally rejects these arguments. Biological evolution does not transmit *culture* between generations: it determines the response of the developing organisms to the environments in which their development takes place. This is as true for Man as for other organisms. Therefore, improvement of the environment in which humans live – through education and accumulated knowledge – is highly important in the cultural evolution of Man. Dobzhansky deals in detail with research on the relative importance of "nature and nurture" (the terms then in use), such as comparisons of "intelligence" (measured by IQ) of twins raised apart[1338]. He wrote:

> A person is what he is because of his nature and his nurture. His genes are his nature. His upbringing is his nurture. The same is true of mankind as a whole[1339].

On Human Races

The problems of racial differentiation occupied much space in Dobzhansky's books. He was accused of being a racist because he insisted that races are biological units, and do exist – in the rest of the biological world as well as in Man.

> Race differences are facts of nature, which can, given sufficient study, be ascertained objectively. Mendelian populations of any kind, from tribes to inhabitants of countries and continents, may differ in frequencies of some genetic variants, or they may not. If they so differ, they are racially distinct[1340].

Dobzhansky recognized races, but opposed the social connotations of different statuses or qualities of the races. He insisted, rather, on creating equal opportunities for all human variants. He reviewed the opinions of different scientists regarding the number of human races (between six and 34), depending on the characters used by the various authors to distinguish races[1341]. Dobzhansky concluded that the differentiation of human races, and their geographical distribution, was not due to adaptation to different climates [by natural selection]:

[1335] Dobzhansky 1962, p. 12. See also Karl Pearson, Chapter 22 above.
[1336] Ibid.
[1337] Ibid. p. 133.
[1338] Ibid. p. 83-85. An important source of data on the subject was faked! See Gould, 1981 (1951).
[1339] Ibid. p. 24.
[1340] Dobzhansky 1962, p. 267.
[1341] Ibid. pp. 254-263.

> Shocking though this may be, solid and conclusive evidence concerning the adaptive significance of racial traits [e.g., skin color or hair characters] in man is scant in the extreme, and the best that can be offered are plausible speculations and surmises[1342]!

The current distribution of human races could be a result of genetic drift. Early human societies were small family groups that roamed the world. An accidental settling of a family in a particular locality or habitat, if that family by chance carried some particular trait, could be the origin of the spread of the trait over a larger area, if such traits were not harmful for their carriers[1343].

On Attempts at Improvement of the Human Race: Dobzhansky strongly objected to the ideas of the Eugenicists (Chapters 21 and 22) for improving human societies by planned assortative mating. He wrote that assortative mating would not yield the anticipated improvement:

> We inherit genes, not genotypes, from our parents, and we transmit our genes, not our genotypes, to our children. A caste recruited from persons of high quality will contain some less-able individuals in the following generations[1344].

Assortative mating – regardless of the good intentions – would reduce variation in the human populations and create castes, according to parental choice (which may be affected by political or other motives). Is this a desirable result? Do we really want to live in a world with millions of Einsteins, Pasteurs, or Lenins?

Dobzhansky suggested that some desirable changes in society may occur if all individuals are given equal opportunities and freedom of choice regarding their careers, and young people are encouraged to select their profession according to their personal inclinations. Thus those interested in music will become musicians, and those tending to mathematics will become mathematicians, etc. The professional tendencies will bring together similarly-inclined individuals, and the mating of persons with similar tendencies may lead to intensification of some of these tendencies for the benefit of society.

[1342] Ibid. p. 272.

[1343] In 1812, a British physician, William Charles Wells, suggested the same scenario as an explanation of the distribution of black populations in Africa. See Eiseley 1961, pp. 119-125.

[1344] Dobzhansky 1962, p. 246.

32

Ernst Mayr: The Biological Species and Speciation

Having reached the rare age of 100 years, I find myself in a unique position: I am the last survivor of the golden age of the evolutionary synthesis[1345].

Most of the European and American evolutionists in the early 20th century believed in the gradual formation of new species by natural selection. The geneticists tended towards de Vries's mutation theory. The greatest achievement of the Theory of Evolution in the 20th century was the incorporation of the science of genetics into the Darwinian Theory of Evolution.

The greatest contributors to this synthesis were Theodosius Dobzhansky and Ernst Mayr. [According to Mayr, the man who consolidated the merging of the two theories was Dobzhansky]. Both scientists had been educated in Europe before migrating to the USA, and were able to discern the issues common to both genetics and evolutionary theory.

Ernst Mayr (1904-2005) was born and educated in Germany and received his PhD in Berlin. Being Jewish, he chose to migrate to the USA in 1932 following the emergence of National Socialism (Nazism) in Germany. He worked as a curator in the American Museum of Natural History in New York, and then, in 1953, was appointed Professor of Zoology and Head of the Museum of Comparative Zoology at Harvard. Mayr's main expertise lay in ornithology and the geographic distribution of bird species in the world. His special interest as an evolutionist lay in the process of speciation. Two of his most important books are "Systematics and the Origin of Species" (1942) and, in particular, "Animal Species and Evolution" (1963).

Mayr reviewed the history of development of species concepts in systematics[1346]. No common definition of species was then available, and each systematist defined "species" to fit his own needs and point of view.

> It may not be an exaggeration to say that there are as many species concepts as there are thinking systematists and students of speciation[1347].

Some 18th -and 19th-century philosophers attached different meanings to the term species[1348]. The main criterion for setting "species" apart from "variety" was the ability or inability to form fertile interspecies hybrids. If two forms did not mate (or their hybrids were sterile) then they belonged to different species – otherwise they were varieties of the same species.

Definitions of "Species"

Species definitions had been based on morphological criteria since Linnaeus's "Systema Naturae". Every newly-collected individual organism that differed from others in some characteristic was

[1345] Mayr 2004.
[1346] Mayr, 1958 Behavior and Systematics.
[1347] Mayr 1942 p. 114.
[1348] See a review in Grant 1995. Lyell 1853, p. 611.

marked as a "type" of a new species. The "typological" system still guides systematic practice in museums around the world to this day.

Mayr, on the other hand, did not think that the "type" necessarily shows the most common or most representative morphology of the species:

> The "Type" is typical of nothing. It is only an indication of which group of individuals must be associated with a particular name ...it is not necessarily the most typical specimen of the species (such probably does not exist)[1349].
>
> The picture of the species as presented by the taxonomist is, however not necessarily a very accurate rendering of the situation as it exists in nature. It is merely due to the need of the museum worker to identify every individual and to place it in a definite pigeonhole[1350].

Mayr (and Dobzhansky) argued that a species is a populational unit, composed of local populations of individuals (demes) which exchange genes freely among themselves, but not with populations belonging to other species: the formulation offered by Mayr is:

> [A species comprises] **a group of populations ...which are actually or potentially interbreeding, and which are reproductively isolated from other such groups[1351].**

This universally-accepted definition is not applicable to large groups of organisms that reproduce asexually – such as bacteria, many invertebrates, and many plants. It is also impractical to test for species identity according to the reproductive isolation of individuals, without relying both on morphological similarity between individuals and on common sense, because the required information on mating success of all individuals cannot be obtained in nature[1352]. Sokal and Sneath argued that the biological species definition is useless as a practical definition, and in practice morphological measurements with the use of high-speed computers could provide a better, statistical definition.

Although well aware of the difficulties inherent in application of the biological species definition, Mayr disagreed with these suggestions:

> Some authors have gone so far as to abandon the biological species concept altogether and return to the morphological species for sexual and asexual organisms. I see nothing that would recommend this solution... Most practicing taxonomists [are] frankly dualistic: they define the term species biologically in sexual organisms and morphologically in asexual ones[1353].

The Founder Principle

Imagine islands in the ocean, oases in the desert, or high mountain peaks. These are habitable places in the midst of a vast area that is not habitable by a particular species. The farther the islands

---

[1349] Mayr 1942, p. 15.

[1350] Ibid. p. 98.

[1351] Although Dobzhansky offered a definition of species from a genetic standpoint in his book in 1937, it was Mayr who coined the final formulation of a biological species, a central concept in biology and evolution. With this definition, it became possible to study the mechanisms of species formation (speciation). Dobzhansky did not have a Chapter on speciation in his book, but Mayr had written his "Systematics and the Origin of Species" at about the same time, and the two books complement one another.

[1352] Sokal & Crovello 1970; Sokal & Sneath 1973; Sneath & Sokal 1963.

[1353] Mayr 1963, p. 28. For over 20 years, Sokal and Sneath's Numerical Taxonomy excited many researchers and produced hundreds of research papers. However, It did not succeed in replacing the taxonomic practice.

are from the source population, the greater the probability that migrant individuals will be lost and the less likely that a small colonizing unit will succeed.

The population size of founding, colonizing populations can be expected to be small, perhaps as small as a single pair or a single fertilized female. Colonizing populations may be expected to vary randomly in composition, even if they come from the same source of a variable population. The fate of the expanding colonizing population – its chance of survival and its competitive ability with other species – may be determined by the genetic composition of the initial chance mixture of genotypes. Mayr coined the term *"the founder principle"* to describe the dependence of the future fate of an island population on the (random) genetic pool of the colonizing group.

The number of species on an island is a function of the balance between the numbers of immigrant species from the mainland, and the rate of extinction of species on the island – both of which depend on island size and the distance from the mainland[1354]. In his early work, Mayr considered island populations as evolutionary traps: the numbers of species on small islands were smaller than on larger islands.

> There is little doubt that, as postulated by Sewall Wright on theoretical grounds, well-isolated islands are evolutionary traps, which in time kill one species after another that settles on them[1355].

A conference in 1965 was dedicated to colonizing species and the properties required for successful colonization[1356]. Mayr contributed a paper on the properties of colonizing species of birds, based on his work on species of birds on distant archipelagos in Asia. His conclusions were that bird species that migrate in flocks tend to be better colonizers than species that migrate individually. Seed-feeders seem to be more successful than insect-feeders, and birds associated with freshwater habitats are good colonizers.

In nature, small population size implies a limited gene pool, but the two factors can be separated experimentally. An ingenious experiment with flour beetles (*Tribolium castaneum*) indicated that the competitive ability of founding populations depended on the size of their initial gene pool, and not on the initial numbers of founders[1357].

The size of the gene pool has become a major concern for the re-introduction of endangered species into natural areas. The sources of animals for such introductions are the small "breeding nuclei" held captive in zoos. A good example of this is the successful re-colonization of Espaniola Island in the Galapagos, by giant tortoises reared at the Darwin Research Station on Santa Cruz island. More than 1,200 individuals were released, and about 300 are known to have survived and grown at the release sites. All of these individuals, however, are the offspring of only twelve females and three males. Only time will tell how likely these tortoises are to establish a long-term population on the island[1358].

Habitat destruction and colonization are going on all the time in nature, and in particular through human activities. Differences among species in their colonizing ability must be important determinants of the structure of natural ecosystems following such destructions.

Speciation

Mayr reviews earlier ideas on speciation. For example, Charles Darwin considered that new species are likely to develop on islands – because isolation is favorable to the process:

1354 Macarthur & Wilson 1967.
1355 Mayr 1942, p. 225.
1356 Lewontin & Stebbins 1965.
1357 Dawson 1970.
1358 Milinkovitch 2004; Gibbs et al. 2008.

> The immigration of a few new forms, or even of a single one, may well cause an entire revolution in the relations of a multitude of the old occupants… I infer that isolation would be eminently favourable to the production, through natural selection, of new forms[1359].

R.A. Fisher[1360] first suggested a genetic model for speciation. He insisted that there must be geographical separation when a population splits into two parts:

> In many cases without doubt, the establishment of complete or almost complete geographic isolation has at once settled the line of fission: the two separate moieties thereafter evolving as separate species …until such time as the morphological differences between them entitle them to "species rank"[1361].

Fisher's model is referred to in the literature as *"allopatric speciation"*. The resulting populations may develop in different directions due to the accumulation of different mutations and the independent effects of selection.

Mayr considered that the issue of speciation deserves more attention, since there is no agreement among scientists about the genetic context of speciation. He dedicated much space in his 1942 book to this subject, and even more so in 1963.

> The fact that an eminent contemporary geneticist [Richard Goldschmidt] can come to conclusions which are diametrically opposed to those of most other geneticists is striking evidence of the extent of our ignorance[1362].

Mayr argued strongly and persistently that new species can only – or almost exclusively – emerge when a geographic barrier prevents gene flow between parts of the population.

> Species descend from groups of individuals which become separated from the other members of the species, through physical or biological barriers, and diverge during this period of isolation. The concept of the isolated population as incipient species is of the greatest importance for the problem of speciation[1363].

Mayr insisted that groups of sexually-reproducing organisms can be considered true species only after reproductive isolation is achieved. Organisms can be classified according to different criteria – but these should not be considered different species unless they are reproductively isolated:

> If we designate an "isolate" any more or less isolated population… we can distinguish in sexually-reproducing organisms between geographical, ecological, or reproductive isolates – of which only the last are species. The unspoken assumption made by certain authors, that the three kinds of isolates coincide, are not supported by the known facts[1364].

According to Mayr, every species has an optimal niche, to which it is adapted by natural selection. The environmental conditions – the niche – determine the limits of the geographical distribution of the species.

> The statement that every species is adapted to its environment is a self-evident platitude. The result is that every species has an optimal environment, presumably somewhere

[1359] Darwin 1856, in Stauffer 1975, p. 271.
[1360] Chapter 28.
[1361] Fisher 1958 (1930) p. 139.
[1362] Mayr 1942, p. 65.
[1363] Ibid. p. 33.
[1364] Mayr 1963, p. 27.

> near the center of its range, and definite limits of tolerance with respect to latitude and longitude[1365].

Speciation and Geographical Clines

Contrary to common thought, geographical clines – the changes in the morphology of populations and species in correlation with geographical, climatic, and other factors – are not conducive to speciation. On the contrary,

> Clines indicate continuity. But since species formation requires discontinuities, we might formulate a rule: the more clines found in a region, the less active is species formation[1366].

Mayr acknowledges the fact that many so-called species reported to have been formed gradually by natural selection were in fact cases of variation within species. Nevertheless he is totally opposed to the suggestion offered by the mutationists that species may be formed spontaneously without geographic barriers[1367].

> A new species develops if a population, which has become geographically isolated from its parent species, acquires during this period of isolation characters which promote or guarantee reproductive isolation when external barriers break down[1368].

> That geographic speciation is the most exclusive mode of speciation among animals... is now quite generally accepted. The theory of geographic speciation is one of the key theories of evolutionary biology[1369].

Mayr believed that polymorphisms observed in nature are largely due to the differentiation of "*sibling species*" – a term Mayr coined for cases of incomplete reproductive isolation with little morphological distinctness ("morphologically similar and closely related, but sympatric species").

> The subspecies is considered the incipient species in much of the evolutionary literature. Goldschmidt ...rightly questions whether the subspecies, being merely ecotypic response to the local environment, have much of a potential as incipient species. As we have shown above... it is the geographical isolate, and not the subspecies, that is the incipient species[1370].

Sympatric Speciation

Another, alternative concept, "*sympatric speciation*", does not require geographic isolation – provided that a reproductive barrier develops in a variable population. Mayr considered that the probability of sympatric speciation is very low. The central process in speciation is geographic isolation, which prevents gene flow and the consequent obliteration of differences between parts of the population:

> It is realized more and more clearly that reproductive isolation is required to make the gap between two sibling species permanent. This is unlikely to happen in one step in a population. Therefore sympatric speciation is unlikely.

[1365] Ibid. p. 60.
[1366] Mayr 1942, p. 97.
[1367] Mayr 1942, pp. 148-155.
[1368] Ibid. p. 151.
[1369] Mayr 1963, p. 481.
[1370] Ibid. p. 491.

> There is no evidence for ecological speciation as a process distinct from geographic speciation. I have been unable to find a single case of so-called ecological speciation in the literature that did not require the spatial isolation of populations[1371].

Mayr admitted that ecological varieties such as host-races feeding on different plants can coexist sympatrically for a long time if their habitats persist, and gene flow between them can be consequently limited – but they will become true separate species only if a genetic reproductive barrier is formed. Such a barrier cannot materialize unless a geographic barrier to gene flow exists[1372]. He was not convinced that it is necessary to invoke hypothetical sympatric speciation to explain strong biological differences between closely-related species; and was frustrated that not everybody agreed that his position on this issue should be accepted:

> It is discouraging to read the perennial controversy because the same old arguments are cited again and again in favour of sympatric speciation, no matter how decisively they have been disproved previously[1373].

Note: *Mayr lived long enough to witness challenges to some of his long-established ideas about evolution. One of these was the insistence on geographic isolation as a required factor of speciation. In particular, Guy L. Bush (1978) studied the theoretical conditions for sympatric speciation. His model requirements were that: 1) there should be two suitable host plants growing within the flight range of the organism in question; 2) that at least two genes will be involved, each with two alleles: one gene affecting the sensory system of the adult, for recognition of the alternative plants as suitable hosts – and the other gene enabling the digestive system of the larval stages of the organism to digest the host tissues; and 3) that the mating of males and females will take place on their natal host.*

[1371] Mayr 1963, p. 458.

[1372] Mayr 1942, p. 193; he admits however that all examples of sympatric speciation are of insects, and that he, an ornithologist, may be unable to evaluate them critically (p. 209).

[1373] Mayr 1963, p. 451.

33

Julian Sorell Huxley

Julian Huxley (1887-1975) was the grandson of Charles Darwin's friend and ally, Thomas Henry Huxley (and the brother of the writer, Aldous Huxley). A zoologist by training, he was interested in ornithology and bird behavior. He was one of the founders of the World Wildlife Fund, and later became the first Director of the UN Organization for Science and Agriculture, UNESCO. During his term of office he reached an agreement with the Government of Ecuador that established the Galapagos Archipelago as a nature preserve. A research station on Santa Cruz Island (Darwin Station) was set up and tasked with the restoration of giant tortoise populations – extinct on most islands – with UN support.

Huxley was very active in promoting public support for the Theory of Evolution, lecturing on the radio and television on the subject, and he had a role in building the "Modern Synthesis of Evolution"[1374]. He was active in the "Eugenics" society in England and served as its president for a number of years. He was knighted in 1958.

Details of Huxley's Career

Julian Huxley was born in Surrey, England. At the age of 13 he was sent to school at Eton College, and then received a scholarship to continue his studies at Oxford. He specialized in ornithology and graduated with honors.

His career was extraordinarily versatile. After a year of research at the Marine Laboratory at Naples, Italy, in 1912 he accepted an offer to establish a Department of Zoology at Rice University, Houston, Texas. Two years later he left and returned to England to join in the war effort (World War I), serving in Intelligence. In 1925 he was appointed Professor of Zoology at King's College, London, but left in order to work with the writer H.G. Wells on their comprehensive book on biology, "The Science of Life" (1929), which emphasized an evolutionary approach to biology.

Sponsored by the British Colonial Office, Huxley visited the then British-held colonies of Kenya, Tanganyika [now Tanzania], and Uganda in 1929, as an advisor for education. He was impressed by the fact that the wild animal populations on the Serengeti plain had not been adversely affected by the colonization by the white man – because white people did not settle in the tse-tse fly [and Malaria] infested areas. He recommended that these areas should be managed as National Parks.

In 1941, while on an invited lecture tour in the USA, he caused some controversy when he declared that America should join Britain in the war. A few weeks later Japan attacked Pearl Harbor, and America indeed joined the war.

Huxley served for seven years as Secretary of the Zoological Society, managing all the zoological parks including London Zoo, which at times led him into conflict with other Fellows of the Society. He received several prestigious awards and medals of honor from the Society. Huxley served for two years as secretary of UNESCO – (the usual six-year term was shortened, perhaps because the Americans did not like his "leftist" attitude). His later publications concerned human society, with emphasis on the role of Man in Nature.

[1374] His famous book, "Evolution - The Modern Synthesis", 1942.

"Problems of Relative Growth"[1375]

Huxley's book, "Problems of Relative Growth", presents a summary of a series of papers – published between 1924 and 1931 – in which he describes the changes in body shape as an animal increases in size during ontogeny. It is well known that as an animal grows, not all its limbs increase in size at the same rate [heterogony]. For example, in human babies the head is larger, relative to the body, than in the adult, because in adult development the body grows much more rapidly than the head. Huxley investigated this phenomenon in detail in a number of organisms and gave it a mathematical formulation.

> I hope to convince the systematist that by a knowledge of the laws of relative growth we are put in possession of new criteria bearing on the validity of species, subspecies, and 'forms'; the nature of certain dwarf forms; and the importance (or the reverse) of size differences in general for systematics.
>
> The physiological mechanism underlying these general rules still remain very obscure, in the absence of experiments specifically directed to the point. But there are some interesting hints and possibilities[1376].

Huxley describes in detail several "case studies" of heterogonic growth (his term). Considered biologically, this is:

> A manifestation of huge macro-evolutionary changes based on small genetical differences, acting on a growth gradient, affecting a large number of parts – many non-adaptive traits, correlated with general growth and not needing natural selection[1377].

Huxley developed a mathematical expression of the law of relative growth:

$$Y = bX^k \quad \text{or} \quad \log Y = \log b + k \log X$$

where Y is the measured size of the limb, X is general body size, and b and k are constants:

> For each part we have a constant differential growth ratio, denoted by the value of k.

on the logarithmic scale, Y is linearly related to X, and this enables comparisons of growth rates of different limbs within the same organism, as well as of the same limb in different organisms.

Many examples are given in the book, with Tables listing the values of k. Examples range from the weight of roe deer (*Capreolus*) antlers relative to body size, to the size of the thorax in polymorphic ants (*Pheidole*), and the relative size of the large chela in male fiddler crabs (*Uca*) – an example treated more intensively than the others.

Expanding on the suggestion by D'Arcy Thompson[1378], Huxley refers to the figure[1379] showing the possible transformation of the fish *Diodon* to that of a different extant fish, the sunfish *Orthagovisus*:

> From the figure it will be immediately obvious that the essence of the transformation, considered biologically, must have been the origin of a very active growth center in the whole of the hind region of the body, whence the intensity of growth diminishes regularly towards the front end[1380].

[1375] Huxley, J. 1932, "Problems of relative growth".
[1376] Huxley, J. 1932, p. 3.
[1377] Huxley 1932.
[1378] D'Arcy Thompson, "on Growth and Form", (1917) 1966.
[1379] Ibid. figs. 154-155.
[1380] Huxley, J. 1932.

Evolution: The Modern Synthesis

Huxley suggested that discussions of evolution should be organized under several main topics, to facilitate a deeper understanding of the processes.

> Evolution must be dealt with under several rather distinct heads. Of these, one is the **origin of species**... another is the origin of **adaptations**. A third is **extinction**. And a fourth, and in many ways the most important, is **the origin and maintenance of long-range evolutionary trends**[1381]**.**

The book opens with an historical review of the main events in the development of the theory since genetics first appeared on the scene.

> In 1894, William Bateson published a large volume called "Materials for the Study of Variation", [which] may have been the prelude to the study of discontinuous, rather than continuous, variation [this was aided by de Vries and Morgan]. Darwinism was on the eclipse, Mendelism on the rise.
>
> The true-blue Darwinian stream... had reached its biometric phase. Tracing its origin to Galton, Biometry blossomed under the guidance of Pearson and Walton... The Biometricians stuck to hypothetical modes of inheritance and genetic variation, on which to exercise their mathematical skills. The Mendelians refused to acknowledge that continuous variation could be genetic, or at any rate dependent on genes, or that a mathematical theory of selection could be of any real service to the evolutionary biologist[1382].

An important insight, engendered by the new science of genetics, was the realization that genetic change in populations may be brought about not only directly by new mutations, but also indirectly by recombination, which may create new genotypes – some of which may be better adapted to the environment than the older ones, and which may be available for natural selection[1383].

On the Species Problem

Huxley suggested that there was perhaps no real need for a formal definition of the term species:

> Logic demands that we should define the term (species). It may be that logic is wrong, and that it would be better to leave it undefined, accepting the fact that all biologists have a pragmatic idea of its meaning at the back of their heads[1384].

The "classical" criterion of infertility, wrote Huxley, is no longer acceptable [as indeed Darwin and Wallace had agreed long ago]. Perhaps species can be defined in different ways – to answer different aspects of biology: "The most important are the geographical, the ecological, and the genetic"[1385].

"Geographical" species may arise due to isolation and the accumulation of small mutational steps; "ecological" species may arise when they occupy different ecological environments – for example when insects occupy different host plants, or plants live on different soils or flower at different seasons; and "Genetic" species may arise when mutations occur in a population, causing a reproductive barrier. All these factors affect the accumulation of genetic changes and may lead to differentiation. In addition, random events in small populations may also lead to speciation. The

[1381] Huxley, J. 1942, p. 153.
[1382] Huxley, J. 1942, p. 24.
[1383] Huxley, J. 1942, p. 124.
[1384] Ibid. pp. 154, 167.
[1385] Ibid. p. 154.

definitions for geographical, ecological, and genetic species may overlap – but need not necessarily coincide!

On Clines

Julian Huxley coined the term "cline" to describe a continuous trend in some biological character that parallels a geographical or climatic trend such as elevation or temperature. These trends are a product of natural selection, adapting local populations to the prevalent conditions along the cline[1386]. Two well-known examples of clines which are often referred to in textbooks and ecological publications, are Bergmann's and Allen's rules:

> **Bergmann's Rule:** Within a polytypic warm-blooded species, the body size of a subspecies usually increases with decreasing mean temperature of the habitat.
>
> **Allen's Rule:** In warm-blooded species, the relative size of exposed portions of the body (limbs, tail, and ears) decreases with the decrease in mean temperature[1387].

The prevalent explanation for both rules applies mostly to homeothermic animals[1388] [which maintain a constant body temperature] as a function of the ratio of body surface to body volume: when an organism of length x increases in size, its surface area increases as $S = X^2$ and its volume as $V = X^3$. Heat loss from the body is a function of surface area relative to volume (S / V). This ratio becomes smaller, the larger the organism. Therefore, larger organisms conserve heat better and can survive further north, or generally under colder average temperatures, than smaller ones. The same principle holds in the explanation of Allen's rule.

On Extinction: Extinction has been a major phenomenon in the evolutionary past, as demonstrated by the fossil record. This was recognized as early as the 1820s by Georges Cuvier[1389]. The Darwinians considered that as species become more and more closely adapted to their environment, they become increasingly specialized, and gradually lose some of their genetic variation. The claim is often made that such reduced genetic variation slows the differentiation process and ultimately leads to extinction of the specialized species. Huxley disagrees with this[1390].

On Long-Term Trends in Evolution

Evolution is thus seen as a series of blind alleys. Some are extremely short – those leading to new genera and species that either remain stable or become extinct. Others are longer – the clines of adaptive radiation within a group such as a class or a subclass, which run for tens of millions of years, before coming up against their terminal blank wall[1391].

But Huxley denied that evolution is heading towards progress. What we observe, he considered, is "progress without a goal".

[1386] Huxley, J. 1942, p. 206.
[1387] Ibid. p. 253.
[1388] Note: This physiological explanation holds true strictly for homeothermic animals. However, clines are known also in organisms which do not maintain constant body temperature, such as some amphibians, reptiles, and invertebrates.
[1389] Chapter 4.
[1390] Huxley, J. 1942, p. 500.
[1391] Huxley, J. 1942, p. 55.

> The ordinary man, or at least the ordinary poet, philosopher and theologian, always was anxious to find purpose in the evolutionary process. I believe this reasoning to be totally false[1392].

In the evolution of the horse – as worked out by his grandfather, Thomas Henry Huxley – ancestors of the present-day horse had first four, then three digits in their hooves – whereas the modern horse runs on single digit hooves. This trend is interpreted as an improvement in running skills [note that the cheetah, nonetheless, has five digits and runs faster]. The reduction in the number of digits is accompanied by other anatomical changes – particularly in the structure of the teeth, facilitated perhaps by the transition from browsing on bushes to grazing on graminaceous grasses. All these trends, claims Huxley, can be explained without the need to assume that evolution is in principle progressive. Different characters in the evolution of the horse diverge in different directions[1393].

Another trend which is often referred to as progressive is that of the increasing volume of the brain in vertebrates, from fishes to mammals, and to Man – with the claim being that a more complex brain allowed more functions to be developed, and was selected for in vertebrate evolution[1394].

Julian Huxley's later book, "Evolution in Action", includes an extended discussion of long-term trends in evolution. One such trend, which was considered unidirectional ("Orthogenetic"), is that of the patterns of change in the sutures of septa in the shell of fossil ammonites (Cephalopoda): older ammonite fossils show a simple sutures between septa, and the pattern becomes more and more convoluted in more "recent" fossils.

Improvement is often expressed as changes that facilitate better utilization of the environment; for example during the establishment of a new life form or the occupation of a new habitat or ecological niche. The process is called "adaptive radiation". Huxley prefers the term "deployment", borrowed from military terminology, meaning the re-distribution of troops – [" to extend the front and reduce the depth". Webster's dictionary]

> The realization of possibilities – that is perhaps the best way to view biological improvement[1395].

Huxley elaborates upon the improvements in the structure and function of the vertebrate nervous system during evolution[1396]. In particular he expands upon the evolution of vision as a response to environmental stimulation of light. "Light, in any proper sense of the word, did not come into existence before there were animals with eyes" ("it is impossible to explain "light" to one born blind")[1397]. All mammals except the primates are color-blind, and even in monkeys red and blue coloring features appear very rarely (the Mandrill is an exception). The improved perception of color culminates in Man, and results from an increase in the number of sensory units [in the eye] and the addition of color-sensitive ones. Other parts of the nervous system evolved in similar ways.

Julian Huxley: on Man

> [Man] has developed a new method of evolution, the transmission of organized experience by way of tradition ...which largely overrides the automatic process of natural selection as the agent of change[1398].

[1392] Ibid. p. 576.
[1393] Morgan 1925.
[1394] Note: This line of thought was greatly elaborated by Dobzhansky (Chapter 31).
[1395] Huxley, J. 1957, p. 71.
[1396] Ibid. p. 74 ff.
[1397] Ibid. p. 79.
[1398] Huxley J., "Evolution in Action". 1953, p. 132.

Huxley, like Dobzhansky, attributed the evolutionary success of Man to the improved capacity of his brain and to the development of cultural evolution[1399].

Evolution in human populations, however, occurs principally by means of changes in the structure of societies, improvements in technology, and in the finding of new uses for existing technologies – and not by an increase in brain capacity, which has not changed since the time that Man dwelt in caves[1400].

> The critical point in the evolution of man ...was when he acquired the use of verbal concepts and could organize his experience in a common pool... Thus, once life had become organized in human form, it was impelled forward, not merely by the blind forces of natural selection, but by mental and spiritual forces as well[1401].

Huxley was active in the British Humanist Society. Human evolution, He wrote, was driven by the behavior and decisions of individual leaders, not by the statistical average behaviors of the populations: the Mongols were an important force in the Middle Ages, but their fate and their effect on Europe was due to one man, Genghis Khan. Similarly, it was the leaders of religious and social movements – and not the common masses – who have been responsible for the fate of humanity in the more recent centuries, such as Napoleon, Stalin, and Hitler.

Huxley thought that the Theory of Evolution – the common origin of life – should become the only unifying basis for mankind, and replace the belief in God (he called this idea "Religious Naturalism").

> For evolution bridges the gap between man and animal, between mental and material, and between the organic and the inorganic. Evolution shatters the pretense of isolationism and sets man squarely in his relation with the cosmos.
>
> Many people assert that this abandonment of the God hypothesis means the abandonment of all religion and of all moral sanctions. This is simply not true. But it does mean, once our relief at jettisoning an outdated piece of ideology furniture is over, that we must construct something to take its place[1402]?
>
> To all people at some time, and to many people much of the time, the world is an unpleasant or even horrible place, and life a trial or even a misery. Little wonder that many ideologies, religious and otherwise, are concerned with providing escapes from the unpleasant reality[1403].

On Eugenics

Huxley was a believer in eugenics (he was one of the founders of the Eugenics society and served first as its vice-president and then as president). He strongly recommended restricting the birthrate, perceiving the rapid increase in the human population size as threatening the future of the world:

> No one doubts the wisdom of managing the germ plasm of agricultural stocks. So why not apply the same concept to human stocks?

[1399] Ibid. p. 108.
[1400] Huxley 1957 p. 21.
[1401] Ibid. pp. 115-116.
[1402] Huxley, J. 1969, "Essays of a humanist".
[1403] Ibid. p. 116.

> If nothing is done to control this flood of people, mankind will drown in its own increase, or… the world economy will burst at the seams, and mankind will become a planetary cancer… The population problem has entered on a new phase. It is no longer primarily a race between population and food production, but between death control and birth control[1404].

But the efforts to limit population growth were not to be applied randomly to human populations: the social structure should be considered. The eugenics ideology is clearly expressed in the following quotations:

> The quality of people, not merely quantity, is what we must aim at. And therefore a concerted policy is required to prevent the present flood of population increase from wrecking all our hopes for a better world[1405].

> The lowest strata are reproducing too fast. They must have no easy access to relief or hospital treatment, lest the removal of the last check on natural selection should make it too easy for children to be produced or to survive. Long unemployment should be a ground for sterilization[1406].

The Role of Man in Future Evolution

Huxley evaluated the central place of Man in nature – as the dominant species with extraordinary destructive means at his disposal. This should make Man aware of his responsibility for maintaining nature – for his own good:

> Though biological science was content to classify him [Man] as just another animal, in his own eyes he was still the Lord of Creation, apart from the rest of nature, and in some unspecified sense, above nature[1407].

> It is as if man had been suddenly appointed managing director of the biggest business of all – the business of evolution – appointed without being asked if he wanted to, and without any proper warning and preparation. What is more, he can't refuse the job[1408].

[1404] Huxley, J. 1957, pp. 178-181.
[1405] Huxley 1957, p. 16.
[1406] Huxley, "Man in the Modern World", 1942.
[1407] Huxley 1957, p. 42.
[1408] Huxley 1957, p. 13.

34

Richard Goldschmidt (1878-1958): Genetics, Evolution, and Hopeful Monsters

Richard B. Goldschmidt – zoologist and geneticist – was born in Frankfurt, Germany, and received his PhD in 1902 from the University of Heidelberg. In his Autobiography, Goldschmidt declared, with obvious pride, that he had contributed greatly to various biological fields:

> My life work is filled with innumerable newly observed facts... in many different fields as cytology, descriptive and experimental embryology, neurology, microscopic anatomy, physiology, taxonomy, parasitology, genetics and evolution[1409].

His career in Germany, which had begun with great hopes, was rudely interrupted by the two World Wars. He taught zoology for ten years at the University of Munich, working on the anatomy and genetics of *Ascaris* worms and *Amphioxus*. He then moved to Berlin to take up a senior position at the Kaizer Institute for Biology. As part of his research on the gypsy moth, *Lymantria dispar*, he received a two-year grant to study this insect in Japan, where its populations are extremely widespread. The First World War broke out when he was on board ship on his way back to Germany and he was unable to return home. The ship docked instead in San Francisco. The USA at that time had not yet entered the war and, using his connections with American scientists, he settled temporarily in Berkeley and then moved to New York. When the USA finally entered the war he was arrested as an enemy alien and spent two years in a detention camp (the atmosphere in the camp was relaxed, and he describes his participation in musical and other cultural events there).

When the war ended, being a proud German he returned to his war-devastated homeland. His home and his laboratory had remained undamaged in his absence and he restarted his research on the gypsy moth. This was interrupted again, in 1936 – when the Nazis came to power in Germany. Being Jewish, he was "encouraged" to quit his job. Under the pretense of being invited to give lectures in the USA, he received permission to leave the country, and with some difficulty succeeded in obtaining a position at Berkeley, California. He immigrated with his family to the USA, where he became a professor at Berkeley. This time he found it more difficult to adjust to American life. He was now required to teach elementary zoology to 600 freshmen students – whereas in Germany he had been his own master, with an ample budget and no teaching obligations. When in 1942 the USA entered into the war with Germany (after Pearl Harbor), Goldschmidt was arrested once more as an enemy alien, leaving his family in financial difficulties and being given the cold shoulder by their neighbors. Eventually, they were granted American citizenship.

Goldschmidt published many papers. His important book, "The Material Basis of Evolution"[1410], created much controversy when his unconventional views on genetics and evolution became known. He was considered a heretic in scientific circles, but his scientific achievements as a geneticist could not be ignored. He was elected to the National Academy of Sciences (1947), became the president

[1409] Goldschmidt 1960, p. 312.
[1410] Goldschmidt 1940.

of the 9th Congress of Genetics (1953), and continued to work at Berkeley until his retirement. He died of a heart condition in 1958.

Genetic Studies on the Gypsy Moth: Intersexes

Goldschmidt carried out intensive studies on the genetics of the gypsy moth, *Lymantria dispar*, a widespread forest pest in the Palearctic region.

The males of *L. dispar* have brown wings while the females have white wings with pale black stripes. In crosses between geographical races of the moth, Goldschmidt obtained males and females with phenotypes intermediate between the male and female morphology, which he named "*intersexes*". In contrast to the known mechanism of sex determination in *Drosophila* and in Man, in which the difference between the sexes is chromosomal (females being XX and males, XY), in the moths there are no "sex chromosomes". Goldschmidt developed a model to explain sex determination and intersexes in the moths. From his crossing results, he inferred that the morphological variation may result from the balance between male and female "factors". Each individual carries two relevant loci, each with two alleles. The females are homozygous for the "female factor" but heterozygous for the "male factor" (symbolically, FFMm), while the males are homozygous at both loci (FFMM). These factors control or regulate the production of a hormone or an enzyme responsible for the morphological appearance of the phenotype. The quantity of the hormone or its production rate can also be affected by environmental factors. Goldschmidt suggested that the model could offer a general explanation of sex determination in the living world.

The Physical Structure of the Genetic Material

The accepted model for the structure of genetic material, as envisioned by Morgan and his group, was that the genes are arranged on the chromosomes in a linear order, like "beads on a string". Each gene was envisioned as a concrete unit responsible for the production of a specific phenotypic character, changeable through mutation that replaces one allele (and phenotype) by another.

The results of genetic studies by Dobzhansky and others, as well as his own observations in *Lymantria*, led Goldschmidt to doubt the validity of this model. Those studies had indicated that mutational and recombinational changes in the phenotype became more frequent when the experimental temperature was raised. Goldschmidt reasoned that if the model was correct, inversions and recombinations should not cause a phenotypic change, because all the "beads" were there – only their order had been changed. Some observations by Dobzhansky supported Goldschmidt's view that the phenotype in general is not determined by individual gene loci, but that all phenotypic changes are due to "position effects" of small chromosomal sections, which affect physiological or hormonal processes in the organism. Goldschmidt indicated that several observed phenomena, which were not effectively explained by the accepted model – such as "incomplete penetrance" and "incomplete expressivity" in *Drosophila* – are well explained by temperature-dependent hormonal activity, affecting the rate of accumulation of certain proteins required for the formation of the observed structure[1411].

Goldschmidt's suggestion that the "beads on a string" model might not be valid set him in opposition to most geneticists, who regarded him as a heretic. He claimed that every phenotypic change is in fact a result of "position effect" – a chromosomal phenomenon – rather than due to a replacement of one allele by another.

> This viewpoint, which in my opinion is daily becoming more probable, [is] that actually no particulate genes exist, but that all mutations are based on very small pattern changes[1412].

[1411] See more recent examples in *Tribolium*: Wool & Mendlinger 1973; Wool 1985.

[1412] Goldschmidt 1940, p. 203.

> The classical atomistic theory of the gene is not indispensable for genetics as well as for evolution. We have already foreshadowed the twilight of the gene[1413].

Population and Geographical Variation

The local strains of the gypsy moth which Goldschmidt had collected on his travels to Asia and reared in his laboratory, were used in a detailed study of the geographic variation in their morphological as well as physiological characters. The timing of the larval diapause was characteristic for each population, in correlation with local environmental clues such as temperature and day length. The differences among populations were heritable and persisted in laboratory rearing. Goldschmidt suggested that each population was adapted to its local environment by natural selection. He termed this process "*micro-evolution*".

By contrast, he detected no barriers to reproduction between the local geographical populations, however genetically different they might be from each other. He concluded that the accumulation of small differences between populations and adaptation to the local environment is insufficient for the formation of a new species. To form a new species a major reorganization of the gene pool was required – a "systemic mutation". The resulting process he termed *macro-evolution.*

> I derived the hypothesis that in addition to small or large mutations of genic loci, there exists a completely different type of mutation. I called this "systemic mutation" meaning that a reshuffling or scrambling of the intimate chromosomal architecture, which might occur rarely, by chance, will act as a macromutational agent. This means that it will produce, suddenly, a huge effect upon a series of developmental processes, leading at once to a new and stable form, widely diverging from the former[1414].

> The systemic pattern mutation – as opposed to gene mutation – appears to be the major process leading to macroevolution, i.e., evolution beyond the blind alleys of microevolution[1415].

Systemic Mutations and Hopeful Monsters

Goldschmidt upset many of his contemporary scientists when he suggested that the formation of new species (and of higher taxonomic categories – genera and families), – macro-evolution – is a different process to that of the adaptation of populations to their environments by natural selection (microevolution). This was a rejection of the traditional Darwinian concept, and the Darwinians reacted – "savagely: this time I was not only mad but almost a criminal"[1416].

Goldschmidt realized that the phenotype resulting from such a "systemic mutation" is likely to be detrimental to the individual organism, which may not survive the test of natural selection. It can be referred to as a "monster" compared with normal, adapted phenotypes. In the long run, however, local conditions may change, and a "monster" occurring by chance may be just the right form to survive and become the prevalent form of a new species.

> The solution was the existence of macromutations, which in rare cases, could affect early embryonic processes so that, through the features of embryonic regulation and integration, at once a major step in evolution could be accomplished and fixed under certain conditions.

[1413] Ibid. pp. 209-219.
[1414] Goldschmidt 1952 p. 96.
[1415] Ibid. p. 245.
[1416] Goldschmidt, Autobiography 1960.

> I spoke half-jokingly of the hopeful monster in my first publication on the subject, a lecture read by invitation in 1922 at the World's Fair in Chicago[1417].

Goldschmidt found many examples of would-be monsters among the mutant strains of *Drosophila* maintained in the genetic laboratories of Morgan, Dobzhansky, and their students. Mutant strains with " homeotic mutations" – deformed wings or duplicate organs (extra pair of legs instead of antennae or wings) – were maintained in the laboratory but were obviously monstrous. Such changes, if they occurred in nature, would be eliminated by natural selection[1418].

Goldschmidt later argued that a "monstrous" phenotype may occur without the need for a major genetic upheaval. A small mutational change, if it occurred *early* on in the embryological development of the individual, may be translated into a large phenotypic change in the adult. In such cases the normal developmental machinery may function without upsetting normal embryological processes, and the individual may be able to survive.

Goldschmidt stated that his explanation of such monsters – and other puzzling evolutionary cases like the eyeless forms of vertebrates living in caves – is preferable to the Darwinian explanation. The Darwinians required the assumption of wide heritable variation in each of the characters, with natural selection then working on a gradual variation from the normal to the mutant form, and selecting for the best fit. This assumption – unsupported by evidence – is not necessary if "monsters" occasionally survive: for example, if a mutation in a cat caused fusion of the tail vertebrae, a monster would be created; but if such a change occurred in the *Archaeopteryx*, for example, it would have an advantage because it increases the maneuverability of the tail in flight.

> The Galapagos cormorant is flightless... A single mutant may produce any degree of wing rudimentation. If such a mutation occurred in a hawk, the resulting monster will not survive. But if it occurs in a bird like the cormorant, which is already organized for catching his food while swimming under water, the monstrosity will not be deleterious, and might even be of the "hopeful" type, if it enhances simultaneously the swimming and diving capacity. The Galapagos cormorant survives successfully without flying[1419].

Goldschmidt found support for his theory of "hopeful monsters" in the mathematical theory of D'Arcy Thompson,[1420] who showed, as early as 1917, that complex variation in shape among organisms (or their parts) can be brought about by changing very few – even one or two – of the parameters controlling the rates of growth in different dimensions. But Goldschmidt admitted sadly,

> I am afraid that both Thompson and I were too far ahead of our times to make an impression on evolutionary thought, and especially upon the reasoning of the geneticists, who were unable to turn their minds from the idea of micro-mutations and their accumulation by selection[1421].

More on Macro-evolution

The separation of macro- from micro-evolution was contrary to the accepted Darwinian doctrine: the separation forced Goldschmidt to abandon the traditional role of geographic isolation in

[1417] Goldschmidt 1960, autobiography, p. 318.
[1418] Goldschmidt 1940, p. 341 ff.
[1419] Goldschmidt 1940, p. 393.
[1420] D'Arcy W. Thompson (1860-1948) was a professor of biology at the University of Dundee, Scotland. He was well-versed in Classical studies (he translated some of Aristotle's' works into English) and had a good background in mathematics. In his book "On Growth and Form" (1917, re-published in America in 1984) he showed mathematically that complex animal forms, like the carapace shapes of different species of crabs, can be derived from simple ones by changing the rate of development along one or a few different axes during larval growth.
[1421] Goldschmidt 1940, p. 316.

speciation, as advocated by Mayr. Since speciation was a strictly genetic phenomenon, isolation was no longer essential for the formation of a new species, and therefore subspecies and geographical races cannot be "incipient species", as the Darwinians maintained, but rather, they are dead "blind alleys" within the species.

> Microevolution by accumulation of micromutations – we may also say neo-Darwinian evolution – is a process which leads to diversification strictly within the species, usually – if not exclusively, for the sake of adaptation... subspecies are actually, therefore, neither incipient species nor models for the origin of species. They are more or less diversified blind alleys within the species.
>
> The decisive step in evolution, the first step toward macroevolution, the step from one species to another, requires another evolutionary method from that of sheer accumulation of micromutations[1422].

Goldschmidt specified how he thought that the higher taxonomic units – above the species level – had evolved. Ernst Mayr, who strongly opposed these ideas, nevertheless cites them in full:

> A phylum [says Goldschmidt] consists of a number of classes, all of which are basically recognizable as belonging to that phylum but, in addition, are different from each other. The same principle is repeated at each taxonomic level. All genera of a family have in common the traits which characterize the family... so it goes down to the level of the species. Can this mean anything but that the type of the phylum was evolved first, and later separated into the types of classes, then into orders and so on down the line? This natural, naive interpretation... actually agrees with the historical facts furnished by palaeontology[1423].

Reactions

In the introduction, entitled "The uses of heresy", to a new [1982] edition of "The Material Basis of Evolution", Steven J. Gould describes the attitude of contemporary biologists to Goldschmidt the heretic. His idea of "hopeful monsters" was ridiculed, and his scientific ideas about evolution were ignored for 40 years – although Goldschmidt was, in Gould's opinion, one of the ten most important geneticists of the 20th century. Goldschmidt himself felt betrayed and insulted by the attitude of his contemporaries.

From a reading of Mayr's books (1942, 1963) the true attitude towards Goldschmidt seems to have been less extreme than Gould's description. Mayr appears to have appreciated Goldschmidt's scientific work, although he certainly disagreed with his ideas[1424].

[1422] Goldschmidt 1982 (1940), p. 183.
[1423] Mayr 1963, p. 600.
[1424] Mayr 1963, p. 312; Ibid. p. 198; Ibid. pp. 437-8.

35

E.B. Ford and H.B.D. Kettlewell: Ecological Genetics

The Theory of Evolution in the 20th century attempted to explain the [phenotypic] changes in terms of changes in allele frequencies or genotypes. These were assumed to be caused by different responses of genotypes to the external conditions – including competitors and predators – or to interactions [competition] between genotypes within populations. Accordingly the formulations offered by Fisher, Haldane, and Wright became the basis of all discussion of evolution. Dobzhansky and his associates applied genetic techniques to the study of living populations of *Drosophila* in nature[1425]. However, data supporting the theory came mainly from laboratory studies.

In the 1930s, a group of British scientists, headed by Edmund B. Ford, initiated quantitative studies aimed at measuring the extent of selection in Nature. They termed this branch of science "*Ecological Genetics*". Their books[1426] contain scientific reports of field experiments carried out by a number of investigators on different organisms, but with the same purpose in mind. In the 1940s-1960s, Ford and his associates established methodologies for conducting field studies and carried out a series of long-term field studies which demonstrated that selection did occur in nature, and they sought to identify the selective force (s).

Despite the efforts of the ecological-geneticists over the years, there are not many cases in which the process of selection was followed in detail over time and the selective agent was identified. In a literature review of adaptive changes in natural populations over the last 200 years, only 47 cases have been recognized[1427]. Some of these cases have become textbook examples of evolution. The majority of them relate to the responses of insect populations to man-made changes in the habitat, or to the acquisition, by insects, of a new host plant. The authors of the review suggest that the reason for the relative scarcity of proven cases of selection in nature may not be that the processes of adaptation are inherently too slow to be noticed – as usually claimed – but the opposite: the process may be quite fast, so that one must be in the right place at the right time in order to observe it[1428].

Industrial Melanism

One of the first, and certainly the most publicized, example of a genetic change in populations in nature is the phenomenon known as industrial melanism[1429].

Early on in the 20th century British entomologists noticed the replacement, by dark or black forms, of common species of moths that were characterized by white or pale wings with thin line markings. By the 1930s the phenomenon had been observed in 80 unrelated species in different families. Examination of the collections in the British Museum and other sources indicated that the dark forms had been rare before 1848, comprising no more than a few percent; whereas 50 years later their frequency was approaching 100% in some areas. A common characteristic of the various moth species was that they spent the daylight hours resting on tree trunks.

[1425] Chapter 28.
[1426] Ford 1965; Creed 1970.
[1427] Reznick and Ghalambor 2005a, b.
[1428] Ibid.
[1429] Ford 1965; Creed 1971.

Simultaneously with the change in the moth populations, which attracted the attention of both professionals and moth fanciers, a more obvious and radical change became noticeable in the countryside: the color of the trees, and even the faces of buildings in the cities, were becoming dark and blackened. These were the days of the Industrial Revolution: the industry was using huge quantities of coal to generate energy, and homes were heated by burning coal. The soot from the chimneys was covering everything, and the prevalent fog in the Thames Valley became the notorious London smog.

Biologists noticed that the yellowish-green lichens that had previously covered the tree trunks in the damp British woods, had died and disappeared, and the trunks and leaves were covered with soot. Experiments in washing leaves revealed that the soot had reached areas downwind of the industrial areas. The coincidence of environmental change with the phenotypic change in the moth populations suggested that the two phenomena might be related. It was initially suggested that some materials in the soot were causing an increased production of the black pigment *melanin* in the insect cuticle [hence the name of the phenomenon]. This suggestion could not be confirmed experimentally. Another suggestion was that mutagenic substances in the soot were causing an increase in recessive black mutations in the moths. Further research showed that the mutations causing melanism were – at least in one thoroughly-investigated geometrid moth, *Biston betularia* – dominant, and the typical pale form was recessive. This explanation too thus had to be rejected.

The British geneticist and mathematician, J.B.S. Haldane[1430], analyzed the spread of melanism in this moth in 1924. Given that the frequency of a black form had been less than 1% in 1850 and more than 95% fifty years later, he calculated that the black genotype must have had a 50% selective advantage over the typical form. Moreover, as Haldane pointed out, the frequency of rare moths in collections is an over-estimate of their real frequency in nature, given that collectors preferentially tend to be attracted to rare forms.

In a later book[1431] Haldane described a case of phenotypic change in a moth population and suggested a plausible mechanism. The pine trees in one half of a wood had been felled by a storm and replaced by birch (*Betula*) trees. Fifteen years later, the frequencies of light and dark moths in the two parts of the wood were observed to differ. In the replanted part, pale forms constituted 95% of the moths and dark forms only 5%. In the old part where the pine trees still stood, the reverse ratio was observed. Haldane noted that in the replanted part of the wood, dark moths were more conspicuous on the pale bark of the birch trees and were probably consumed by birds and bats more often than pale forms, as could be verified by counting the remains of wings on the forest floor.

E.B. Ford suggested that the melanic forms had not arisen *de novo* by some mutagen in the polluted environment: the melanic allele was present throughout, albeit at very low frequencies. He considered that the reason why the dominant black forms were not more frequent was that dark moths were very conspicuous when they settled on the lichen-covered trunks of trees. They were picked out more easily than pale forms by predators searching for prey, and the dark alleles were thus selected against. The dark forms gained an advantage when pollution caused the death of the lichens and the environment was blackened by soot: the pale moths were then much more conspicuous against the black background and the selection pressures were reversed. The dark allele, being dominant, quickly increased in frequency when selected for[1432].

A series of large-scale experiments were carried out by H.B.D. Kettlewell[1433] in an attempt to validate the hypothesis that natural selection really works. The first study established the protocol for the studies that followed.

A large sample of moths was collected in a pollution-free wood. The majority of the moths were typical [=pale-colored], with melanics comprising less than 5% of the sample. Several hundred

[1430] Chapter 27; Haldane 1924.
[1431] Haldane 1993 (1932).
[1432] Ford 1965; Creed 1971.
[1433] Kettlewell 1958; 1961.

laboratory-reared moths, marked by paint for identification, were released into a wood clearing before sunset. About half of the released moths were dark, the other half the typical, pale form. Cages with female moths were placed in the clearing to attract the moths to the area. Early the next morning, the researchers positioned themselves around the clearing, armed with binoculars and cameras, and identified and photographed every moth they saw captured by birds. Of the total 190 captures recorded that morning, 164 were melanics and only 26 were typical pale.

The experimental protocol was repeated twice in the polluted woods near Birmingham. The results were reversed: most of the moths captured by birds were typical, and dark forms were a minority. Similar experiments were carried out in other locations in Britain in the following years.

The results supported the hypothesis that bird predation accounted for the change in the frequencies of the melanic moths: the predators were attracted to the more conspicuous prey. When the background was light and not polluted, they thus selected against the dark moths. When the background became dark, the paler typicals were more easily eliminated.

Supporting evidence for selective predation as a cause of "industrial melanism" was obtained many years later. In 1956, the British parliament passed an anti-pollution law banning the use of coal for heating in homes and factories, and coal was gradually replaced by oil products[1434]. Air pollution was considerably reduced in the following years. Consequently, the frequency of the melanic moths declined with time[1435].

Geographic Variation in Melanism and Air Pollution

In the Liverpool area, a clear pattern of decreasing melanic frequencies was obtained in samples: from 85-90% near the city to 5-10% 50 km farther away. To test the predation hypothesis, freshly-killed (frozen) melanic and typical moths (*Biston betularia*) were glued to tree trunks in a "natural" position in a number of woods along a transect from the urban to the surrounding rural area of south Wales. The moths were counted daily, and missing ones – presumably removed by predators – were replaced with similar phenotypes. The numbers of missing melanic moths were highest in the rural areas and lowest near the city. The results qualitatively supported the suggestion that bird predation intensity depended on the appearance of the habitat (i.e., the amount of pollution). However, when the melanic frequencies at the observation sites along the transect were used in computer simulations, predation alone was insufficient to explain the pattern[1436]. To obtain a better fit of the observed to the simulated patterns, it was necessary to postulate the effects of other variables on melanic frequencies. Surveys of melanic frequencies were carried out in Denmark, Sweden, and Finland. Although the findings supported qualitatively the correlation of melanism with environmental pollution, in none of these surveys could the pattern be fully explained by selective predation alone. Other mechanisms were then suggested to explain the departure of the observed geographical patterns of melanic frequencies from predictions based on predation[1437].

Experiments were carried out to determine which of the components of the polluted air were responsible for the disappearance of the lichens from the tree trunks. Six variables of lichen cover and eight environmental variables were measured on oak trees at 104 sites in England. The correlations of these variables with melanic frequencies (and with each other) were estimated statistically. Melanic frequencies were negatively correlated with the concentration of sulfur dioxide (SO_2) in the air, but not with the amount of soot, the component intuitively responsible for the color of the habitat[1438].

[1434] The arguments for enacting the anti-pollution law were not based on scientific or public health considerations. They were a response to the death of some valuable prize cows in an agricultural exhibition and the cancellation of opera performances because the audience could not see the stage for the smog. (Berry 1990, p. 308).

[1435] Cook et al. 1986.

[1436] Mani 1980.

[1437] Liebert and Breakfield 1987.

[1438] Cook et al. 1986.

Thus the predation hypothesis, though appealing, does not provide a complete solution for the spread of melanism. It cannot be denied that predation plays a role, but local conditions too may determine the relative importance of the environmental variables responsible for the phenomenon.

Polymorphism in the European Land Snail

The European land snail (*Cepaea nemoralis*) seemed a promising candidate for field studies of selection. Snails are common and easily collected. The genetics of phenotypic variation in snail shells revealed genes determining basic shell color (yellow or pink) and the presence of five stripes on the shell[1439].

Sheppard and Cain[1440] reported that certain birds – (*Turdus ericitorum*) – were unable to break the shells of the snails they collected with their beaks, but instead slammed them against stones ("anvils") that occurred in their territory, in order to crush them. The snail's shell fragments remained near the anvils. The frequencies of the snail phenotypes reconstructed from the fragments were compared with random samples of snails from the field. The findings suggested that the birds collected predominantly those phenotypes that were most conspicuous against the background color of the surrounding grass. These results were considered evidence that selective predation affected the genetics of the snail population.

Land snails are very common in Britain, and geographic polymorphism in snails was studied intensively over forty-five years, yielding more than 300 scientific papers. The conclusion was that bird predation was insufficient to account for the geographic distribution of the polymorphic snail phenotypes. The birds only infrequently preyed on the snails, and the role of background color was not confirmed[1441]. One suggestion was that snail colonies were established by a few founder individuals carrying a particular genotype, which happened to arrive at a new suitable site and then expanded their range by dispersal to create a "mosaic" geographical pattern of patches of snails of similar phenotypes ("area effects"[1442]). No convincing correlation was found between snail morphology or electrophoretic variation, and habitat characteristics. More puzzling was the absence of reproductive barriers between snail populations from very distant localities: although dispersal distances of snails are typically very limited, laboratory crosses were fertile even between individuals from different countries[1443].

Melanism in the Two-Spot Ladybird

Another organism that seemed promising for the study of natural selection in the field was the two-spot ladybird, *Adalia bipunctata* (Coccinellidae). This polymorphic beetle is widespread in England and the Netherlands. While the typical form has red elytra with two black spots – one on each wing (hence the Latin and English names), forms with four or six black spots are also found, and in some individuals the spots coalesce and cover the entire wing[1444]. Such individuals are commonly referred to as melanics. The genetic basis of the polymorphism in *A. bipunctata* is complex – up to 12 dominant alleles were found to affect the numbers of spots on the elytra.

In several parts of England, particularly in industrial areas, up to 97% of these beetles were melanics. This similarity with industrial melanism was suggestive of a similar mechanism: the involvement of natural selection via predation. However, this interpretation seems to be incorrect. Coccinellids exude a yellow repellent (and toxic?) fluid from their coxae when caught or disturbed. They are generally rejected by captive birds when offered as food, although frozen beetles given

---

[1439] Cain 1971.
[1440] Sheppard & Cain (in Ford 1965).
[1441] Review in Jones et al., 1997; Goodhart 1987.
[1442] Cameron 1992.
[1443] Johnson et al. 1984.
[1444] Creed 1971.

to hungry birds were sometimes eaten regardless of their color. Red and black colors are both considered warning colors in insects. An alternative interpretation was thus called for.

Research on melanic polymorphism in *A. bipunctata* continued intermittently for many years. A strong correlation was reported with the amount of smoke particles in the air. Moreover, following air-pollution control regulations in Birmingham, between 1960 and 1969 melanic frequencies were reduced.

In beetle pairs collected *in copula,* the frequencies of melanics were higher than in randomly-collected beetles: although this suggested possibly non-random mating, this could not be confirmed in controlled experiments.

The beetles aggregate in large numbers in crevices (in tree bark and even in buildings) in the Netherlands. Seasonal differences in morph frequencies suggested that melanics were selectively favored in early spring and selected against in summer. The advantage in spring was that the melanics absorb more of the sun's radiation on cool days and warm up more quickly than the typical red beetles, enabling them to start feeding and mating earlier in the day and the season[1445]. This suggestion was supported in extensive ecological studies in the Netherlands[1446, 1447].

Insecticide Resistance

The resistance of insects to insecticides poses an alarming threat for agriculture and modern food production. (Similar, and even more threatening for disease control and epidemiology, is the resistance of microorganisms to antibiotic drugs). Examples of pest control failure due to resistance were already being reported in the early 20th century[1448], but the scale of the phenomenon increased following World War Two with the introduction of DDT and the increasing use of pesticides to protect crops[1449]. Thousands of research papers were published over the years, recording cases of resistance and suggesting alternative ways to protect the crops. A few early examples are listed here.

As Dobzhansky had recognized in 1937, insecticide resistance is one of the best examples of an evolutionary process. As in the case of industrial melanism, the genetic alleles conferring resistance to some individuals are present in low frequencies. When the pesticides are applied, the frequencies of resistant individuals increase, because all the other individuals are selected against.

The progress of this evolutionary process is measurable by the time – often very short – between the first application of a new pesticide and the first report of control failure[1450].

It is not clear how many genes are involved in conferring resistance. Some authors have suggested that resistance may be polygenic[1451]. In the field, selection is applied to very large populations and the agricultural control treatment aims at killing all of the pests: the rare ones that do survive may be those that are extremely resistant. Laboratory populations, by contrast, are small and the range of genetic variants available in any one population is inherently limited. Therefore, if resistance is discovered in a laboratory population, it is likely to be due to the cumulative effects of several genes[1452].

Physiologically, resistance may result when enzymes (coded for by resistance alleles) break down the poison and convert it to harmless products. In aphids and mosquitoes, organophosphorus

1445 O'Donoald & Muggleton 1979.

1446 Brakefield 1984.

1447 In a comparative study in the Sinai highlands (near the monastery of Santa Catherina), a northern black goat, a white goat, and a native mountain goat (ibex) were held together and examined both behaviorally and physiologically. It turned out that the black goat started feeding earlier than the white one in the early morning. On bright days in winter, the black color enabled the goat to save up to 28% of the energy required by the white goat to heat its body (Dmi'el et al. 1980).

1448 Forgash 1984.

1449 In Huxley 1942.

1450 Forgash 1984.

1451 Firko & Hayes 1990; Roush & McKenzie 1984.

1452 Macnair 1991.

insecticides that penetrate through the insect's cuticle are metabolized by esterases and phosphatases. The quantity of relevant enzymes is several times higher in resistant than in susceptible individuals. The production of large numbers of the enzyme molecules, which facilitate survival of these individuals, is due to multiple copies of the gene coding for the relevant allele – a process of gene amplification[1453].

Evolution of Heavy-Metal Tolerance in Plants

Evolutionary processes in plant populations are slower than in animals, due to the plant's longer generation times and slower rates of reproduction (e.g., flowering only once a year; it may take years for a seed of a tree to reach maturity). Cases have nonetheless been documented in England of rather rapid evolution in plants, over a span of less than 100 years. Those studies concerned plants growing on heaps of earth (called tailings) near abandoned copper and zinc mines in Wales. Both metals are toxic to plants: experiments have shown that a concentration of 5 ppm (parts per million) of copper is lethal for many plants. The mines were abandoned when the quantity of ore recovered did not justify continuing their operation (the dates are on record). However, the tailings – heaps of waste earth excavated while the mine was active – still contained metal concentrations several hundred times higher than the lethal concentration to plants. Nevertheless, after the mines were abandoned, plants from neighboring fields succeeded in colonizing the tailings[1454].

Tolerance of the plants to copper or zinc (in different studies) was estimated according to the length of the longest root of seedlings grown in nutritive solutions with known concentrations of metals added. Seedlings from the mine plants were tolerant to metal concentrations that were lethal to plants from the neighboring fields. Moreover, copper-tolerant plants spread away from the mine in the direction of the prevailing winds. This suggested that although a variety of seeds from the fields were blown in that direction, only the tolerant seeds were able to establish at the newly-colonized sites[1455]. A case illustrating this process was observed near a metal refining factory in Wales. The factory, built in 1900, caused severe environmental pollution that killed the vegetation around it. The most potent components of the polluted air from the chimneys were lead, copper, and zinc. Grass was planted in the foreground using seeds from nearby grasses, and after 70 years the grass lawn appeared continuous. Tolerance analysis revealed that seedlings from the "old lawn" were far more copper-tolerant than seedlings from the field[1456].

Predation in Natural Fish Populations

Fishermen in different countries are reporting an alarming trend of decreasing yield – fewer and smaller fish of commercial species, such as tuna and salmon. It was suggested that overfishing by man is an extreme form of predation, which affects not only the abundance of the fish populations but also their genetics.

Populations of guppies (*Poecilla reticulata*) – a popular, colorful aquarium fish – were studied to provide support for the suggestion that predation can select for smaller fish. The guppy can be bred inexpensively in the laboratory and the females produce live young which can easily be isolated and counted. Natural populations of guppies were manipulated at their native sites, several small, fast-flowing mountain streams in Trinidad[1457]. The streams run through a number of waterfalls, which create ecological barriers between guppies living above or below the waterfalls. Only one predatory fish is found above the waterfalls, while below them several natural predators of guppies exist. The scientists postulated that predation pressure was stronger in the ponds below the waterfalls.

[1453] Bunting and van Emden 1978; Devomnshire & Field 1991; Pasteur & Raymond 1996).
[1454] Bradshaw 1971, in Creed 1971.
[1455] Bradshaw 1971.
[1456] Wu et al. 1975.
[1457] Reznichk and Ghalambor 2005 a, b.

To test how predation affected the guppies, fish were transferred from below to above the waterfalls. Several different streams were used as independent replicates. Fish were sampled above and below the waterfall in each stream, four, seven and 11 years after the transfer. The sampled fish were brought to the laboratory, where fish size, female fecundity, and offspring weight were measured.

The transferred females – released from the heavy predation – reproduced more frequently, but produced smaller offspring compared with the original fish from above the falls. The results thus confirmed that a genetic (evolutionary) change had occurred[1458].

[1458] Reznick and ghalmbor 2005b

36

Motoo Kimura and the Neutrality Hypothesis

Motoo Kimura (1924-1994) studied biology in Japan and was interested in botany, but was attracted to the mathematical theories of population genetics and evolution, in particular those of Sewall Wright and J.B.S. Haldane. To comprehend these theories he studied mathematics informally, and developed an important mathematical model of populations interconnected by migration (the "stepping-stone" model) as an extension of Wright's island model. Kimura received a Fulbright scholarship for advanced study in the USA, where he worked with the American geneticist James Crow at the University of Wisconsin, receiving his PhD in 1956. His association with Crow lasted several years and they co-authored a book[1459]. Kimura returned to Japan, and continued his theoretical work in population genetics and evolution at the Institute of Genetics in Mishima.

The enormous extent of heritable protein (enzyme) variation in natural populations then discovered (1960-1970s) by electrophoretic methods, was difficult to explain by natural selection. This led Kimura to suggest that molecular variation is mostly selectively neutral: its magnitude is maintained by random processes – mutation rate, genetic drift, and random fixation. This hypothesis attracted many very active supporters. Kimura's second book [1460]established his position as the originator of this theory.

The neutrality hypothesis became one of the most important contributions to evolutionary biology in the 20th century, also becoming the center of controversy in 1969 with the publication of an article with the provocative title, "Non-Darwinian Evolution"[1461]. The controversy split the biologists into two camps: "neutralists" and "selectionists". It stimulated theoretical research in population genetics and much experimental research in evolutionary ecology of natural populations for the following 20 years or more. Thousands of research papers on genetic variation in hundreds of animal and plant species were published world-wide. The dispute was generally decided by the end of the 20th century (in favor of the neutralists, at least at the molecular level), but continued to stimulate thinking about evolution.

Modeling Individual Genomes

In the early 20th century, two conceptual models of the genetic structure of an individual were discussed. The classical or "typological" model assumed that most of the loci in the genome of a diploid individual were in a homozygous condition. Rarely, a mutation will occur, creating a heterozygote at some locus by changing one allele for another. Most of these mutants are deleterious, and when they are exposed as homozygotes by recombination, they tend to be removed by natural selection. This model agreed with the observed scarcity of phenotypic mutants in any population and with the general concept of the "wild type"[1462].

The alternative, "neo-Darwinian", model suggested that most of the loci in an individual genome are *heterozygous*. Each new mutation creates a different allele, differing from those previously

[1459] Crow & Kimura 1970 "Introduction to Population Genetics Theory".
[1460] Kimura 1983a, "The Neutral Theory of Molecular Evolution".
[1461] King & Jukes 1969.
[1462] Chapter 28.

existing (the "infinite-allele" model). Each mutation affects individual fitness – either positively or negatively. According to this model, the variation in the population is maintained by some form of balancing natural selection.

The Neutrality Hypothesis

The "neutralists" rejected this last assumption. Following Kimura, they considered that most of the mutations in the DNA do not affect the phenotypic fitness of the individuals at all [are "neutral" as regards natural selection]. Consequently, natural selection is not the major driving force of evolution, unlike the Darwinians' assumption. Evolution depends on the mutation rate and proceeds by random processes – genetic drift and random fixation – that were formerly considered of minor effect[1463].

Supporting Evidence for the Neutrality Hypothesis

The genetic code: the genetic evidence accumulated in the middle of the 20th century – following the discovery of the structure of DNA by Watson and Crick – confirmed that natural selection does not work on the genotype, encoded by the nucleotides in the DNA, but on its phenotypic expression in the proteins. Simple calculations showed that the code for a given amino acid must contain more than two nucleotides ("letters"): combinations of two "letters" yield only 16 codes (4^2) for the 20 most common amino acids. A three-letter code yields 64 combinations (4^3) – far too many. When the codes for the 20 amino acids were determined[1464] ("The Universal Code"), it became clear that the genetic code is "degenerate", in the sense that some amino acids have more than one codon, creating ambiguity. For example, the four codons CGU, CGC, CGA, CGG all code for arginine. Such codons are referred to as "*synonymous*". A synonymous mutation will not change the amino-acid composition of the protein coded for by the mutated DNA. It is therefore inherently neutral, undetectable by natural selection.

Sixty-one of 64 codons code for amino acids (three are "stop" codons). Each codon can theoretically be changed by point mutation into nine codons. Therefore 549 codons (61 x 9) can be formed by point mutations. It was calculated that 134 of these (24.4%) are synonymous.

Homologous Proteins: Homologous proteins are similar in molecular amino-acid sequence and structure and fulfil the same function in different species of organisms.

One of the first proteins to be analyzed was Cytochrome C, a co-enzyme in the metabolic pathway. The protein molecule is 104 amino acids long and has a central heme group with oxygen-carrying capability. The protein was isolated and sequenced in 38 species of organisms – among them yeast, insects, amphibians, reptiles, mammals and Man[1465]. It turned out that at 35 of the 104 side-chain sites, the same amino acid was located in all 38 organisms tested. At the other 69 sites, 2 to 9 different amino acids were found to occupy each site. The replacement of one amino acid by another at these sites does not affect the essential properties of the protein, and such replacements are therefore "neutral". Many other examples of this kind were reported and discussed[1466].

Gene Duplication and Pseudo-genes: A mutation in an active gene may impair or inhibit the normal function of the resulting protein, and may be harmful for the organism. However, if a given DNA sequence (gene) is duplicated, the genome will contain two (or more) copies of that sequence. If a mutation occurs in *one* of the copies, and the other copy still produces an intact product, the organism may not be negatively affected. Moreover, the duplicate copy of the gene is free

1463 "The Sewall-Wright Effect". See Chapter 30.
1464 Strickberger 1968, Graur & Li 2000.
1465 Dickerson 1972.
1466 Ohta 1974; Kimura & Ohta 1971.

to accumulate more deleterious mutations without affecting the fitness of its carrier, since it is in any case "neutralized". Nuclear DNA contains many repetitive sequences[1467]. The functional significance of this mass of DNA is not clear[1468], but mutations in repetitive DNA sequences are unlikely to harm the individual so long as some of the sequences retain the original functionality.

To the category of sequences in which mutations may accumulate without causing harm one may add *introns* – non-functional spacers between functional genes (exons); and *pseudo-genes* – sequences of previously functional genes rendered non-functional by a harmful mutation (pseudo-genes are similar in sequence to functional genes except at the site of the disabling mutation). The replacement rate of nucleotides in pseudo-genes is higher than in functional genes[1469]. These facts support the suggestion that mutations in inactive parts of the genome are selectively neutral.

The Principles of the Neutrality Hypothesis:[1470] A population may contain large amounts of genetic variation, which is maintained without the involvement of natural selection.

1. This variation is of two kinds: (a) *neutral* (or equivalent) mutations in which the replacement of one allele by another has no effect on the fitness of the individual; and (b) *"transient"* mutations, which do have effects on fitness but persist in the population either "on their way" to extinction or, alternatively to fixation, and are detected when the same population is sampled sequentially. In very large populations, the fate of mutations of type b is determined by their selective values; but in finite, small natural populations their fate is determined by genetic drift and random fixation.
2. In a population of size N, per site, $2N\mu$ mutations are expected on average (when μ is the mutation rate at the site). Of these, $1/2N$ will be fixed, and the rest will be lost.
3. The time to fixation of a positive (favorable) mutation is very long – in the order of 4N generations. The larger the population, the longer the time required for fixation.
4. Genetic variation accumulates by past mutations and is simply an historical result of the processes of evolution, either selective or random, that the population has undergone in the past. There is no way of estimating the value of this variation for the persistence and future of the population[1471].

Neutrality of Variation in Evolutionary Thinking

As stated above, the neutrality hypothesis caused a major split among evolutionists. The strict neo-Darwinians of the 20th century, who strongly opposed the idea that not all characters of an organism affected fitness, ignored the fact that the issue was not new. Darwin had considered and did not reject the idea that some characters are "neutral":

> Variations neither useful not injurious would not be affected by natural selection and would be left either a fluctuating element, as perhaps we see in certain polymorphic species, or would ultimately become fixed, owing to the nature of the organism and the nature of the conditions[1472].

In his "Descent of Man", Darwin reflected on this idea and seems to have had second thoughts:

> I am convinced… that very many structures which may now appear to us useless, will hereafter be proved to be useful, and will therefore come within the range of natural selection. Nevertheless, I did not formerly consider sufficiently the existence of structures

1467 Graur & Li 2000.
1468 Orgel & Crick 1980.
1469 Graur & Li 2000
1470 Nei 1975.
1471 Nei 1982.
1472 Darwin, Origin of species, p. 58.

> which, as far as we can judge, are neither beneficial nor injurious. And this I believe to be one of the greatest oversights as yet detected in my work[1473].

The geneticist Richard Goldschmidt suggested that small, even somewhat deleterious mutations with no lethal effects, may accumulate in the individual until sufficient change is induced to cause the appearance of a "hopeful" monster[1474].

Even Ernst Mayr, in his early book, expressed the idea that some morphological variants may have no selective value. As examples he considered the banding patterns on snail shells:

> There is no reason to believe that the presence or absence of a band on a snail's shell would be a noticeable selective advantage or disadvantage... It should not be assumed that all the differences between populations and species are purely adaptational and that they owe their existence to their superior selective qualities. Many combinations of color patterns, spots and bands, as well as extra bristles and wing veins, are probably largely accidental[1475].

However, in his later (1963) book, Mayr reconsidered the possibility that neutral *genetic* changes are unlikely, contending that different genotypes may be selectively equivalent[1476].

The Selectionists' Arguments

It was not incidental that the most active supporters of the neutrality hypothesis were statisticians or geneticists with a good background in mathematics. If the genetic variation is merely random, then models based on probability theory are applicable [and many such models were in fact published]. If selection is involved, mathematical modeling becomes much more complicated.

The attack on neutralism was led by ecologists. Hundreds of research papers were published. The attack took place from two directions. One was to conduct surveys on electrophoretic variation in large numbers of natural populations, while in parallel measuring important environmental parameters (like temperature or precipitation) at the sampled localities. If genetic variation was correlated with an environmental parameter, so the argument went, it could be the result of adaptation by natural selection.

The other approach was to conduct laboratory studies of specific enzyme systems in which electrophoretically-different strains could be bred and compared [this naturally limited the number of organisms that could be used – most of these studies used *Drosophila* strains]. The aim was to show that individuals carrying different alleles (detected by electrophoresis) were also different in fitness, especially in specific environments, indicating that these differences were not "neutral".

> The analysis suggests that the amounts of genetic polymorphism and heterozygosity vary non-randomly between loci, species, habitats and life zones, and are correlated with ecological heterogeneity. Natural selection, in some form, may often be the major determinant of genetic structure and differentiation[1477].

The second approach to a counter-attack on neutralism is illustrated by the analysis of the alcohol-dehydrogenase (ADH) system in laboratory populations of *Drosophila*[1478]. *Drosophila* larvae inhabit ripe and rotting fruit and seepage pools in wineries, where the sugar in the fruit is converted to alcohol by yeast fermentation. Alcohol is toxic to insects in general (and to bacteria, which makes

1473 Darwin 1874, p. 76.
1474 Goldschmidt 1933.
1475 Mayr 1942, p. 75; Ibid. p. 86.
1476 Mayr 1963, pp. 213-214.
1477 Nevo 1978.
1478 e.g., Kamping and Van Delden 1978, 1979, 1983.

it useful as a disinfectant). In *Drosophila*, ethanol is converted to acet-aldehyde [which is even more toxic] by ADH, and further by aldehyde-oxidase (AO) into the harmless CO_2 and water.

Genetic variation in these two enzyme system was investigated in detail. Two alleles of ADH were studied in particular: ADH – f and ADH – s [the two proteins differ in the migration distance on the gel. The fast (f) allele is the more active]. The same alleles were found in different populations of *Drosophila*, and strains homozygous at these loci were bred for laboratory experimentation. When the two strains were introduced in competition (as larvae), on experimental media containing different concentrations of various alcohols, the frequency of individuals carrying the ADH – f alleles increased with time. No change was observed on media without alcohol. In the presence of alcohol, the f allele had a selective advantage[1479].

Another enzyme system that was used to demonstrate fitness value for electrophoretic variants in *Drosophila* was amylase. In the regular digestive process of starch-containing foods, amylases break down starch (a poly-saccharide) into the component di-saccharide maltose, which is further digested into CO_2 and water. [This system is not essential for *Drosophila*, which can be bred on media lacking starch – as long as they contain maltose].

On gels containing starch, the enzyme is visualized after staining with iodine – the migrating proteins digest the gel and leave "holes" while the undigested gel stains blue-black. In particular, two alleles were selected for study and bred as homozygous strains – $AMY^{4\text{-}6}$ and AMY^{1} (the former is more active than the latter). When placed in competition with AMY^{1} on media containing starch, larvae carrying $AMY^{4\text{-}6}$ had an advantage and their frequency increased. On maltose media, no change was observed[1480].

In one exceptional case, an AMY-"null" strain of *Drosophila* – which had no amylase activity on the gels but was viable on maltose media – was introduced in competition with a normal strain on starch media. Surprisingly, the "null" strain did quite well in competition. Further investigation revealed that the normal larvae produced excessive amounts of amylase, and the surplus was excreted into the medium. The enzyme digested some of the starch into maltose, which the "null" larvae could use to survive.

In conclusion, these laboratory experiments demonstrated that some electrophoretic variants do affect the fitness of their carriers. This variation cannot be considered "neutral".

The amount of published material on the issue of "neutrality" is far too great to do it justice in this book.

[1479] Review in Van Delden 1982.
[1480] DeJong & Scharloo, 1980.

Closing Comments

The methodology and technical developments in the late 20th century enabled detailed studies of variation and evolution in the DNA of many organisms, including Man. The DNA methods quickly replaced electrophoresis, and the neutralist-selectionist controversy was settled in favor of the former.

Thus led to the development of a new field of research: Molecular Evolution (see Graur 2016[1481] for an up-to-date review).

The Theory of Evolution evolved with time, absorbing some ideas – and rejecting others. The methods of research on evolution turned from philosophical ("armchair") discussions to experimental work, which varied as technology developed. Throughout it all, however, for nearly 170 years, the Theory of Evolution has been the best explanation for the origin and development of the biological world. The concept of natural selection – suggested by Darwin – has remained valid and central to biological thinking, even when rival mechanisms of population change have gained prominence. It seems that the Theory is here to stay.

[January 14, 2020]

[1481] Graur, D., 2016, Molecular Genome Evolution; Sinauer.

References

Agassiz, L. 1850. The diversity of origins of the human races. Christian Observer 49: 1-36.

Agassiz, L. 2004 (1857). Essay on Classification. Dover, New York.

Agassiz, L. and Gould, A.A. 1875 (1848). Principles of Zoology, Part I: Comparative Physiology. Sheldon & Co., New York.

Arbuthnot, J. 1710. An argument for divine providence, taken from the constant regularity observed in the births of both sexes. Transactions of the Royal Society, London.

Ayala, F.J. and Kiger, J.A. 1980. Modern Genetics. The Benjamin/Cummings publ., California. p. 1.

Ayala, F.J., Powell, J.R. and Dobzhansky, T. 1971. Enzyme variability in the *Drosophila willistoni* group. II: Polymorphisms in continental and island populations of *Drosophila willistoni*. Proceedings of the National Academy of Science, USA, 68: 2480-2483.

Baker, H.G. 1986. In the introduction to Darwin's "different forms of flowers…". University of Chicago Press.

Barlow, N. (ed.). 1958. Autobiography of Charles Darwin, 1809-1882. Collins, London.

Benson, K.R. 2001. T.H. Morgan's resistance to the chromosome theory. Nature Reviews, 2: 469-474.

Berry, R.J. 1990. Industrial melanism and peppered moths (*Biston betularia* (L.)). Biological Journal of the Linnaean Society, 39: 301-322.

Bibby, C. 1960. T.H. Huxley: Scientist, Humanist, and Educator. Horizon Press, New York.

Blyth, E. 1835. The variety of animals. *In*: Eiseley, L. 1981. Darwin and the Mysterious Mr. X. Harvest-HBJ, New York, pp. 97-111.

Blyth, E. 1837. Seasonal and other changes in birds. *In*: Eiseley, L. 1981. Darwin and the Mysterious Mr. X. Harvest-HBJ, New York, pp. 112-140.

Bodmer, W 2003. R.A. Fisher, statistician and geneticist extraordinary: a personal view. International Journal of Epidemiology, 32: 938-942.

Bonnet, C. 1782. Abridgement of the contemplations of Nature, p. 7.

Brackman, A.C. 1980. A Delicate Arrangement. Times Books, New York.

Bradshaw, A.D. 1971. Plant evolution in extreme environments. *In*: Creed, R. (ed.). Ecological Genetics and Evolution. Blackwell, Oxford, pp. 20-50.

Brakefield, P.M. 1984. Selection among clines in the ladybird *Adalia bipunctata* in the Netherlands: a general mating advantage to melanics and its consequences. Heredity, 53: 37-49.

Brumby, M. 1979. Problems in learning the concept of natural selection. Journal of Biological Education, 13: 119-122.

Brumby, M. 1984. Misconceptions about the concept of natural selection by medical biology students. Science Education, 68: 493-503.

Bunting, S.W. and Van Emden, H.F. 1980. Rapid response to selection for increased esterase activity in small populations of an apomictic clone of *Myzus persicae*. Nature, 285: 502-503.

Bush, G.L. 1978. Modes of animal speciation. Annual Review of Ecology and Systematics, 6: 339-364.

Cain, A.J. 1971. Colour and banding morphs in subfossil samples of the snail *Cepaea*. *In*: Creed, R. (ed.). Ecological Genetics and Evolution. Blackwell, Oxford, pp. 65-92.

Cameron, R.A.D. 1992. Change and stability in *Capaea* populations over 25 years: a case of climatic selection. Proceedings of the Royal Society, London, B, 248: 181-187.

Chambers, R. 1994 (1844). Vestiges of the Natural History of Creation. University of Chicago Press, Chicago.

Connolly, T. 2016. Flower porn: form and desire in Erasmus Darwin "the loves of plants". Literature Compass, 13 (10): 604-616.

Cook, L.M., Mani, G.S. and Varley, M.E. 1986. Post-industrial melanism in the peppered moth. Science, 231: 611-613.

Cope, E.D. 1894. The energy of evolution. American Naturalist 28: 205.

Costa, J.T. and Ross, K.G. 1994. Hierarchical genetic structure and gene flow in macrogeographic populations of the eastern tent caterpillar (*Malacosoma americanum*). Evolution, 48: 1158-1167.

Creed, R. (ed.) 1971. Ecological Genetics and Evolution. Blackwell, Oxford.

Crick, F. 1981. Life Itself: Its Origin and Nature. Simon and Schuster, New York.
Crow, J.F. and Kimura, M. 1970. An Introduction to Population Genetics Theory. Harper & Row, New York.
Cuvier, G. 1995. Historical Portrait of the Progress of Ichthyology, from its origin to our own time. *In*: Pietsch, T.W. (ed.). Johns Hopkins University Press. Baltimore.
Darwin, C. 1844. Darwin's "draft", In van Wyhe 2005.
Darwin, C. 1859. First edition of the Origin. Murray London (facsimile).
Darwin, C. 1873. The Expression of Emotions in Man and Animals. Appleton, New York.
Darwin, C. 1874. The Descent of Man, and Selection in Relation to Sex. 2nd ed. Hurst & Co., New York.
Darwin, C. 1875. Insectivorous Plants. Murray, London.
Darwin, C. 1875. Climbing Plants. Murray, London.
Darwin, C. 1890 (1842). On the Structure and Distribution of Coral Reefs and Geological Observations on the Volcanic Islands and Parts of South America. Minerva Library, Ward, Lock and Bowden, London.
Darwin, C. 1891 (1877). The Effects of Cross- and Self-Fertilisation in the Vegetable Kingdom. 3rd ed. Murray, London.
Darwin, C. 1898a (1868). The Variation of Animals and Plants Under Domestication. Vols. I, II. Appleton, New York.
Darwin, C. 1898b (1872). The Origin of Species by Means of Natural Selection, or the Preservation of Favoured Races in the Struggle for Life. 6th ed. Murray, London.
Darwin, C. 1904 (1862). The Various Contrivances by which Orchids are Fertilized by Insects. 2nd ed. Murray, London.
Darwin, C. 1912 (1845). Journal of Researches into the Natural History and Geology of the Countries Visited during the Voyage Around the World of H.M.S. Beagle, under the Command of Captain FitzRoy Esq., R.N. Murray, London.
Darwin, C. 1937 (1845). The Voyage of the Beagle. 2nd ed. Harvard Classics, F.J. Elliott & sons, New York.
Darwin, C. 1952. The Origin of Species, The Descent of Man, and Selection in Relation to Sex. The Modern Library, New York.
Darwin, C. 1985 (1881). The Formation of Vegetable Mould, through the Action of Worms, with Observations on Their Habits. Facsimile. University of Chicago Press, Chicago.
Darwin, C. 1986 (1877). The Different Forms of Flowers on Plants of the Same Species. University of Chicago Press, Chicago.
Darwin, C. 1966 (1981). The Power of Movement in Plants. (Facsimile of Appleton & co.) De Capo press, New York.
Darwin, C. 2003 (1879). The Life of Erasmus Darwin. Cambridge University Press, Cambridge.
Darwin, E. 1803. The Temple of Nature or the Origin of Society: a poem with philosophical notes. T. Bensley, London.
Darwin, F. 1887. The Life and Letters of Charles Darwin, Vols. I-III. 3rd ed. Murray, London.
Darwin, F. 1904. The Life and Letters of Charles Darwin. Appleton, New York.
Dawson, P.S. 1970. A further assessment of the role of founder effects in the outcome of *Tribolium* competition experiments. Proceedings of the National Academy of Science, USA, 66: 1112-1118.
De Jong, G. and Scharloo, W. 1980 (1976). Environmental determination of selective significance or neutrality of amylase variants in *Drosophila melanogaster*. Genetica 84: 77-94.
De Vries, H. 1889. Intracellular Pangenesis. [on internet: www.esp.org/ books/diaries/pangenesis]
De Vries, H. 1912 (1904). Species and Varieties: Their Origin by Mutation. The Open Court Publ., Chicago.
Demarest, B. and Wolfe, C.T. 2017. The organism as reality or as fiction: Buffon and beyond. History (HLPS) 39: 1-16
Devonshire, A.L. and Field, L.M. 1991. Gene amplification and insecticide resistance. Annual Review of Entomology, 36: 1-23.
Dewbury, A. 2007. The American school and scientific racism in American anthropology. Histories of Anthropology Annual, 3: 121-147.
Dickerson, R.E. 1972. The structure and history of an ancient protein. Scientific American, 226: 58-72.
Dietrich, M.R. 2003. Richard Goldschmidt: hopeful monsters and other "heresies". Nature Reviews, 4: 68-74.
Dobzhansky, T. 1955. Evolution, Genetics, and Man. J. Wiley & Sons, New York.
Dobzhansky, T. 1962. Mankind Evolving. Yale University Press, New Haven.
Dobzhansky, T. 1970. Genetics of the Evolutionary Process. Columbia University Press, New York.
Dobzhansky, T. and Pavlovsky, O. 1957. An experimental study of interaction between genetic drift and natural selection. Evolution, 11: 311-319.
Dow, B.D. and Ashley M.V. 1998. High levels of gene flow in Bur oak revealed by paternity analysis using microsatellites. Journal of Heredity, 89: 62-70.

Drummond, H. 1896. The Ascent of Man. Hodder and Sloughton, London.
Edwards, A.W.F. 1974. The fundamental theorem of natural selection. Biological Reviews, 69: 443-474.
Eiseley, L. 1981. Darwin and the Mysterious Mr. X. Harvest-HBJ, New York.
Eldredge, N. 2005. Darwin's other books: Red and Transmutation notebooks, Sketch, Essay, and natural selection. PLoS Biology, 3: 1864-1867.
Falconer, C.D. 1960 (new edition 1981). Introduction to Quantitative Genetics. Ronald Press, New York.
Firko, M.J. and Hayes, J.L. 1990. Tools for insecticide resistance risk assessment: estimating the heritability of resistance. Journal of Economic Entomology, 83: 647-654.
Fisher, R.A. 1936. Has Mendel's work been rediscovered? Annals of Science, 1: 115-137.
Fisher, R.A. 1958 (1930). The Genetical Theory of Natural Selection. Dover, New York.
Ford, E.B. 1965. Ecological Genetics. Methuen, London.
Forgash, A.J. 1984. History, evolution, and consequences of insecticide resistance. Pesticide Biochemistry and Physiology, 22: 178-186.
Futuyma, D.J. 1983. Science on Trial: The Case for Evolution. Panthcon Book, New York.
Galton, F. 1889. Natural Inheritance. Richard Clay, London.
Galton, F. 1894. Discontinuity in evolution. *In*: Materials for the Study of Variation, treated with special regard to discontinuity in the origin of species. Macmillan, London, pp. 362-372.
Galton, F. 1962 (1869). Hereditary Genius. The World Publ., New York.
Galton, F. 1971 (1872). The Art of Travel. Stockpole Books, Harrisburg, Pennsylvania.
Galton, F. 2006 (1892). Finger Prints. Prometheus Books, Amherst, New York.
Gardner, M. 1957. Fads and Fallacies in the Name of Science. 2nd ed. Dover, New York.
Glass, B., Temkin, O. and Straus, W.L. (eds.) 1968. Forerunners of Darwin, 1745-1859. Johns Hopkins University Press, Baltimore.
Goldschmidt, R. 1933. Some aspects of evolution. Science, 78: 539-547.
Goldschmidt, R. 1940. The Material Basis of Evolution. Yale University Press, New Haven.
Goldschmidt, R.B. 1960. In and Out of the Ivory Tower. University of Washington Press, Washington.
Goldschmidt, R.B. 1982 (1940). The Material Basis of Evolution. Harvard University Press, Cambridge.
Goodhart, C.B. 1987. Why are some snails polymorphic and others not? Biological Journal of the Linnaean Society, 31: 35-58.
Gosse, P.H. 1998 (1857). Omphalos, an attempt to untie the geological knot. Ox Bow Press, Connecticut.
Gould, S.J. 1977. Ontogeny and Phylogeny. Belknap Press, Harvard.
Gould, S.J. 1981. The Mismeasure of Man. Norton, New York.
Gould, S.J. 1982. The Panda's Thumb: More Reflections in Natural History. Norton, New York.
Gould, S.J. 1982a. Piltdown revisited. *In*: The Panda's Thumb. Norton, New York, pp. 117-124.
Gould, S.J. 1982b. The uses of heresy. Introduction to: Goldschmidt, R. 1940. The Material Basis of Evolution. Harvard University Press, Cambridge.
Gould, S.J. 1985. The Flamingo Smile. Norton, New York.
Gould, S.J. 1994. Knight Takes Bishop. *In*: Bully for Brontosaurus. (in Hebrew, Dvir Publications) pp. 397-414.
Gould, S.J. and Eldredge, N. 1977. Punctuated Equilibria: the tempo and mode of evolution rediscovered. Paleobiology, 3: 115-151.
Grant, P.R. and Grant, B.R. 1995. The founding of a new population of Darwin's Finches. Evolution, 49: 229-240.
Graur, D. and Li, W-H. 2000. Fundamentals of Molecular Evolution. Sinauer, Amsterdam.
Graur, D., Gouy, M., and Wool, D. 2009. In retrospect: Lamarck's treatise at 200. Nature, 460: 688-689.
Haeckel, E. 1876. The History of Creation, or the Development of the Earth and its Inhabitants by the Action of Natural Causes. Appleton, New York.
Haeckel, E. 1902. The Riddle of the Universe, at the Close of the Nineteenth Century. Harper, New York.
Haldane, J.B.S. 1923. Daedalus, or science and the future. Paper read at the university of Cambridge.
Haldane, J.B.S. 1924. A mathematical theory of natural and artificial selection. Transactions of the Royal Society of Cambridge, 23: 3-41.
Haldane, J.B.S. 1929 (1927). On being the right size. *In*: Pritchard, F.H. (ed.). The World's Best Essays. Methuen, London, pp. 346-351.
Haldane, J.B.S. 1937. The effect of variation on fitness. American Naturalist, 71: 337-349.
Haldane, J.B.S. 1940a. Science and Everyday Life. Macmillan, New York.
Haldane, J.B.S. 1940b. A great Soviet Biologist. *In*: Science and Everyday Life. Macmillan, New York, pp. 134-138.
Haldane, J.B.S. 1941. The faking of genetical results. Eureka, 27: 21-24.
Haldane, J.B.S. 1993 (1932). The Causes of Evolution. Princeton University Press, Princeton.

Hoyle, F., and Wickramasinghe, N.C. 1978. Lifecloud, the Origin of Life in the Universe. Harper & Row, New York.

Hubby, J.L. and Lewontin, R.C. 1966. A molecular approach to the study of genic heterozygosity in natural populations. I. The number of alleles at different loci in *Drosophila pseudoobscura*. Genetics, 54: 557-594.

Huff, D. 1954. How to Lie with Statistics. Norton, New York.

Hutton, J. 1998 (1788). The Theory of the Earth. Kessinger Publishing, USA. p. 6.

Huxley, J. 1932. Problems of relative growth. Methuen, London.

Huxley, J. 1942. Evolution, The Modern Synthesis. George Allen & Unwin, London.

Huxley, J. 1957a. Transhumanism. *In*: Huxley, J. (ed.). New Bottles for New Wine. Harper, New York, pp. 41-59.

Huxley, J. 1957b. Man's place and role in nature. *In*: Huxley, J. (ed.). New Bottles for New Wine. Harper, New York, pp. 41-59.

Huxley, L. 1900. Life and Letters of Thomas Henry Huxley. Vols. I, II. Macmillan, London.

Huxley, T.H. 1858. On the agamic reproduction and morphology of Aphis. Transactions of the Linnaean Society, 22: 193-241.

Huxley, T.H. 1887. On the reception of the origin of species. *In*: Darwin, F. 1904, The Life and Letters of Charles Darwin. Appleton, New York. Vol. I: 1-11.

Huxley, T.H. 1890 (1870). Biogenesis and abiogenesis. *In*: Critiques and Addresses. Macmillan, London, pp. 218-250.

Huxley, T.H. 1890a (1871). Administrative Nihilism: *In*: Critiques and Addresses. Macmillan, London, pp. 3-32.

Huxley, T.H. 1890b (1870). Palaeontology and the doctrine of evolution. *In*: Huxley, T.H. 1890a. Critiques and Addresses. Macmillan, New York. pp. 181-217.

Huxley, T.H. 1890c (1870). On the formation of coal. *In*: Huxley, T.H. 1890a. Critiques and Addresses. Macmillan, London, pp. 92-110.

Huxley, T.H. 1890d (1870). Yeast. *In*: Huxley, T.H. 1890a. Critiques and Addresses. Macmillan, London, pp. 71-90.

Huxley, T.H. 1890e (1874). Universities: actual and ideal. *In*: Science and Culture, and Other Essays. Appleton, New York, pp. 31-72.

Huxley, T.H. 1890f. The School Boards. in Huxley, T.H. 1890a. Critiques and Addresses. Macmillan, London, pp. 33-55 (esp. p. 47).

Huxley, T.H. 1894. Discourses, Biological and Geological. Popular lectures and addresses II. Macmillan, New York.

Huxley, T.H. 1900 (1863). Man's Place in Nature, and Other Essays. Macmillan, London.

Huxley, T.H. 1910 (1874). Joseph Priestley. *In*: Science and Culture, and Other Essays. Appleton, New York, pp. 102-134.

Huxley, T.H. 1989 (1894). Evolution and Ethics. *In*: Paradis, J. and Williams, G.C. (eds.). Evolution and Ethics. Princeton University Press, Princeton.

Huxley, T.H. 1897 (1876). Lectures on Evolution, I-III. *In*: Science and the Hebrew Tradition. Appleton, New York, pp. 46-138.

Huxley, T.H. 1938a (1868). On a piece of chalk. *In*: Lectures and Lay Sermons. Everyman's Library, J.M. Dent & Sons, London, pp. 1-21.

Huxley, T.H. 1938b. Lectures on the structure of the skull. I-VIII. *In*: Lectures and Lay Sermons, E.P. Dutton & co., New York, pp. 121-272.

Imbert, E., LeFevere, F. 2003. Dispersal and gene flow of *Populous nigra* (Salicaceae) along a dynamic river system. Journal of Ecology, 91: 447-456.

Jenkin, F. 1867. Review of the origin of species. The North British Review, 46: 277-318.

Johnson, M.S., Stine, O.C. and Murray, J. 1984. Reproductive compatibility despite large scale genetic divergence in *Cepaea nemoralis*. Heredity, 53: 655-665.

Jones, C.G., Lawton, J.H. and Shachak, M. 1997. Positive and negative effects of organisms as physical ecosystem engineers. Ecology, 78: 1946-1957.

Joravsky, D. 1962, 1971. The Lysenko Affair. Scientific American, 207: 41-49.

Joravsky, D. 1986. The Lysenko Affair. University of Chicago Press, Chicago.

Kammerer, P. 1924. The Inheritance of Acquired Characters. Boni & Liveright, New York.

Kamping, A., and Van Delden W. 1978. Alcohol dehydrogenase polymorphism in populations of Drosophila melanogaster. Biochemical Genetics 16: 541-551.

Kargbo, D.B. 1980. Children's beliefs about inherited characteristics. Journal of Biological Education, 14: 137-146.

Kerr, W.E. and Wright, S. 1954. Experimental studies of gene frequencies in very small populations of *Drosophila melanogaster*. I-III. Evolution 8: 172-177; 293-307.

Kerkut, G.A. 1960. Implications of Evolution. Pergamon Press, London.

Kettlewell, H.B.D. 1958. A survey of the frequencies of - (L.) and its melanic forms in Great Britain. Heredity, 12: 51-72.

Kettlewell, H.B.D. 1961. The phenomenon of industrial melanism in the Lepidoptera. Annual Review of Entomology, 6: 245-262.

Kimura, M. 1983a. The Neutral Theory of Molecular Evolution. Cambridge University Press, Cambridge.

Kimura, M. 1983b. The neutral theory of molecular evolution. *In*: Nei, M., and Koehn, R.K. (eds.). Evolution of Genes and Proteins. Sinauer, Sunderland, pp. 208-233.

Kimura, M., and Ohta, T. 1971. Protein polymorphism as a phase of molecular evolution. Nature, 229: 467-469.

King, J.L. and Jukes, J.H. 1969. Non-Darwinian Evolution. Science, 164: 788-798.

King-Hele, D. 1977. Doctor of Revolution: The life of Erasmus Darwin. Faber & Faber, London.

Koestler, A. 1973. The Case of the Midwife Toad. Vintage Books, Random House, New York.

Kropotkin, P. 1902. Mutual Aid: A Factor of Evolution. Anarchist Archives.

Kutchera, U. 2009. Charles Darwin Origin of Species, directional selection and the evolutionary science today. Naturwissenschaften, 96: 1247-1263;

Kutchera, U., and Niklas, K.J. 2009. Evolutionary plant physiology: Charles Darwin forgotten hypothesis. Naturwissenschaften 96: 1339-1357.

Lack, D. 1947. Darwin's Finches. Cambridge University Press, Cambridge.

Lahav, N. 1991. Pre-biotic coevolution of self-replication and translation, or RNA world? Journal of Theoretical Biology, 151: 531-539.

Lahav, N. 1993. The RNA-world and co-evolution hypotheses and the origin of life: implications, research strategies and perspectives. Origin of Life and Evolution of the Biosphere, 23: 329-344.

Lamarck, J.B. 1914 (1809). Zoological Philosophy, an Exposition with Regard to the Natural History of Animals. Macmillan, London.

Lamarck, J.B. 1984 (1809). Zoological Philosophy. University of Chicago Press, Chicago.

Le Guyader, H. 2004. Etienne Geoffroy Saint-Hilaire, a Visionary Naturalist. University of Chicago Press, Chicago.

Leigh, E.G. 1993. Afterwards. *In*: Haldane 1993 (1932). The Causes of Evolution. Princeton University Press, Princeton, pp. 130-212.

Lewontin, R.C. 1965. Selection for colonizing ability. *In*: Baker, H.G. and Stebbins G.L.(eds.). The Genetics of Colonizing Species. Academic Press, New York, pp. 70-92.

Lewontin, R.C. 1974. The Genetic Basis of Evolutionary Change. Columbia University Press, New York.

Lewontin, R.C. and Hubby, J.L. 1966. A molecular approach to the study of genic heterozygosity in natural populations. II. Amount of variation and degree of heterozygosity in natural populations of *Drosophila pseudoobscura*. Genetics, 54: 595-609.

Li, C.C. 1966. Population Genetics. University of Chicago Press.

Li, W-H. and Graur, D. 1991. Fundamentals of Molecular Evolution. Sinauer, Amsterdam.

Liebert, T.G. and Brakefield, P.M. 1987. Behavioral studies on the peppered moth *Biston betularia*, and a discussion of the role of pollution and lichens in industrial melanism. Biological Journal of the Linnaean Society, 31: 129-150.

Lyell, C. 1853. Principles of Geology. 9th ed. Murray, London.

Lyell, C. 1873 (1863). Geological Evidences of the Antiquity of Man. 4th ed. Murray, London.

Lyell, C. 1874. The Student's Elements of Geology. 2nd ed. Murray, London.

Lyell, K.M. (ed.) 1881. Life, Letters and Journals of Sir Charles Lyell, Vols. I, II. Murray, London.

MacArthur, R.H. and Wilson E.O. 1967. The Theory of Island Biogeography. Princeton University Press, New Jersey.

Macnair, M.C. 1991. Why the evolution of resistance to anthropogenic toxins normally involves major gene changes: the limits to natural selection. Genetica, 84: 213-219.

Malo, A.F., Roldan, E.R.S. and Garde, J. 2005. Antlers honestly advertise sperm production and quality. Proceedings of the Royal Society, London B272: 149-157.

Malthus, T. 1798. An Essay on the Principle of Population. J. Johnson, London.

Mani, G.S. 1980. A theoretical study of morph ratio clines with special reference to melanism. Proceedings of the Royal Society of London, B, 210: 299-316.

Martin, W., Baross, J., Kelly, D. and Russell, M.J. 2008. Hydrothermal vents and the origin of life. Nature Reviews 6: 805-814.

Maynard-Smith, J. 1989. Evolutionary Genetics. Oxford University Press, Oxford, UK.

Mayo, O. 2008. A century of Hardy-Weinberg equilibrium. Twin Research and Human Genetics, 11: 249-256.

Mayr, E. 1942. Systematics and the Origin of Species. Columbia University Press, New York.

Mayr, E. 1963. Animal Species and Evolution. Harvard University Press, Cambridge.

Mayr, E. 2004. 80 years of watching the evolutionary scenery. Science, 305: 46-47.

Milinkovitch, M.C. 2012. Recovery of a near-extinct Galapagos tortoise despite minimal genetic variation. Evolutionary Applications 6: 377-383. Also: Proceedings, National Academy of Sciences, USA, Feb. 2012, E821-838.

Milinkovitch, M.C., Monteyne, D., Gibbs, J.P., Fritts, T.H., Tepia, W., Snell, H.L., Tiedemann, R., Caccone, A., and Powell, J.R. 2004. Genetic analysis of a successful repatriation programme: giant Galapagos tortoises. Proceedings of the Royal Society of London, B, 271: 341-345.

Miller, S. 1953. A production of amino acids under possible primitive earth conditions. Science, 117: 528-529.

Miller, S.L. 1974. The atmosphere of primitive earth and the prebiotic synthesis of amino acids. Origins Life, 5: 139-151.

Mitchell, G., Sittert S.J. and Skinner, J.D. 2009. Sexual selection is not the origin of long necks in giraffes. Journal of Zoology, 2009, 53: 1-6.

Mivart, St. J.J. 1871. On the Genesis of Species. 2nd ed. Macmillan, New York.

Moczek, A.P. and Emlen, D.J. 2000. Male horn dimorphism in the scarab beetle *Onthophagus taurus*: do alternative reproductive tactics favour alternative phenotypes? Animal Behaviour, 59: 459-466.

Moore, J.A. 1986a. Evolution by creeps and evolution by jerks. Science as a way of knowing, III: Genetics. American Zoologist, 26: 583-747.

Moore, J.A. 1986b. Science as a way of knowing, III: Genetics. American Society of Zoologists, Baltimore.

Morgan, T.H. 1903. Evolution and Adaptation. Macmillan, New York.

Morgan, T.H. 1909. Chromosomes and Heredity. American Naturalist, 44: 449-496.

Morgan, T.H. 1910. Sex – limited inheritance in Drosophila. Science, 32: 120-122.

Morgan, T.H. 1925. Evolution and Genetics. Princeton University Press, Princeton.

Nei, M. 1972. Genetic distance between populations. American Naturalist, 106: 283-292.

Nei, M. 1975. Molecular Population Genetics and Evolution. American Elsevier, New York.

Nei, M. 1983. Genetic polymorphism and the role of mutation in evolution. *In*: Nei, M. and Koehn, R.K. (eds.). Evolution of Genes and Proteins. Sinauer, Sunderland, pp. 165-190.

Nevo, E. 1978. Genetic variation in natural populations: pattern and theory. Theoretical Population Biology, 13: 121-177.

O'Donald, P. and Muggleton, J. 1979. Melanic polymorphism in ladybirds maintained by sexual selection. Heredity, 43: 143-146.

Ohta, T. 1974. Mutational pressure as the main cause of molecular evolution and polymorphism. Nature, 252: 351-354.

Oparin, A.I. 1957. The Origin of Life on Earth. Oliver & Boyd, Edinburgh.

Oparin, A.I. 1968. Genesis and Evolutionary Development of Life. Academic Press, New York.

Opler, P.A. 2016. Theodore Boveri (1862-1915), to commemorate the centenary of his death and contribution the Sutton-Boveri hypothesis. American Journal of Medical Science 1704A: 2803-2829.

Orgel, L.E. and Crick, F.H.C. 1980. Selfish DNA: the ultimate parasite. Nature, 284: 604-607.

Osborn, H.F. 1910. in Huxley, J., 1942: Evolution, the modern synthesis. p. 486.

Owen, R. 1863. On the Archeopteryx of von Meyer, with a description of fossil.... Philosophical Transactions of the Royal Society, London. 153: 33-47.

Paley, W. 1844 (1802). Natural Theology, or evidences of the existence and attributes of the Deity, collected from the appearances of Nature. *In*: Paley's Works, H.G. Bohn, London.

Pasteur, N., and Raymond, M. 1996. Insecticide resistance genes in mosquitoes: their mutations, migration, and selection in field populations. Journal of Heredity, 87: 444-449.

Pearson, K. 1900. The Grammar of Science. 2nd ed. Adam and Charles Black, London.

Pearson, K. 1949 (1892). The Grammar of Science. Everyman, New York.

Portin, P. 2009. The elusive concept of the gene. Hereditas 146: 112-117.

Price, P.W. 1996a. Biological Evolution. Saunders College Publishing. New York.

Price, P.W. 1996b. Empirical research and factual based theory. American Entomologist 42: 219.

Provine, W. 1986. Geneticists and race. *In*: Science as a Way of Knowing. III. Genetics. American Society of Zoologists, Baltimore. pp. 857-887.

Quammen, D. 2006. The Reluctant Mr. Darwin. Norton, New York.

Raby, P. 2001. Alfred Russel Wallace: ALife. Princeton University Press, Princeton.

Rak, Y., and Arensburg, B. 1987. Kebara 2 Neanderthal pelvis, a first look at a complete inlet. American Journal of Physical Anthropology 73: 227-231.

Remoff, H. 2016. Malthus, Darwin, and the descent of economics. American Journal of Economics and Sociology 75 (4): 862-905.

Reznick, D.N., and Ghalambor, C.K. 2005a. Can commercial fishing cause evolution? Answers from guppies (*Poecilla reticulata*). Canadian Journal of Fishery and Aquatic Science, 62: 791-801.

Reznick, D.N., and Ghalambor, C.K. 2005b. Selection in nature: experimental manipulation of natural populations. Integrative and Comparative Biology, 45: 456-462.

Rohlf, F.J., and Schnell, G.D. 1971. An investigation of the Isolation by Distance model. American Naturalist, 105: 295-324.

Romanes, G.J. 1899. An examination of Weismannism. Open Court Publ., Chicago.

Roush, R.T. and Mckenzie, J.A. 1987. Ecological genetics of insecticide and acaricide resistance. Annual Review of Entomology, 32: 361-380.

Rudwick, M.J.S. 1997. Georges Cuvier, Fossil Bones, and Geological Catastrophes. University of Chicago Press, Chicago.

Sander, K., and Schmidt-Ott, U. 2004. Evo-Devo aspects of classical and molecular data in a historical perspective. Journal of Experimental Zoology (Mol. Dev. Evol.), 302, B: 69-91.

Sandler, J., and Sandler, L. 1986. On the origin of Mendelian genetics. Science as a Way of Knowing, III: Genetics. American Zoologist, 26: 753-768.

Sapp, J. 2003. Genesis – The Evolution of Biology. Oxford University Press, Oxford.

Schilthuizen, M., and Lombaerts, M. 1994. Population structure and levels of gene flow in the Mediterrranean land snail *Albinaria corrugata* (Pulmonata, Clausilidae). Evolution, 48: 577-586.

Sedgwick, A. 1850. A Discourse of the Studies, University of Cambridge. Preface to the 5th edition.

Selander, R.K., Kaufman, D.W. 1975. Genetic structure of populations of the brown snail (*Helix aspera*) I. Microgeographic variation. Evolution 29: 385-401.

Simmons, R.E., and Scheepers, L. 1996. Winning by the neck: sexual selection in the evolution of giraffe. American Naturalist, 148: 771-786.

Sinnott, E.W., Dunn, L.C. and Dobzhansky, T. 1958. Principles of Genetics. 5th ed., McGraw-Hill, New York.

Sneath, P.H.A. and Sokal, R.R. 1973. Numerical Taxonomy. 2nd ed., Freeman, New York.

Sokal, R.R. and Crovello, T.J. 1970. The Biological Species concept: a critical review. American Naturalist, 104: 127-155.

Sokal, R.R. and Sneath, P.H.A. 1963. Numerical Taxonomy. Freeman, New York.

Sokal, R.R. and Rohlf, F.J. 1995. Biometry. 5th ed. Freeman, New York.

Spencer, H. 1862. First Principles. W.J. Johnson, printer. London.

Spiess, E.B. 1977. Genes in Populations. Wiley, New York.

Stauffer, R.C. 1975. Charles Darwin's Natural Selection. Edinburgh University Press.

Stott, R. 2012. Darwin's Ghosts: The Secret History of Evolution. Spiegel & Graw, New York.

Sulloway, F.J. 1984. Darwin and the Galapagos. Biological Journal of the Linnaean Society 21: 29-59.

Sulloway, F.J. 2009. Tantalizing tortoises and the Darwin-Galapagos legend. Journal of the History of Biology 42: 3-31.

Sulloway, F.J. 2009. Why Darwin rejected intelligent design. Journal of Bioscience 34: 173-183.

Sutton, W.S. 1903. Chromosomes in heredity. Biological Bulletin 4: 231-251.

Takahashi, M., Arita, H., Hiraiwa-Hasegawa, M., and Hasegawa, T. 2008. Peahens do not prefer peacocks with more elaborate trains. Animal Behavior, 75: 1209-1219.

Thompson, D'Arcy W., 1992, 1984 (1917). On Growth and Form. Cambridge University Press, Cambridge.

Vallery-Radot, R. 1930? The Life of Pasteur. Garden City, New York.

Van Delden, W. 1982. The Alcohol dehydrogenase polymorphism in *Drosophila melanogaster*: selection at an enzyme locus. Evolutionary Biology, 15: 187-222.

Van Delden, W. and Kamping, A.K. 1983. Adaptation to alcohols in relation to the Alcohol dehydrogenase locus in *Drosophila melanogaster*. Entomologia experimentalis et Applicata, 33: 97-102.

Van Delden, W., Boerema, A.C. and Kamping A. 1978. The Alcohol dehydrogenase polymorphism in *Drosophila melanogaster*. I: selection in different environments. Genetics, 90: 161-191.

Van Wyhe, J. (ed.) 2005. The writings of Charles Darwin on the web: Francis Darwin (1909). The Foundation of the Origin of Species. The 1844 Essay.

Wade, M.J., 1977. An experimental study of group selection. Evolution 31: 134-153.

Wade, M.J. 1978. A critical review of the models of group selection. Quarterly Review of Biology, 53: 101-114.

Wallace, A.R. 1858. A narrative of travels in the Amazon. Reeves et al. London.

Wallace, A.R. 1864. The origin of human races and the antiquity of man, deduced from the theory of natural selection. Journal of the Anthropological Society of London, 2: 1-25.
Wallace, A.R. 1889. Darwinism. Macmillan, London.
Wallace, A.R. 1962 (1869). The Malay Archipelago. 10th ed. Dover, New York.
Wallace, A.R. 1874. A defense of modern spiritualism. Fortnightly Review 15: 1-38.
Wallace, A.R. 1891. Natural Selection and Tropical Nature. Macmillan, London.
Wallace, A.R. 1893a. The non-inheritance of acquired characters. A letter to the editor, Nature, July 20th 1893.
Wallace, A.R. 1893b. Are individually-acquired characters inherited? Fortnightly Review, 53: 1-17.
Wallace, A.R. 1895. The method of organic evolution. Fortnightly Review, February 1st, March 1st 1895. *In*: Smith, C.H. (ed.), List of Wallace's Writings, S-510.
Wallace, A.R. 2003 (1876). The Geographic Distribution of Animals. Elibron Classics, Macmillan, London.
Wallace, B. 1969. Topics in Population Genetics. Norton, New York.
Weismann, A. 1889. The duration of Life. *In*: Essays upon Heredity, p. 35ff.. Clarendon Press, Oxford.
Weismann, A. 1892. Essays Upon Heredity and Kindred Biological Problems. Vol. II. Oxford University Press, Oxford.
Weismann, A. 1893. The Germ Plasm: A Theory of Heredity. Walter Scott, London.
Weismann, A. 1893. The supposed transmission of acquired characters. *In*: The Germ Plasm: A Theory of Heredity. Walter Scott, London, pp. 392-409.
Weismann, A. 1891-1893. Essays Upon Heredity, and Kindred Biological Problems. Vols. I, II. Clarendon Press, Oxford.
Wells, K.D. 1818. William Charles Wells and the races of man. Journals.uchicago.edu isis 1973.
Whewell, W. 1866. History of the Inductive Sciences. II. History of Geology, p. 388ff. IV. History of Botany, p. 564ff.
Wilberforce, S. 1860. Review of Darwin's The Origin of Species. Quarterly Review, 108: 225-264.
Williams, J.E. 1925. In Search of Reality. I: Evolution. Duckworth, London.
Winchester, S. 2003. Krakatoa. Penguin Books, London.
Wolfe, A.J. 2016. Cold-war contest of the golden jubilee. Journal of the History of Biology 45: 389-414.
Wood, R.J., and Orel, V. 2001. Genetic Prehistory in Selective Breeding: APrelude to Mendel. Oxford University Press, Oxford.
Wool, D. 1985. Blind flour beetles: variable eye phenotypes in the microcephalic (mc) mutant of *Tribolium castaneum* (Coleoptera: Tenebrionidae). Israel Journal of Entomology, 19: 201-210.
Wool, D. 1987. Differentiation of island populations: a laboratory model. American Naturalist, 129: 188-202.
Wool, D. 2001. Charles Lyell – "The father of Geology" – as a forerunner of modern ecology. Oikos, 94: 385-391.
Wool, D. 2006. The Driving Forces of Evolution. Science Publishers, Enfield.
Wool, D., and Mendlinger, S. 1973. The eu mutant of *Tribolium castaneum*: environmental and genetic effects on penetrance. Genetica, 44: 496-504.
Wright, S. 1931. Evolution in Mendelian Populations. The Iowa State College Press, Ames, Iowa.
Wright, S. 1921-1958. Systems of Mating and other papers. The Iowa State College Press, Ames, Iowa.
Wright, S. 1968-1984. Evolution and the Genetics of Populations. Vols. I-III. University of Chicago Press, Chicago.
Wright, S., and Kerr, W.E. 1954. Experimental studies of the distribution of gene frequencies in very small populations of *Drosophila melanogaster*. I. Forked Evolution, 8: 172-177.
Wu, L., Bradshaw, A.D., and Thurman, D.A. 1975. The potential for evolution of heavy-metal tolerance in plants. III: The rapid evolution of copper tolerance in *Agrostis stolonifera*. Heredity, 34: 165-187.

Author Index

Subject Index